Lecture Notes in Electrical Engineering

Volume 679

The book series *Lecture Notes in Electrical Engineering* (LNEE) publishes the latest developments in Electrical Engineering—quickly, informally and in high quality. While original research reported in proceedings and monographs has traditionally formed the core of LNEE, we also encourage authors to submit books devoted to supporting student education and professional training in the various fields and applications areas of electrical engineering. The series cover classical and emerging topics concerning:

- Communication Engineering, Information Theory and Networks
- Electronics Engineering and Microelectronics
- Signal, Image and Speech Processing
- Wireless and Mobile Communication
- Circuits and Systems
- Energy Systems, Power Electronics and Electrical Machines
- Electro-optical Engineering
- Instrumentation Engineering
- Avionics Engineering
- Control Systems
- Internet-of-Things and Cybersecurity
- Biomedical Devices, MEMS and NEMS

For general information about this book series, comments or suggestions, please contact leontina.dicecco@springer.com.

To submit a proposal or request further information, please contact the Publishing Editor in your country:

China

Jasmine Dou, Associate Editor (jasmine.dou@springer.com)

India, Japan, Rest of Asia

Swati Meherishi, Executive Editor (Swati.Meherishi@springer.com)

Southeast Asia, Australia, New Zealand

Ramesh Nath Premnath, Editor (ramesh.premnath@springernature.com)

USA, Canada:

Michael Luby, Senior Editor (michael.luby@springer.com)

All other Countries:

Leontina Di Cecco, Senior Editor (leontina.dicecco@springer.com)

**** Indexing: The books of this series are submitted to ISI Proceedings, EI-Compendex, SCOPUS, MetaPress, Web of Science and Springerlink ****

More information about this series at http://www.springer.com/series/7818

Wynand Lambrechts · Saurabh Sinha

Millimeter-wave Integrated Technologies in the Era of the Fourth Industrial Revolution

Wynand Lambrechts
University of Johannesburg
Johannesburg, Gauteng, South Africa

Saurabh Sinha
University of Johannesburg
Johannesburg, Gauteng, South Africa

ISSN 1876-1100 ISSN 1876-1119 (electronic)
Lecture Notes in Electrical Engineering
ISBN 978-3-030-50474-8 ISBN 978-3-030-50472-4 (eBook)
https://doi.org/10.1007/978-3-030-50472-4

This Springer imprint is published by the registered company Springer Nature Switzerland AG
The registered company address is: Gewerbestrasse 11, 6330 Cham, Switzerland

Preface

The Fourth Industrial Revolution (Industry 4.0) will take advantage of ubiquitous connectivity to grow economies. There is a vast difference between the telecommunications infrastructure of developed countries and that of emerging markets; emerging markets often lack broadband Internet infrastructure. A sustainable solution is required to bridge the digital divide and enable emerging markets to participate in Industry 4.0. Such a solution, however, has numerous facets and requires introducing technology that is sustainable and future-proof and ensures scalable, feasible, and affordable access. Unfortunately, globally, the cost of accessing broadband Internet in unequal markets remains high, with current solutions reaching their thresholds in terms of available bandwidth on the mass consumer level.

Spectrum demand and allocation in emerging markets often lag behind what is happening in developed countries and expensive to (unfairly) maximize profit for local regulators and frequently for local governments as well. Arguments typically refer to the limited amount of available spectrum, which is true of current-generation networks. However, millimeter-wave (mm-Wave) frequencies (30—300 GHz) offer unprecedented amounts of bandwidth for wireless broadband connectivity. Built on the premise of mm-wave connectivity is the fifth-generation (5G) networks, a clear contender to spearhead connectivity for Industry 4.0, also in emerging markets.

This book focuses on researching the capabilities of mm-wave-based 5G networks, specifically for emerging markets. The technology is researched from several perspectives, including technological advantages and disadvantages, as well as its unique characteristics to offer scalable and affordable connectivity to grow emerging economies. In the current information age, bringing connectivity to as many people as possible (ideally everyone) boosts socioeconomic benefits through innovation in sciencc and technology, with the common goal of bringing about positive change in the lives of individuals. The importance of connectivity has recently been further emphasized with the global pandemic, COVID-19, forcing millions of people to adopt an online approach to conduct business and communicate with one another.

The research presented in this book is structured to investigate not only the economic benefits of mm-wave and 5G. It also gives a strong theoretical background on the underlying technologies required to realize transmitters and receivers that are capable of operating at extremely high frequencies. The technical contribution of this book is in its presentation of electronic subsystem analysis and review of high-frequency active circuits that are capable of operating at mm-wave frequencies. These subsystems include frequency mixers, oscillators, low-noise amplifiers, and power amplifiers. To implement feasible future-generation technologies and ensure future-proof infrastructure that is easily expandable, it is of the utmost importance to understand the principles that influence the integrity of transmitted information. This book researches the importance of identifying, describing, and analyzing technology from a purely technological standpoint, but equally so, acknowledges and investigates the challenges in introducing such technologies into emerging markets.

A detailed investigation of techniques to introduce mm-wave 5G networks to emerging markets is presented in the latter part of this book. As the digital service sector evolves, certain key characteristics remain universal, such as adequate digital infrastructure, technology-literate end users, innovative skilled entrepreneurs, and business environments that encourage creative thinking. A core characteristic and major advantage of 5G are its potentially dynamic pricing and low dependence on the infrastructure of previous-generation wireless networks. This book aims to investigate these features thoroughly and provide potential solutions to introduce 5G, possibly in a staggered approach, to emerging markets. In emerging markets, the complexity and challenges of distributing critical products and services in significantly unequal markets are much higher. Governments have been forced to innovate policies and strategies to achieve the successful distribution of products and services to poor households and rural areas. 5G is a future-proof and sustainable technology with numerous advantages not associated with previous-generation mobile technologies that offer governments and local entrepreneurs a means to innovate modern policies and strategies to bridge the digital divide.

The primary audience of this book is learners in the fields of engineering and information technology who want to identify and act upon ways of advancing connectivity in emerging markets and rural areas by bringing high-bandwidth, low-latency mobile connectivity to individuals through future-proof technologies. The audience is presented with an introduction to the role of mm-wave and 5G in Industry 4.0, followed by a theoretical background on mm-wave circuit design, and finally, an in-depth investigation into potential solutions to prepare emerging markets to participate in Industry 4.0 through mm-wave broadband connectivity.

Acknowledging the Technical Peer-Review Process

The authors would like to recognize the support of the numerous technical reviewers, as well as language and graphics editors, who have participated in the development of this research contribution. We value the system of scholarly peer review and the perspective that this adds to the production of research text that augments the body of scientific knowledge.

Johannesburg, South Africa

Wynand Lambrechts
Saurabh Sinha

Contents

About the Authors

Dr. Wynand Lambrechts, SMIEEE, obtained his B.Eng., M.Eng., and Ph.D. degrees in electronic engineering from the University of Pretoria (UP), South Africa. He achieved his M.Eng. with distinction. He has authored two publications in peer-reviewed journals and has presented at various local and international conferences. Wynand is the lead author on four books; in the fields on sustainable energy and in microelectronic engineering, published by international publishers. He has co-authored four contributing chapters in other books in the fields of green energy and technology and the Fourth Industrial Revolution. He previously held a position as an electronic engineer at Denel Dynamics, a state-owned company in South Africa. He is currently employed by SAAB Grintek Defence (SGD) and is also serving as a part-time research associate at the University of Johannesburg (UJ), South Africa.

Prof. Saurabh Sinha, Ph.D.(Eng), Pr. Eng., SMIEEE, FSAIEE, FSAAE, MASSAf. Prof. Sinha obtained his B. Eng. (with distinction), M. Eng. (with distinction), and Ph.D. degrees in electronic engineering from the University of Pretoria (UP). As an established researcher, rated by the National Research Foundation (NRF), he has authored or co-authored over 130 publications in peer-reviewed journals and at international conferences. He served UP for over a decade; his last service being as the director of the Carl and Emily Fuchs Institute for Microelectronics, Department of Electrical, Electronic and Computer Engineering. On October 1, 2013, he was appointed as an executive dean of the Faculty of Engineering and the Built Environment (FEBE) at the University of Johannesburg (UJ). As of December 1, 2017, he is the UJ deputy vice-chancellor: Research and Internationalization. Among other leading roles, he also served the IEEE as a board of director and IEEE vice-president: Educational Activities.

Chapter 1
The Role of Millimeter-Wave and 5G in the Fourth Industrial Revolution

Abstract In digital production, an important sector in the fourth industrial revolution, data communication, is an imperative function of numerous applications in the current and modern era. Big data analytics, preventative maintenance and remote maintenance are among the initially targeted sectors where new generation technologies will play a crucial role. Storage of large quantities of data has become relatively contained in recent years; however, the transfer of data, especially audio and visual media, requires high-bandwidth communication channels that are reliable, safe and stable. Wireless broadband is a convenient method of transferring data between two or more points and enables critical applications in the fourth industrial revolution. In this chapter, the significance of wireless-broadband-enabling technologies, such as the fifth-generation (5G) mobile network with underlying millimeter-wave carrier frequencies, is researched. Wired implementations such as fiber have allowed gigabits per second data transfer for quite some time, but broadening access through wireless technologies requires wireless infrastructure that matches the transfer capabilities of fiber. The principles of millimeter-wave frequencies as enabling technologies for future wireless communication in the era of the fourth industrial revolution are critically investigated in this book and introduced in this chapter.

1.1 Introduction

The incipient fourth industrial revolution (Industry 4.0 or 4IR) and the rising need for communication, not only between humans but also between ubiquitous devices, require the broadband capabilities of wireless technologies. The fifth-generation (5G) is necessary to realize the new demands of flexible production, massive multiple-input multiple-output (MIMO) systems and the internet of things (IoT), virtual reality (VR), augmented reality (AR), intelligent automation and autonomous driving [20]. The estimated gains of smart manufacturing are forecast at US $1.5 trillion [52]. This includes improved efficiency, cost and energy savings and new services accompanied

W. Lambrechts and S. Sinha, *Millimeter-wave Integrated Technologies in the Era of the Fourth Industrial Revolution*, Lecture Notes in Electrical Engineering 679,
https://doi.org/10.1007/978-3-030-50472-4_1

by revenue growth. Successful positioning of 5G, which will lead to a competitive advantage, depends on collaboration between research institutions and industrial partners. Industry 4.0 will evolve at an exponential pace as opposed to the relatively linear growth of earlier industrial revolutions. Industry 4.0 brings about the grouping of technologies, data algorithms, data quality and data integrity as well as necessitating ethical approaches to data sharing. At the core of 5G, one of the enabling technologies of Industry 4.0 is millimeter-wave (mm-Wave) technology. Referring to information carrier signals with frequencies (up to 100 GHz or 100 billion cycles per second) having wavelengths in the millimeter range, mm-Wave permits transmitting and receiving information across wide pipelines of bandwidth.

This book will research and review the roles that mm-Wave technology and 5G play in Industry 4.0 from a technical and economic perspective. Rather than only reviewing the potential applications of mm-Wave in 5G, this book distinguishes itself by also researching the capabilities of the underlying technology and describing its potential and limitations. Maturity of any technology is crucial for widespread deployment and mm-Wave-based 5G communication is yet to reach maturity, albeit already implemented in numerous applications. This could be cause for concern, and the aim of this book is to highlight these concerns, placed in a technical context. Furthermore, this book will highlight the efforts of higher education in the developing world and in emerging markets to plan successfully for Industry 4.0 and intelligent systems and align policies to effect a shift in business and industry, ethics and the future of work.

Leading to mm-Wave communications, ancient methods of interacting and collaborating looked vastly different. To compare the period between sending and receiving a message of modern technology to ancient ones is rather prejudicial, but it remains informative to acknowledge how the earliest communication took place. Equally important is defining what communication is—a relatively recent formalization of information theory, what constitutes information and how (and if) the correct information reaches its destination. The following paragraph reviews information theory and available definitions of communication, followed by a brief overview of the evolution of communication.

1.2 Communication and Information Theory

If communication had been absent from our lives, there would be very little progress in any form of information transfer. There would be no inventions, no problems solved, no books or novels, no intimate relations among people and no way of expressing ourselves. Luckily, communication has always been possible and has always found ways of evolving. Prior to exploring the evolution of communication in the following paragraph, the term should be defined as accurately as possible. Hauser [21] follows an encompassing and explanatory approach to communication in the animal kingdom, which includes human language, categorized into four primary approaches:

1. Mechanistic, the ability to understand the mechanisms that underlie the expression of a trait, which includes neural, physiological and psychological ones).
2. Ontogenetic, identifying the genetic and environmental influences that direct the development of a trait.
3. Functional, considering the effectiveness and effects of a trait on survival and reproduction.
4. Phylogenetic, identifying the evolutionary history of a trait, and isolating and evaluating its ancestral features.

The definition of communication has elicited argument and it is no easy task to define it Mellor [36], especially since definitions of communication *should* be valid across disciplines. This means that a common definition should be agreed upon by researchers in, for example, sociobiology, behavioral ecology, sensory ecology, neuropsychology, cognitive psychology and of course linguistics [21]. Various researchers agree that the notions of *information* and *signal* form essential components of many of the definitions of communication [21]. Information is an aspect of interaction concerning a sender and a perceiver (calling it a receiver would assume that the transmitted information is understood exactly as it was meant to be). The *signal* is capable of carrying various types of *informational content* and *context* and can be manipulated by both the sender and the perceiver. Hauser [21] additionally highlights that a signal is designed to serve specific purposes that should be evaluated in terms of both the production and the perception constraints.

Information, like communication, is a difficult term to define, and some commonly accepted multidisciplinary frameworks exist that aim to identify the common characteristics of information. Information theory is the study of the quantification, storage and communication of information that satisfies both modern general communication systems and ancient ones. One framework (from the perspective of an electronic engineer) suggested by Shannon [50] is essentially a mathematical (statistical) theory of information. This publication, titled "*A mathematical theory of communication*" has been broadly discussed (and accepted) by engineers, biologists, psychologists as well as philosophers and has become a popular technique to theoreticize information. Figure 1.1 is an adaptation of a schematic diagram of a generalized communication structure as presented in Shannon [50].

As shown in Fig. 1.1, a general communication system, in any discipline and from any era, must consist of a source of information, the message that is to be conveyed, and a transmitter that will send the message to its destination. The message is passed on from one point to another using a signal, where unwanted noise is typically added to the information (which ideally needs to be removed to ensure accurate perception of the transmitted message). The received signal, as it passes through a medium (channel), is acknowledged by the receiver (perceiver) and the message reaches its final destination, with its integrity dependent on multiple factors along its path.

Shannon [50] maintained that the ultimate problem of communication is that of replicating at one end either precisely or nearly a message selected at another end. The message often has meaning; therefore, it denotes or is interrelated agreeing to some system with specific physical or abstract objects. Shannon [50] also reasoned

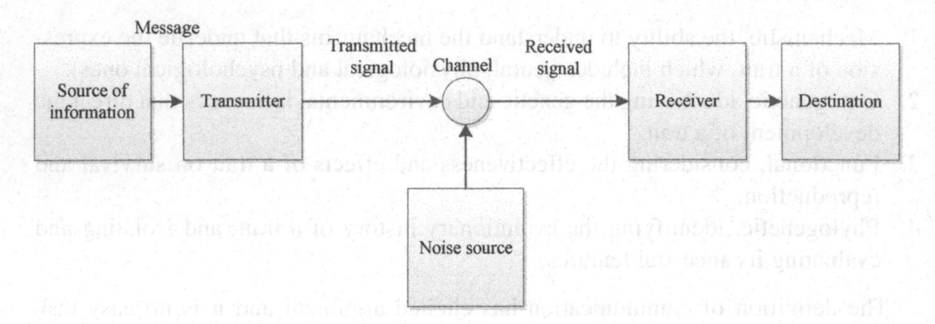

Fig. 1.1 The Shannon-Weaver model of communication, a schematic diagram of a generalized communication structure. Adapted from Shannon [50]

that this diagram of features of communication is inapt to a specific engineering problem and that the system ought to be designed to function for any message in a set of possible messages. If, for example, the quantity of messages in a group is determinate and one message is selected from the group where each message is equally probable to be selected, the number or any monotonic function of this number can be considered as a quantity of the information produced [50] and is a function of *information entropy*.[1] Shannon [50] formalized this theory of information and this contribution to information entropy follows the mathematical approach. According to Shannon [50],

$$H(X) = -\sum_{i=1}^{n} p_i \log_b p_i \tag{1.1}$$

where *H(X)* is Shannon entropy, *n* is the number of events, *X* designates a set or field of events, *b* is the logarithmic base (typically used values of *b* are 2 if binary information is decoded, Euler's number *e*, or 10) and p_i is the probability that the *i*th event of set *X* will occur. Using (1.1), it is theoretically possible to

- assess the extent to which the uncertainty at a receiver is reduced through the transmission of a signal by the sender, or
- determine if communication has occurred between two or more interacting bodies [50].

Entropy is defined in terms of a probabilistic statistical model, where high uncertainty corresponds to high entropy. If there is no ambiguity (or uncertainty), the entropy is nil/zero. Entropy can also be normalized by dividing it by the length of the information. The normalized entropy is a measure of randomness of information and is called metric entropy. Information theory, especially Shannon's concept of information entropy, has numerous applications in communications and in assessing previous generations of communication techniques. Shannon's information theory

[1] Information entropy is the median rate that a random source of data produces information.

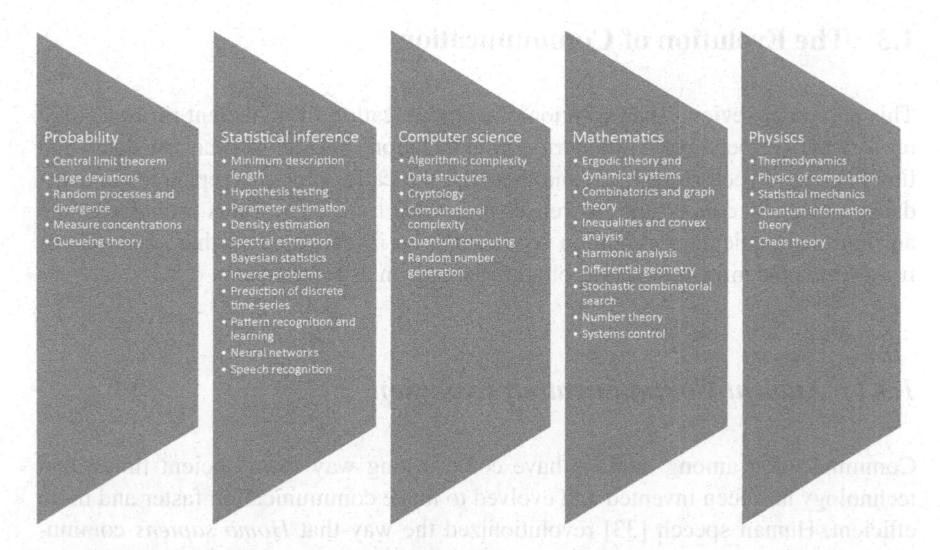

Fig. 1.2 Scientific disciplines and fields in which Shannon's theory on information has contributed to the particular field, adapted from Verdú [54]. Modern disciplines and fields, especially in computer science, would have numerous additional contributions

aids studies of linguistics in communications (using letter frequency [19] as shown in Appendix 2) and other technical applications such as data compression and channel coding. It has been linked to the origination of the optical disc (compact disc), the viability of mobile phones and the advance of the internet as we know it today, displaying the importance and the reputation of information theory and the work presented in Shannon [51]. Verdú [54], on the fiftieth anniversary of the publication of Shannon's information theory, listed numerous scientific disciplines that have adapted the theory in some way. These scientific disciplines and the fields to which information theory have contributed, according to Verdú [54], are summarized in Fig. 1.2.

Additional to the disciplines and fields summarized in Fig. 1.2, Verdú [54] also referred to works on information theory in economics, biology and chemistry. As a modern example, in Larsson [31], entropy is used to model 5G mobile network traffic aggregation as a single parameter characteristic to measure the memory and the behavior of a given data traffic trace in a 5G system. This book will also refer to Shannon's theory on information and entropy in analyzing modern communication techniques, especially enabling technologies for the Industry 4.0, such as mm-Wave-based 5G and the IoT where big data are present. To understand the importance of modern communication (leading to 5G), it is useful to review its historical development and the following paragraph briefly reviews the evolution of communication and gives perspective on how modern transferring of information has changed, how we communicate and what has stayed the same.

1.3 The Evolution of Communication

This paragraph reviews the evolution of communication from ancient times leading up to what we perceive as modern communication. Various sources of dates and timelines have been researched and there are typically some discrepancies in exact dates, especially concerning references to ancient times. The dates provided in the ancient communication evolution section can be investigated further; the sources from which the information was obtained are given as reference.

1.3.1 Ancient Communication Evolution

Communication among humans have come a long way from ancient times, and technology has been invented and evolved to make communication faster and more efficient. Human speech [33] revolutionized the way that *Homo sapiens* communicates and the timing around its origins are unclear [28]. Following the oldest forms of long-distance communication from ancient China, Egypt and Greece, which were smoke signals and drum beats, forms of *written* symbols were developed *only* 35,000 years ago and the earliest discovered type of visual communication is an ancient cave painting dating back 35,600 years, discovered in the northern part of Spain. Note that these theories have been challenged and traces of the pigment ocher used 164,000 years ago in South Africa have been found [34]. These visual symbols (typically carvings in rocks, known as petroglyphs) were primarily used to record small sections of history and to tell stories. Petroglyphs evolved into pictograms, a symbol that represents a concept, object, activity, place or event. Essentially, petroglyphs showed certain events and pictograms told the story behind the event. The evolution of pictograms led to ideograms, a symbol that showed an idea rather than an event that occurred.

The first consonantal alphabet (a combination of the words *Aleph* and *Beth*, the first two letters in the Phoenician[2] alphabet) was created by the ancient Egyptians between 2700 BC and 2000 BC. It consisted of a set of 22 hieroglyphs to represent combinations of syllables and consonants to pronounce ideograms and foreign names [48]. Modern-day written language in the Roman alphabet (also known as the Latin alphabet) evolved from the Greek alphabet and has become the official script in numerous countries worldwide. As languages and script evolved, transferring messages and communicating over longer distances became possible. Messages sent by pigeons, likely starting with the Persians, were used by the Romans to aid their military over 2 000 years ago. The use of carrier pigeons lasted through World Wars I and II to warn armies of impeding enemy attacks [1]. In 1844, the first electronic way to send messages through a wire was recorded, when Samuel Morse (1791–1872) sent a telegraph in Morse code from Washington DC to Baltimore, Maryland (about

[2]Phoenician city-states were the first to make extensive use of an alphabet.

40 miles/64 km). This laid the groundwork for the telephone, fax machine and the internet.

1.3.2 Evolution of Modern Communication

Refraining from delving into too much detail on the evolution of modern communication, it is useful to highlight some of the defining technologies of the last one-and-a-half centuries, stemming from the first electronically sent telegraph. In 1876, on March 7, Alexander Graham Bell (1847–1922) patented the first electronic telephonic device (the telephone as we know it today) (patent #174,465), capable of transmitting voice over a cable over a distance. The telephone, or *landline*, gained popularity rapidly and was commonly used by the 1950s. Almost a century after the telephone patent, information transmission would change drastically when the ability to transmit data over a wire led to the invention of the internet. Officially dated back to when ARPANET[3] used the transmission control protocol/internet protocol (TCP/IP) on 1 January 1983 based on earlier generation prototypes of the internet experimented with from the 1960s to the 1980s. Moreover, born in 1955, Tim Berners-Lee conceived the World Wide Web (www) by around 1990. The first iterations of physical connections to the www used dial-up modulator-demodulators (modems) to place information on a digital carrier, transmit the information over a wire, and pack the data bits at the receiver to represent the original information. Wireless communication started to gain traction from the 1990s through mobile phones and the short messaging service allowed people to send digitized texts to each other from wireless devices. Skipping forward, the next three decades brought about quick development, and today we are accustomed to

- smart phones, tablets and computers capable of making phone calls, messaging, and accessing the internet,
- social media allowing us to connect to theoretically anyone, anywhere in the world, and
- numerous online services that cater for virtually all our cyber needs.

Having so much communication technology at our disposal has also drastically changed the way we communicate and share information over the last couple of decades. The amount of data that is generated and stored daily is staggering. Forbes estimates that "*2.5 quintillion (10^{18}) bytes of data are created each day*" (only three times fewer bytes per day compared to all the sand grains (2.5 quintillion) on earth [4]) and the pace is accelerating, especially owing to the IoT [35]. The term big data is therefore not an overcompensation for the quantities of data we *consume*. It is therefore also relevant to review the quality and the integrity of the generated data.

[3]The Advanced Research Projects Agency Network.

According to Rui Carvalho,[4] managing director of Enterprise Solutions for S&P Capital IQ, there is a distinct, albeit connected, difference between data quality and data integrity. According to Carvalho, data quality concerns the completeness, accuracy, timeliness and consistency of information contained in the data. Data integrity, as described by Carvalho, pertains to the validity of data and therefore also to the precision and uniformity of saved (stored) data. Generating quality data leads to an end-product that has integrity in its data.

The face of communication has been affected during the course of traditional media. Marketing has evolved from traditional print to digital platforms, industry relations have become instant and high-priority for competitiveness in the digital marketplace, communication devices are becoming more powerful and cost-effective to allow more people access to the internet and the workplace is transforming as the internet allows employees to work off-site (the future of work). Much of the transformation of communication has given rise to Industry 4.0 and the evolution thereof is rapidly changing. Apart from the virtually infinite technological advances occurring annually in multiple sectors, one technology is bound to play a large part in defining the future of communications, which is 5G. 5G wireless communication built on mm-Wave technology is receiving a lot of attention from mobile network operators (MNOs) in the telecommunications industry as well as from industry and academic institutions that develop technologies in line with the Industry 4.0 market. Prior to reviewing mm-Wave and 5G, the terms fixed wireless access (FWA) and mobile broadband are briefly defined. The telecommunication industry often refers to these technologies independently and MNOs make separate reports on the progress of each of these terms, especially when stating business cases for 5G development. These are two different concepts, both important for MNOs to pursue; clearly highlighting the differences is useful and this is presented below.

1.4 Fixed Wireless Access Versus Mobile Broadband

FWA is a means to provide internet access to fixed locations using wireless mobile technology such as 4G and (preferably) 5G, rather than fixed lines such as fiber or digital subscriber line (DSL). Using 5G as the backbone of FWA allows for access speeds comparable to those of many fiber offerings. The basic premise of an FWA system is presented in Fig. 1.3.

As shown in Fig. 1.3, a dedicated active antenna system is used to serve fixed locations with wireless broadband internet access. 5G, with its additional spectrum availability when compared to 4G, will play a crucial role in FWA systems to ensure that the bandwidth and latency offered to consumers match those of fixed line alternatives. FWA is therefore a more cost-effective and resourceful means to deploy broadband in regions with partial access to fixed broadband services (DSL or fiber). 5G enables FWA on scales significantly larger than those of which 4G is capable. 5G

[4]Interview publicly available at http://www.argylejournal.com.

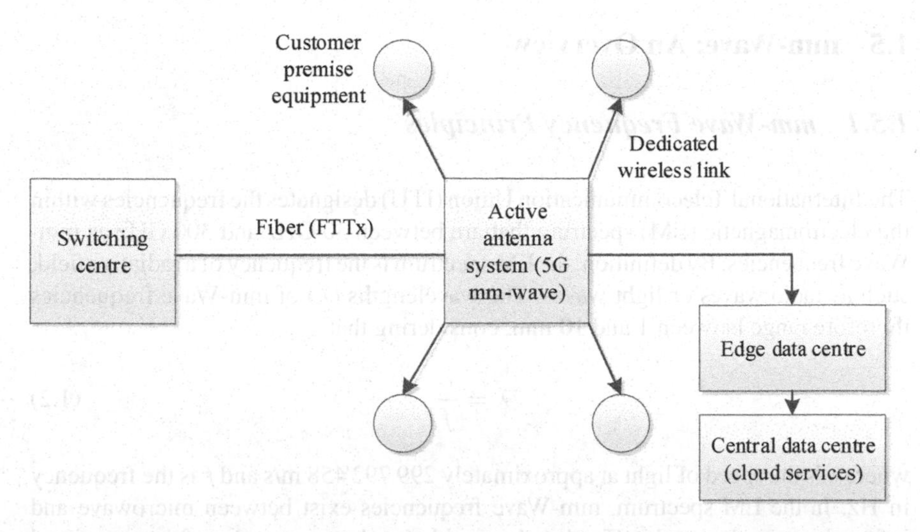

Fig. 1.3 The basic premise of an FWA system showing the dedicated active antenna system serving multiple fixed locations with broadband internet access and the edge data centre and cloud services are also served from a central location

FWA eliminates the necessity for expensive deep-fiber fixed access infrastructure [6] and still provides peak transfer rates that outperform most other wireless broadband technologies. [This requirement is estimated at US $130–150 billion [6] investment needed in the United States of America (USA); see Appendix 3 for a brief overview of deep-fiber architecture.]

Mobile broadband refers to a more traditional approach where mobile devices that have a modem on board access existing mobile networks to access the internet, irrespective of their location (assuming it is within the coverage area of the MNO's cellular tower). Coverage could include several generations of mobile technology within the global system for mobile communication family[5] (GPRS, EDGE, WCDMA, HSPA, LTE and 5G).

This book will refer to mm-Wave integrated technologies, 5G communications and Industry 4.0 throughout. Therefore, the following paragraphs briefly review mm-Wave and 5G, providing the reader with a convenient reference point as these topics are discussed in further detail in later chapters. Industry 4.0 as a concept is not explicitly reviewed in this book, but will rather it be referred to throughout in terms of the technologies discussed in relevant sections and chapters.

[5]These acronyms are not defined in this section; it is assumed that the reader is broadly familiar with these terms.

1.5 mm-Wave: An Overview

1.5.1 mm-Wave Frequency Principles

The International Telecommunication Union (ITU) designates the frequencies within the electromagnetic (EM) spectrum that are between 30 GHz and 300 GHz as mm-Wave frequencies. By definition, the EM spectrum is the frequency of a radiative field, such as radio waves or light waves. The wavelengths (λ) of mm-Wave frequencies therefore range between 1 and 10 mm, considering that

$$\lambda = \frac{c}{f} \tag{1.2}$$

where c is the speed of light at approximately 299 792 458 m/s and f is the frequency in Hz. In the EM spectrum, mm-Wave frequencies exist between microwave and infrared waves, depicted in Fig. 1.4 along with the photon energies of the associated wavelengths [29].

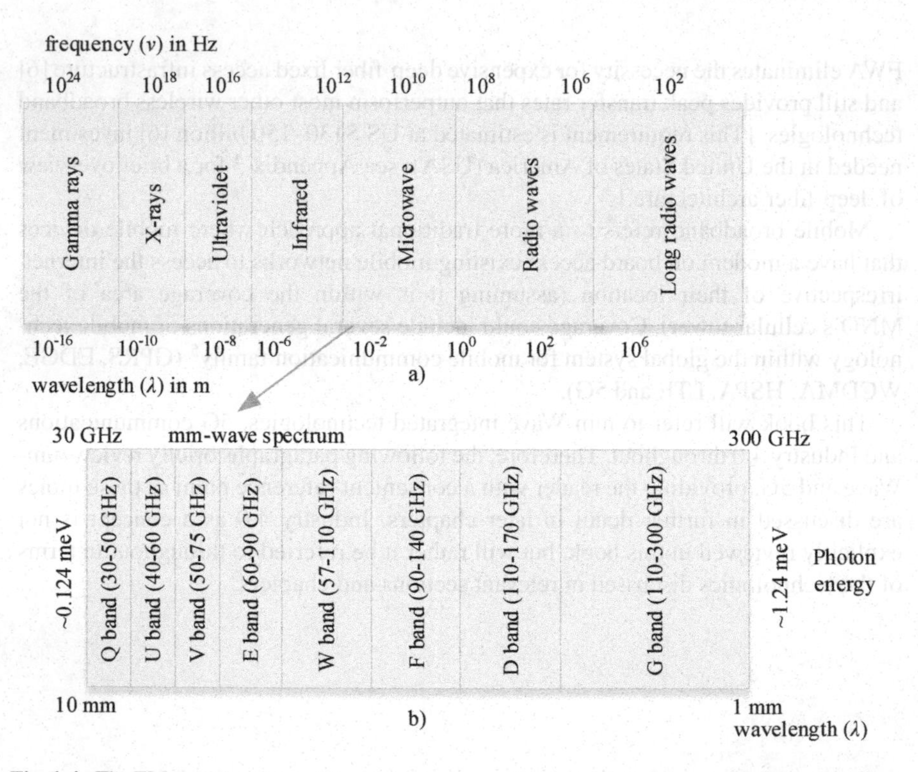

Fig. 1.4 The EM spectrum (**a**) and the wavelengths and frequencies that define mm-Wave (**b**). The approximate photon energies are also shown in **b** with respect to the wavelengths of the mm-Wave band

The propagation characteristics of mm-Wave radio signals are directly linked to the effective range over which a transmitter and a receiver (transceiver) can communicate (a rule of thumb is that the higher the frequency, the shorter the transmission range, assuming a constant power source). Several factors at higher frequencies are not applicable at lower frequencies (or at least these factors can be assumed negligible at lower frequencies) and can limit the transmission range of an mm-Wave signal. As some 5G frequencies are within the mm-Wave spectrum, or close to it when considering the 26 GHz and 28 GHz allocations, which are just outside the 30 GHz and above designation, many of the considerations with respect to propagation and the electronic/optical circuitry required to generate the signals of mm-Waves are also valid for 5G communication. The following section reviews some important and distinctive (compared to lower frequencies) characteristics of mm-Wave propagation. Detailed analysis of signals propagating with carrier frequencies in the mm-Wave band is presented in Chap. 2 of this book.

1.5.2 mm-Wave Propagation

Atmospheric propagation of mm-Wave frequencies experiences inherent attenuation from constituents in the propagation medium/channel (typically air). Figure 1.5 depicts these atmospheric losses (in dB/km) as a function of frequency (in GHz)

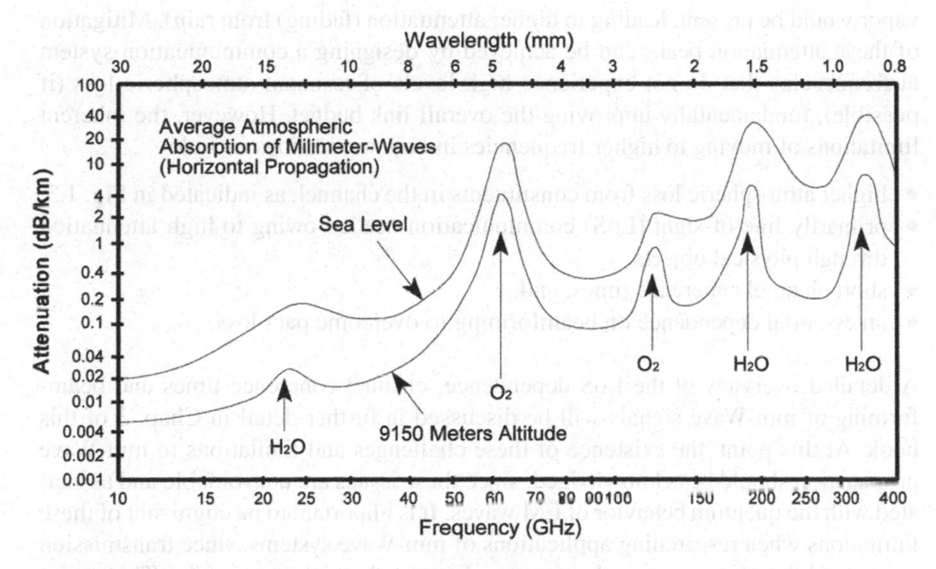

Fig. 1.5 The atmospheric absorption of electromagnetic radiation (radio waves) between 10 and 400 GHz (0.4 THz) as a function of attenuation in dB/km versus frequency in GHz and wavelength in mm. This figure was obtained from http://www.everythingrf.com

across the frequency bands, starting at approximately 10 GHz up to approximately 400 GHz (0.4 THz).

As shown in Fig. 1.5, the atmospheric absorption of mm-Waves experiences several local minima and maxima over the spectrum of 10 GHz and 400 GHz (equivalent wavelengths of 30–0.8 mm). These atmospheric losses are primarily due to constituents in the air such as water vapor (H_2O), oxygen (O_2) as well as ozone (O_3), apart from their inherent path loss as defined by the Friis free-space equation [11] as given in the following equation:

$$P_r(d) = \frac{P_{RAD}\lambda^2}{(4\pi)^2 d^2} \tag{1.3}$$

where "*$P_r(d)$ is the power received by the receiving antenna (in watt) at distance d (in m), P_{RAD} is the equivalent isotropic radiated power from the transmitter and λ is the wavelength of the transmitted signal (in m)*" [11]. The Friis free-space equation will be referenced in this book to highlight path losses at relatively high frequencies, for example mm-Wave) and the effects that the physical distance that separates the transmitter and receiver has on the integrity of the signal. The high losses that a mm-Wave signal experiences through free space are also indicative of the number of repeaters (small cells) required in a 5G architecture, further reviewed in Chap. 3 of this book. Referring back to the constituents in the air, as a function of frequency, resonance occurs with these constituents and causes an increase in attenuation, increasing if the constituent(s) are present in abundance (for example during rain, excess water vapor would be present, leading to higher attenuation (fading) from rain). Mitigation of these attenuation peaks can be achieved by designing a communication system at frequencies that do not experience high levels of resonant atmospheric loss (if possible), fundamentally improving the overall link budget. However, the inherent limitations of moving to higher frequencies in communications include

- higher atmospheric loss from constituents in the channel, as indicated in Fig. 1.3,
- primarily line-of-sight (LoS) communication needed owing to high attenuation through physical objects,
- short channel coherence times, and
- an essential dependence on beamforming to overcome path loss.

A detailed overview of the LoS dependence, channel coherence times and beamforming of mm-Wave signals will be discussed in further detail in Chap. 3 of this book. At this point, the existence of these challenges and limitations to mm-Wave propagation should be acknowledged, since these issues are unavoidable and associated with the quantum behavior of EM waves. It is important to be cognizant of these limitations when researching applications of mm-Wave systems, since transmission range (and therefore repeater density) is reliant on these characteristics. The propagation of mm-Waves is not the only challenge in employing a technology where it is utilized, such as in 5G wireless communications. The integrated electronic and/or optical circuitry required to generate these signals is complex, challenging and incurs additional limitations on the entire system. The following section identifies some of these challenges and an in-depth review is presented in Chap. 4 of this book.

1.5.3 mm-Wave Integrated Technologies

Signal generation for mm-Wave transceivers requires multiple building blocks, as in the case of any general communication system. Integrated circuits (ICs) are frequently used to create the building blocks of mm-Wave transceivers and these circuits are most commonly realized in complementary metal-oxide semiconductors (CMOS). At mm-Wave frequencies, however, the maximum operating speed of the electronic circuits is often limited by the achievable quality factor (Q-factor) of passive components such as inductors, capacitors, and transmission lines within the specific technology, as well as the switching speeds of active devices such as transistors [44]. Design challenges of mm-Wave ICs include providing

- high-power and high-linearity signal outputs from power amplifiers (PAs) and low-noise amplifiers (LNAs),
- wide tuning range of integrated voltage-controlled oscillators (typically limited by low obtainable Q-factors of passive components), as well as
- dealing with high levels of parasitic interference.

Device modeling is a crucial step when designing mm-Wave building blocks. At higher frequencies, active and passive components in a monolithic microwave integrated circuit (MMIC) behave differently compared to the same components at lower frequencies. Active devices such as transistors experience parasitic effects from internally generated capacitances, and passive components such as inductors and capacitors behave differently owing to interference from adjacent components, substrate skin effects, parasitic inductance and capacitance generated by (copper) tracks, as well as various other effects from interconnects and external components. MMIC design and manufacturing (requiring several iterations of prototyping) are expensive and it is essential to take into account the parasitic effects of high-frequency operation that require rigorous efforts in device modeling.

As CMOS technology scales and gets physically smaller, the maximum operating frequency of active components is improving and enabling mm-Wave (and even THz circuits in certain configurations) systems, although the supply voltages of these smaller nodes are also decreasing. Therefore, it becomes more difficult to generate high-power transmitters that also operate linearly, another aspect where device modeling is crucial when designing mm-Wave circuits. In Chap. 4 of this book, mm-Wave lumped device modeling of components, both active and passive, and subsystems of the generic front end of a transceiver, as shown in Figs. 1.6 and 1.7, are reconsidered in order to provide techniques to achieve successful mm-Wave transceivers for 5G communication.

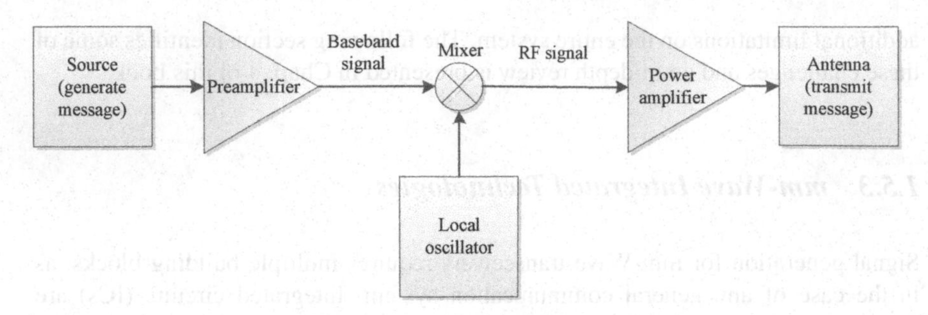

Fig. 1.6 The generic front end of a transmitter. Each of these subsystems should be modeled using lumped equivalent circuit models of active and passive components for mm-Wave operation to predict (simulate) real-world operation

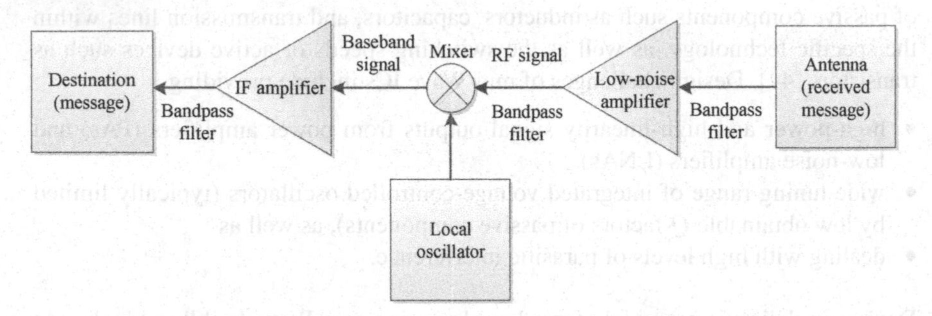

Fig. 1.7 The generic front end of a receiver. Each of these subsystems should be modeled using lumped equivalent circuit models of active and passive components for mm-Wave operation to predict (simulate) real-world operation

As shown in Figs. 1.6 and 1.7, the subsystems of a generic transceiver front end include an LNA, several oscillators, mixers, a PA, intermediate-frequency (IF) amplifiers, frequency division and in certain scenarios, multiple antenna systems. Modeling each subsystem requires each active and passive component to be individually modeled by a lumped equivalent circuit, presented in Chap. 4 of this book. Chap. 4 also aims not to repeat already existing modeling approaches that are available in literature, but to apply these principles specifically to 5G-related systems and the IoT with its unique limitations in terms of power availability and transmission distance. These equivalent circuits are important, since "*typical parasitic extraction tools that are used in post-layout simulations can prove to be inadequate*" [32]. These tools do not always account for all real-world frequency requirements or dispersed effects at mm-Wave [32]. Real-world simulations can also be realized (and ideally should) through full-wave EM analysis [25]. However, this typically requires additional development time and specialized knowledge and skills that increase overall development cost significantly. Accurate and effective device modeling for mm-Wave components and circuits could lead to more effective 5G systems with higher reliability, a longer range and lower development cost.

1.5.4 5G Technical Specifications and Architecture

The complete 5G system architecture technical specification, the 3GPP TS 23.501 version 15.2.0 release 15 (hereafter referred to as *3GPP release 15*) will be referred to in this book. Specific reference will also be made to the functional layers ("*service, management and orchestration, control, multi-domain network operating system facilities and data layers*") of 5G. A brief overview of the functional layers of 5G is listed below, adapted from 5GPPP [12]. A full description of the 5G architecture is publicly available in the 3GPP *release 15*, with certain sections highlighted throughout this book.

1.5.5 Service Layer

The service layer in 5G contains systems related to business support, policy at business-level and the "*decision function*", including applications and services executed by the client. End-to-end orchestration is also included in the service layer.

1.5.6 Management and Orchestration Layer

From *3GPP release 15*, the management and orchestration layer consists of an inter-slice dealer that takes responsibility for cross-slice resource distribution and allocation and collaborates with the "*service management function*". Service management is therefore an intermediary amid the service layer and the inter-slice dealer. Service management converts end-user-facing service description to resource-facing service description and the other way round, as outlined in *3GPP release 15*.

1.5.7 Control Layer

From *3GPP release 15*, the two primary controls, the "*software-defined network coordinator*" and the "*software-defined mobile network controller*", are accommodated in the control layer. These controllers handle dedicated as well as the common network functions, correspondingly, and translate decisions of the control requests into instructions to "*virtual network functions*" (VNFs) and "*physical network functions*" (PNFs).

1.5.8 Multi-domain Network Operating System Facilities

The multi-domain network operating system facilities include multiple adaptors and network abstractions. This layer is responsible for allocation of virtual network resources and maintaining the reliability of the network in multi-domain environments.

1.5.9 Data Layer

The data layer contains the VNFs and PNFs required to transport and process the user data traffic, according to *3GPP release 15*.

Apart from the architecture and functional layers of 5G, of technical interest in this book are the characteristics of the 5G new radio, network slicing, spectrum considerations, and the data and network aggregation that 5G offers. These characteristics are identified as crucial for the modular implementation of technologies based on mm-Wave communications and systems in Industry 4.0, such as mm-Wave massive IoT systems (reviewed in this chapter and critically reviewed in Chap. 3 of this book). The following section refers to the 5G new radio and the benefits it offers in deploying 5G architectures. It is essential to consider that initial 5G roll-out (starting in 2018 and 2019) is initially using (in 2020) lower-band frequencies (sub-6 GHz); the 5G technology (architecture and technical specifications) allows one to increase the carrier frequencies to mm-Wave with low impact on the current deployments.

1.6 5G New Radio

The 5G of mobile networking uses a new radio access technology (RAT), developed by 3GPP *release 15*, termed 5G new radio (NR). In its first phase, published in 3GPP *release 15*, 5G NR has certain distinct characteristics and enhancements in terms of flexibility, scalability and efficiency in its use of power and spectrum. The fundamental characteristics of the new RAT include

- new broadband spectrum allocations as opposed to solely relying on spectrum efficiency of current frequency allocations,
- optimized orthogonal frequency-division multiplexing (OFDM) modulation techniques, specifically cyclic-prefix OFDM adopted for the downlink signal (similar to 4G long-term evolution (LTE)) that effectively optimizes multiple user access,
- beamforming to enable the base station signal beam to be directed towards the receiving system, additionally allowing (small) programmable high-directivity beam-steering antennas,
- multi-user MIMO data streams at higher carrier frequencies that utilize the distributed and uncorrelated spatial location of multiple users,

- spectrum sharing (network slicing) to improve efficiency,
- a unified design across the spectrum with a common flexible framework, and
- the use of small cells for network densification.

Unlike previous generations of mobile networks, 5G NR does not require that "*both the access and core network of the same generation be deployed*" (3GPP *release 15*), for example in 4G, an evolved packet core (EPC) and LTE together formed a functional 4G system. Although EPC could be deliberated an evolution of former generation core systems, 5G with its cloud native and virtualization features makes it conceivable to incorporate features of diverse generations in various arrangements, namely

- standalone (SA)—using a single RAT, and
- non-standalone (NSA)—combining multiple RATs [13].

In the SA configuration, the core network is operated independently from the 5G NR or the evolved LTE cells and these cells can be used for both the control plane as well as the user plane. MNOs are able to deploy this configuration and offer seamless handover between 4G and 5G technologies, dependent on the immediate environment and receiver devices used. There are also variations of SA that are being finalized by 3GPP *release 15* (which should be final in the 3GPP *release 16*); these include

- operation as per current 4G LTE networks that use EPC and LTE evolved node B (eNB) access (Option 1),
- using the 5G core network (5GC) together with NR 5G node B (gNB) access (Option 2), or
- using 5GC and LTE ng-eNB access (Option 5—options 3 and 4 are within the NSA configurations).

The NSA configuration differs in that the 5G NR and the LTE radio cells are shared to deliver access to the network. Additionally, the core network could be either EPC or 5GC, depending on the decision made by the MNO. The MNO therefore has the option to leverage an existing 4G deployment and provide additional functionality through a 5G NR architecture and the user experience of such a configuration depends on the RAT(s) used. Similar to SA, NSA has three distinct variations, namely

- usage of EPC as well as LTE eNB that acts as a master connection, with 5G NR en-gNB acting as the secondary network (Option 3),
- using 5GC as well as NR gNB as the master network with LTE ng-eNB operating as a secondary network (Option 4), or
- 5GC and LTE ng-eNB acting as the master network with 5G NR gNB acting as a secondary network.

The strategies/variations of network deployments that 5G can adopt (SA and NSA configurations) are summarized in Fig. 1.8. Figure 1.9 is a representation of the combinations of migration strategies as outlined in 3GPP *release 15* of SA and NSA configurations.

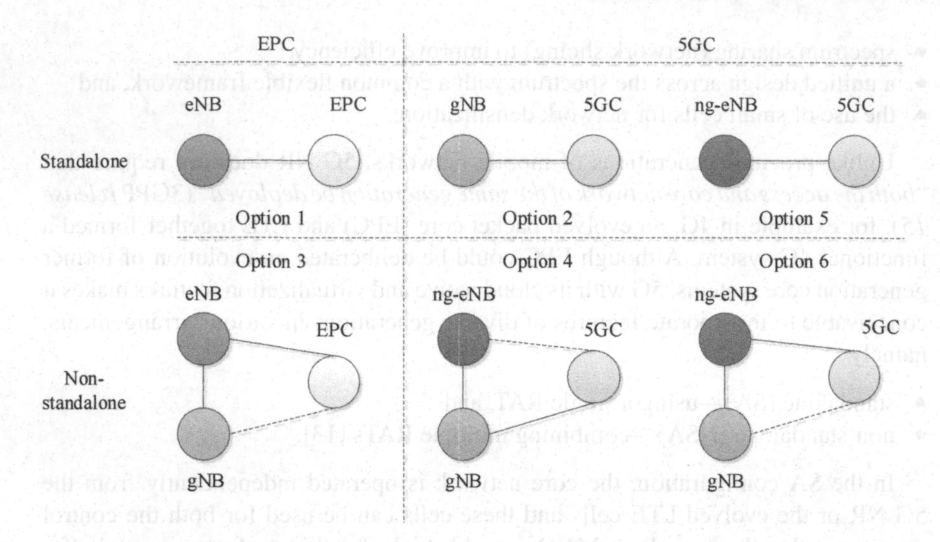

Fig. 1.8 Mobile network operators' 5G network deployment variations of SA and NSA configurations, Options 1 through 6

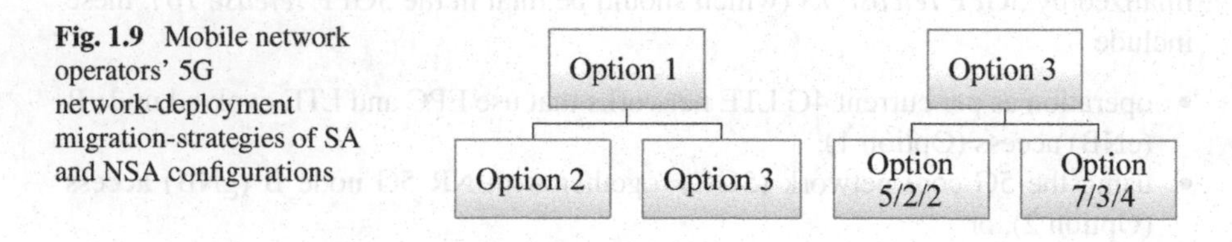

Fig. 1.9 Mobile network operators' 5G network-deployment migration-strategies of SA and NSA configurations

In Figs. 1.8 and 1.9, the alternative network deployment variations as well as migration strategies are summarized (as described earlier in this section). A technical review on the connectivity throughput and scalability of these options is presented in Chap. 4 of this book. Furthermore, in terms of characteristics that separate 5G from earlier generations, 5G has at its core design superior network slicing capabilities, a feature that allows scalability, as described in the following section.

1.7 5G Network Slicing

Network slicing effectively refers to the operation from the viewpoint of the MNO, of creating an "*independent end-to-end logical network that runs on a shared, but physical, network infrastructure*" [14]. These slices can negotiate quality of service (QoS). A network slice can potentially be deployed across multiple MNOs and can span several fragments of the physical network, including the

- terminals,
- access-,
- core-, and
- transport network.

A network slice consists of either dedicated resources, or ones that are shared among MNOs and/or the physical network. These dedicated or shared resources include power allocation and consumption, storage capabilities as well as spectrum (bandwidth) and are, importantly, isolated from each other. The MNOs should define the network slices as either a

- network function, or
- functional behavior.

The scalability of these characteristics allows MNOs to deploy network slices to multiple clients either with similar requirements (behavior), or as more than one network slice type for specific applications (functions). For example, an MNO might decide to assign a high-bandwidth slice to a customer requiring a high-definition video feed, while at the same time providing the (same) customer with a reliable slice to encrypt sensitive metadata of the feed. This therefore opens up additional bandwidth for a second customer, as the first customer is not assigned *two* equal and high-bandwidth slices. Essentially, network slices could be categorized as

- IoT slices requiring reliable and always-on connectivity,
- broadband (>10 Gbps) slices for audio and visual feeds, or
- low-latency (<10 ms) slices for instant communication, for instance in mission-critical military applications.

In GSMA [14] a use case of network slicing customization is provided, where network capability parameters of the network such as

- "*latency,*
- *data security,*
- *energy efficiency,*
- *mobility,*
- *massive connectivity,*
- *reachability,*
- *guaranteed QoS, and*
- *throughput*" [14]

are customized to serve the needs of a user or an application. Furthermore, not only are network capabilities prioritized, but also network services, which include

- "*big data*
- *partner integration,*
- *localization,*
- *edge computing,*
- *cloud storage,*

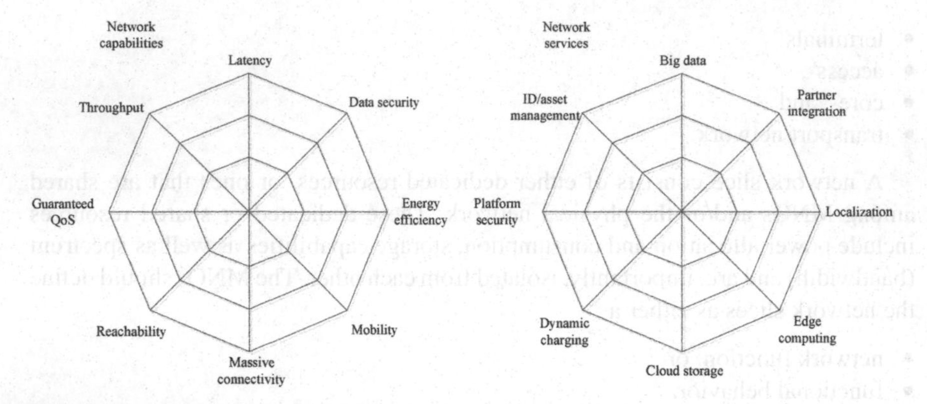

Fig. 1.10 A diagram that can be used to identify where certain network capabilities and services are prioritized based on the user or application requirements, effectively allowing the client to reduce overall cost by only selecting capabilities and services required per use case

- *dynamic charging,*
- *platform security, and*
- *identification management*" [14].

These precise levels of slicing can therefore uniquely serve users and applications and allow scalability through their ability to make resources not required by a particular service available. Adapted from GSMA [14], Fig. 1.10 shows a diagram that can be used to identify requirements from a 5G network in terms of network capabilities and services.

Diagrams such as the ones presented in Fig. 1.10 provide users with a quick and easy reference to the services that could be requested from the MNO, and can give an indication of the costs involved in requesting the unique capabilities and services.

Additional features of which 5G can take advantage are for example data aggregation and traffic aggregation, primarily owing to the modularity of the network offered by network slicing, as reviewed in the following section.

1.8 Aggregation in 5G

1.8.1 Data Aggregation

Fundamentally, data aggregation involves collecting information (data) and expressing it in a meaningful and interpretable, summarized form for statistical analysis. Modern sensing technology enables various institutions, companies and governments to collect data, although the morality of data collection is a separate topic, but owing to the volumes of collected data, in many cases it is not effectively used. A dataset can be aggregated and grouped to enable a human interpreter to determine a trend based on specific attributes of the data. In Chap. 3, an example of data aggregation is presented to display some of the uses and queries that can be performed on aggregated data. There are two kinds of data aggregation, "*time*" and "*spatial*" aggregation [2]. In time aggregation, all of the data points for a *single* resource are gathered over a quantified time (the granularity), whereas in spatial aggregation, all data points for a *group* of resources are gathered over a quantified time. Furthermore, the time recesses for collecting and aggregating data are reported in the context of a reporting period, granularity and a polling period. The reporting period is the total time over which all the data are collected. The granularity is the interval over which "*data points for a given resource (or set of resources) are collected for aggregation*" [2]. The polling period governs how frequently resources are sampled for available data. In certain scenarios, the granularity and polling period are equal. Figure 1.11 summarizes the concepts of reporting period, granularity and polling period for arbitrary data points collected from a group of three resources (Resources I, II and III). In this example, the collection system has a granularity of 10 min with a polling period of five minutes, and the reporting period is 40 min.

As shown in Fig. 1.11, the data points collected during each granularity period are aggregated into a single statistical value. This value could, for example, be the median value of the data gathered by each resource, although a vast amount of statistical analysis can be performed on the data and grouped as aggregated points of data. These points gathered in the five-minute polling period at the end of the 40-min reporting time are not included in the final aggregation dataset owing to the 10-min granularity margin.

Data aggregation, if done efficiently and effectively, eliminates collection of redundant data [42]. As a result, less power is consumed by sensors, actuators and transceivers, leading to a reduction in energy expenditure, which in the modern world

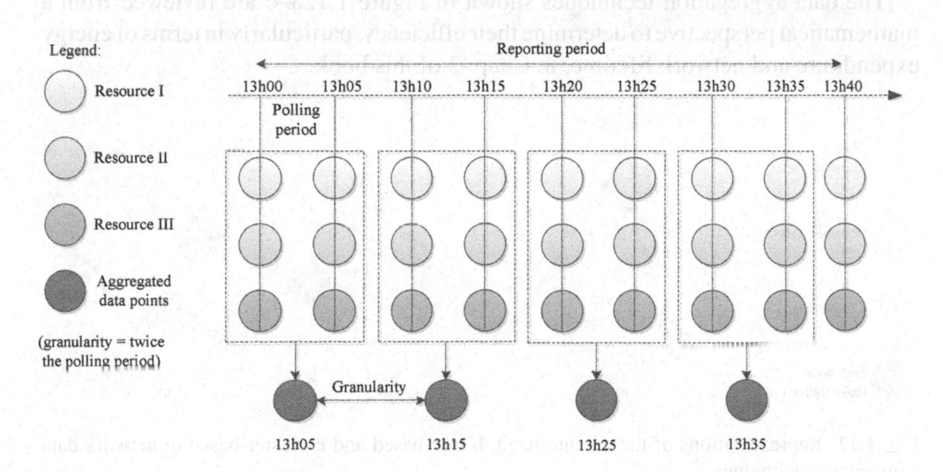

Fig. 1.11 An arbitrary data collection and aggregation system with a reporting period of 40 min, granularity of 10 min and a polling period of five minutes

has become a major challenge for researchers. Atoui et al. [2] also acknowledge the effectiveness of data aggregation to reduce energy consumption of ubiquitous sensor nodes through *filtering* similar data and avoiding data redundancy.

The IoT and WSNs are in essence a collection of sensor nodes and transmitting and receiving devices that collectively transmit information (collected data) to a base station. Four primary techniques/algorithms of data aggregation are typically implemented and can be used in practical sensor node architectures [26]. These techniques are referred to as

- *"tree-based,*
- *cluster-based,*
- *centralized, and*
- *in-network aggregation techniques*" [26].

The descriptions exclude hybrid approaches of the listed techniques.

The tree-based data aggregation [2] technique transfers *raw* data between interconnected intermediate member (*leaf*) nodes towards a central *sink/root* node and data aggregation is performed by this node (also referred to as the base station). A cluster-based data aggregation technique transmits raw data from a cluster of sensors to a nearby aggregator or *cluster head*. This cluster head at that point transfers a short digest of data to the *sink/root* node. The centralized data aggregation technique is simply used where all sensor nodes transmit raw data directly via the shortest path to the base station that is tasked with aggregating all incoming data. Finally, in-network data aggregation, the most complex and arguably the most efficient form of data aggregation, is a comprehensive procedure of collecting and directing data using a network in a multi-hop manner. This technique controls data in transitional nodes to reduce resource usage and increase the lifetime of the network. These techniques of data aggregation are summarized in Fig. 1.12a–c.

The data aggregation techniques shown in Figure 1.12a–c are reviewed from a mathematical perspective to determine their efficiency, particularly in terms of energy expenditure and network lifetime, in Chap. 2 of this book.

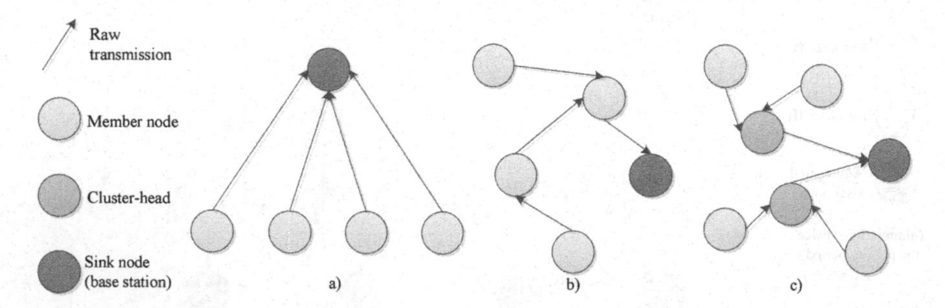

Fig. 1.12 Representations of the **a** centralized, **b** tree-based and **c** cluster-based in-network data aggregation techniques

1.8.2 Network Traffic Aggregation

Network traffic is an extremely heterogeneous system and exhibits variations on numerous time scales [31]. Various traffic types are long-range dependent and could cause congestion when resources are allocated statically. Improving network performance can be achieved through dynamic resource allocation based on the instantaneous traffic. Furthermore, if autocorrelation is employed, it is possible to estimate traffic load that serves as a control variable for dynamic traffic aggregation. Entropy [51] on the traffic traces can be used as aggregation strategies as single measures of randomness. Larsson [31], for example, uses an efficiency measure to aggregate network traffic to determine the statistical performance in throughput, packet loss and delay. This and other similar approaches are technically reviewed in Chap. 3 of this book.

Network slicing in 5G will have numerous other uses, especially considering emerging markets that require access to only specific 5G capabilities and services. Apart from these advantages of 5G and the foreseeable use cases, spectrum allocation and licensing for 5G are progressing slowly worldwide. There are various reasons for the slow response from allocation authorities (not denying that weak business cases from MNOs play a role, which is also discussed in this chapter), but there is agreement on the developing spectrum considerations and allocations for 5G, reviewed in the following section.

1.9 5G Spectrum Considerations

5G will "*support significantly faster mobile broadband speeds and enable the full potential of the IoT*" (GSMA 2019). Importantly, 5G is not a single-frequency standard and mobile devices will have multiple radios that support 5G NR frequency range 1 (FR1) (450 MHz to 6 GHz) and frequency range 2 (FR2) (24.25 GHz to 52.60 GHz) bands. For 5G to reach its full potential, MNOs should invest in more widely harmonized spectrum across three primary frequency ranges: (i) sub-1 GHz, (ii) 1–6 GHz and (iii) above 6 GHz. The 3GPP 5G standard, *release 15*, submitted as a applicant for International Mobile Communications-2020 (IMT-2020), comprises several technologies, including 5G NR that supports wider 5G frequency bands. Channel proportions extending from 5 MHz up to 100 MHz for bands beneath 6 GHz are supported, as well as channel proportions between 50 and 400 MHz in the frequency allocations above 24 GHz (which include true mm-Wave frequencies above 30 GHz and up to 71 GHz). Further development of the 3GPP standard towards *release 16* (which should be finalized by March 2020) is proceeding rapidly to support timelines defined by leading markets, such as the USA, several European countries and China.

Of particular interest, apart from the prime 5G mid-bands such as the 3.5 GHz band, are near-mm-Wave and true mm-Wave frequency allocations for 5G as

Table 1.1 3GPP *release 15* NR 5G frequency band allocations of near-mm-Wave and true mm-Wave frequencies

Designation	Frequency band (GHz)	Comment(s)
n77	3.30–4.20	Sub-6 GHz
n78	3.30–3.80	Sub-6 GHz
n79	4.40–5.00	Sub-6 GHz
n257	26.50–29.50	Commonly called 28 GHz
n258	24.25–27.50	Commonly called 26 GHz
n260	37.00–40.00	Commonly called 40 GHz, true mm-Wave
n261	27.50–28.35	Added band in *release 15*

proposed by the 3GPP *release 15*. Table 1.1 summarizes new bands for the introduction of NR specified by 3GPP, with developments and changes expected in *release 16*. A breakdown of the 3GPP *release 15* frequency bands for 5G NR is given in Appendix 1.

Table 1.1 shows the frequencies that have materialized as the most probable candidates that would facilitate the ultra-high bandwidth capabilities visualization for 5G, designated as n257, n258, n260 and n261. The mid-band spectrum (sub-6 GHz) will also play a vital part in building 5G to be a conventional reality, since the 3.5 GHz spectrum has gained traction globally with, importantly, licensing already being achieved in many countries and with more being registered (GSMA 2019).

In March 2019, the USA Federal Communications Commission (FCC) started the auctioning[6] process of blocks of the 24 GHz spectrum (between 24.25–24.45 and 24.75–25.25 GHz, with each block allocated a bandwidth of 100 MHz) dedicated to future 5G use [9]. However, in May 2019, the US Navy released a memorandum warning against potential interference of the 24 GHz 5G bands with current weather satellites, urging the FCC to halt spectrum allocation specifically between 23.6 GHz and 24 GHz, which could experience bleed-over from adjacent auctioned 5G bands. This band is used by weather satellites to monitor water vapor and could limit the National Aeronautics and Space Administration and the National Oceanic and Atmospheric Administration in their critical satellite-based measurements needed to predict the weather.

Another important frequency band in the unlicensed spectrum is the 66–71 GHz band within the designated V-band and E-band (V-band frequencies are allocated between 50 and 75 GHz and E-band frequencies are between—albeit overlapping—60 and 90 GHz, as designated by the IEEE[7]). This band has some distinct advantages, apart from not being heavily used at present; its vast bandwidth capacity is another key advantage for early adopters. Device miniaturization related to smaller antennas is another key characteristic of this mm-Wave band. The significantly higher data speeds offered in the V-band have use cases in support of expansion of fixed broadband

[6]The auction was posted on 14 March 2019 on http://www.fcc.gov/auction/102/factsheet.
[7]Institute of Electrical and Electronic Engineers.

internet in urban areas as well as backhaul [23] for mobile broadband in last mile connections [3]. Distances of communication in these mm-Wave frequency bands are typically limited to the lower-tier sub-10 km range, because of various factors, including the distance-frequency relationship and attenuation peaks from resonant constituents in the channel.

Apart from the technical characteristics of 5G that classify it beneficial to implement, uses that have significant uses and take advantage of the technologies such as 5G and mm-Wave broadband are also reviewed in this book. The next section reasons mm-Wave massive IoT systems as a use case of 5G and mm-Wave broadband.

1.9.1 mm-Wave Massive IoT Systems

The ITU Radiocommunications Sector has classified mobile network connectivity supported by 5G capabilities into three primary categories, as part of the *Role of IMT*[8] *for 2020 and Beyond* campaign (IMT-2020). These categories are

- enhanced mobile broadband (eMBB) that aims to deliver greater data bandwidth of up to 10 Gbps peak throughput, "*1 Gbps throughput in high-mobility and up to 10 times total network traffic*" at moderate-to-low latency, particularly aimed at VR, AR and high-definition video streaming applications,
- massive machine-type communications (mMTC) that are the foundation for connectivity in IoT and allow low-power wide area (LPWA) tools like infrastructure managing, environmental observing and healthcare uses, and
- ultra-reliable and low-latency communications (URLLC) aimed at delivering dependable and low-latency data in mission-critical applications such as military communications, natural disaster monitoring or industrial and automotive applications [49].

As IMT classifications assist as a communication instrument for societies as well as a implementer that supports in the growth of commerce segments, the vision of IMT-2020 is to contribute to

- placing a similar level of priority on global broadband availability as access to electricity and having a wireless infrastructure that connects the world,
- promoting an integrated information and communications technology industry aimed at constituting a driver for economies globally,
- promoting affordable and sustainable mobile and wireless technologies in emerging markets to bridge the digital divide,
- encouraging cross-platform and essentially unlimited content sharing,
- developing new forms of education that are digitally driven,
- promoting efficient use of energy sources to support ubiquitous and always-on monitoring, communicating and other smart applications through the IoT,

[8] The International Mobile Telecommunications (IMT) are requirements issued by the ITU-R.

- connecting people both socially and politically through the ability to exchange information anytime and anywhere, and
- supporting new arts and a culture, for example through virtual attendance and online collaboration, that encourage cross-culture partnerships,

primarily through use cases that are expected to emerge from its modular approach to eMBB, mMTC and URLLC. IoT systems, supported by mm-Wave-enabled 5G connectivity infrastructure, back a diverse assortment of future uses, and effective and efficient use cases are being developed to support the shift in infrastructure from earlier generations. 3GPP predicts an assorted group of usage situations as well as specialized applications that are driven by several factors, including

- massive numbers of connected devices,
- low-cost devices capable of accessing advanced networks (albeit not yet a reality in 2019, especially in emerging markets),
- low-energy consumption of each connected device,
- high-speed connectivity, upwards of 10 Gbps paired with very low latency (below 10 ms),
- high reliability, and
- high levels of availability [46].

As in most emerging technologies, particularly in mm-Wave communications driving massive IoT networks, several challenges arise that must be overcome before widespread adoption becomes feasible. The challenges lie not only in the technology itself, 5G at millimeter-wave frequencies, but also in the 5G-capable devices, MNO adoption and distribution of 5G, backward compatibility with legacy devices, protocols and software applications and various additional factors such as spectrum allocation and licensing. As a result, immediate 5G adoption is unlikely, but a phased approach and infrastructure development of 5G IoT systems that complement Industry 4.0 are feasible in the near future.

Larger corporations, such as Sony Semiconductor Solutions Corporation, have recently launched an integrated IoT system-on-chip that allows transmission of data on their proprietary ELTRES LPWA networks, enabling transmission up to approximately 60 miles (almost 97 km) and work in noisy (spectrally) urban environments. The wireless transmission occurs at sub-1 GHz frequencies (920 MHz) and can obtain location and time data [8]. It enables many new innovations and applications in security, monitoring and tracking, although it is not designed for mm-Wave transmission, but achieves a long range by utilizing a lower transmit frequency. This book will highlight several of these challenges from a technical and economic perspective in Chap. 5, and will provide a holistic review on the feasibility of 5G mm-Wave-enabled IoT networks in various markets (developed and emerging).

As mentioned in this chapter, the business case of 5G from MNOs to attract investors and build infrastructure has been criticized and is an ongoing challenge, especially in emerging markets where financial backing from investors is crucial. The following section reviews the importance of a strong business case for 5G and MNOs' perception of this space.

1.10 5G Business Cases and Investor Appetite

The global economy is at a pivotal point with Industry 4.0 approaching; artificial intelligence, IoT, VR and AR are becoming more than just buzzwords, but offer concrete long-term economic potential. 5G has been identified as the catalyst for the growth in economies, as it is not just an extension of 4G LTE but rather an enabler of *billions* of devices collecting and sharing data. Countries are spending large amounts of money on wireless communications infrastructure to build new sites to enable seamless transition towards 5G in the near future to meet the technical needs of vertical industries such as healthcare, energy and automotive services. According to Deloitte [7], the USA, Japan and South Korea have all prepared noteworthy steps towards 5G readiness, but none of these countries' commitment to 5G is comparable to that of China. Network densification is one of the primary characteristics of 5G deployments, requiring the accumulation of base station towers and small cells in the network. Deloitte [7] estimates that MNOs need to add three to 10 times the number of existing sites to their networks for ubiquitous 5G connectivity. Fortunately, a large portion of these additional sites are created on existing infrastructure such as lampposts, utility phones or other structures that are capable of hosting these (less) obtrusive sites. China, the current leader in 5G infrastructure development, already has an estimated 1.9 million wireless locations, compared to the 200 000 of the USA [7]. Deloitte [7] compared the USA, China, Germany and Japan in terms of the number of 5G wireless sites per 10,000 people and the number of sites per 10 square miles. These results are adapted and presented in Table 1.2.

As shown in Table 1.2, Japan has most 5G wireless sites per 10 000 people (17.4) and has most 5G wireless sites per 10 square miles (15.2). Importantly, Japan's population in 2017 was 127 million spread over a geographical area of 377 972 km^2, whereas China's population is over 1.3 billion spread over a geographical area of 9.6 million km^2, indicating that the size of its investments to achieve its 5G readiness is large and determined. Germany has shown commitment to investments in 5G wireless technology and the USA has shown that it is committed to facilitating the technology.

Although 5G is the successor of 4G LTE and the technology presents major improvements in capacity and latency, MNOs need to define their business case(s) to stakeholders as to why this service should be invested in and distributed globally. In emerging markets where investments in telecommunications are typically lower

Table 1.2 The number of 5G wireless sites per 10,000 people and the number of sites per 10 square miles in the USA, China, Germany and Japan, adapted from Deloitte [7]

Country	Sites per 10 000 people	Sites per 10 square miles
Japan	17.4	15.2
China	14.1	5.3
Germany	8.7	5.1
USA	4.7	0.4

compared to those in developed countries, the business case must be strong enough to convince stakeholders such as industry investors and governments of its necessity. Simply referring to improvements in bandwidth (capacity) and efficiency are not always convincing arguments for regulators to open up the 5G spectrum, since earlier generations of mobile networks such as 3G and 2G are still underutilized in many markets. According to Graham [18], the focus on 5G applications has shifted from commercial applications such as ultra-high definition video streaming and VR towards manufacturing and Industry 4.0. Unlike video streaming and VR, systems such as industrial IoT do not require high data rates, but widespread coverage is crucial.

According to Gartner [15], most 5G deployments will firstly focus on islands of distribution as opposed to continuous, ubiquitous national coverage around the world. It would therefore be up to enterprises and businesses to adapt digital business initiatives to take advantage of the available mobile services. It is recommended [15] that early adopters

- incorporate realistic networking assumptions and identify the availability of services of mobile 5G networks, and
- specifically address current use case requirements to bring down the cost through only paying for these required services.

Not all 5G services are necessarily required or beneficial to early adopters; some 4G LTE services may suffice in many instances and could be combined with edge computing and network slicing to tailor individual 5G solutions. Gartner [15] estimated that 90% 5G coverage (in a geographical area) would only be achieved by

- 2023 in North America and Canada, Australia, New Zealand, China and Japan,
- 2026 in the United Kingdom and
- 2026–2029 elsewhere in the world.

These estimates can change if the global interest in 5G increases drastically, but the ubiquitous availability of 5G will be, and is currently, hampered by the

- slow roll-out and allocation of spectrum,
- successes of 4G LTE,
- deployment of new physical radios for 5G NR, and
- a general lack of *killer applications* and business cases by enterprises and MNOs to justify investments in 5G.

Apart from an increase in bandwidth, low latency and support for IoT devices, Gartner [15] also published the findings of a survey on the expected uses for 5G networks, adapted for this book and presented in Fig. 1.13.

As shown in Fig. 1.13, IoT communication tops the list of expected 5G uses for 5G networks, with a 59% interest level from respondents. Following IoT communications are video (high-definition 4 K and 8 K streaming) with 53% interest, control and automation at 45%, fixed wireless access at 44%, and high-performance edge

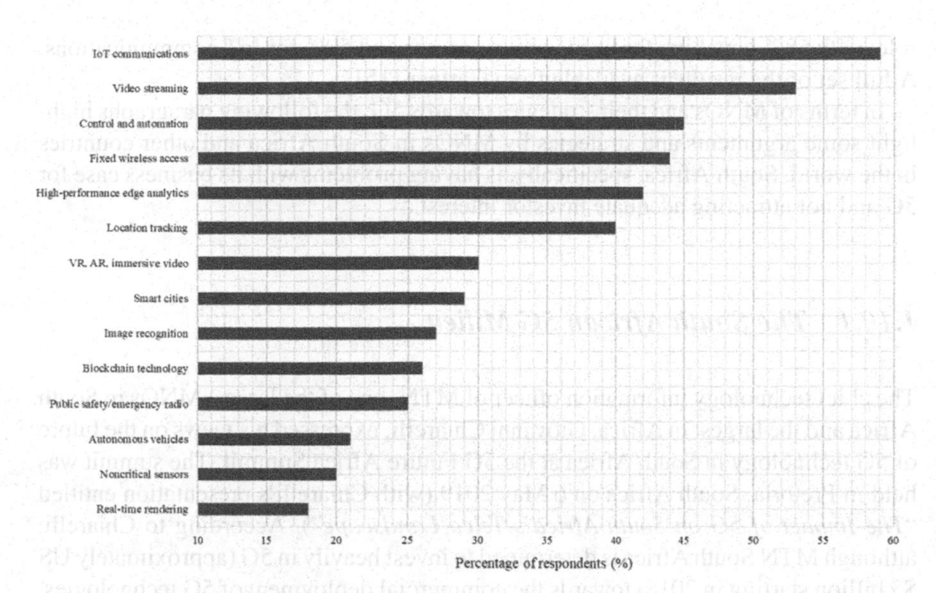

Fig. 1.13 The expected use-cases from respondents for 5G networks according to a 2018 Gartner® survey

analytics at 42%. At 40% interest is location tracking, before interest in 5G technologies drops to 30% and lower. Autonomous vehicles, which have become popular research and development projects for various large corporations, such as Google, Facebook and Uber, received 21% interest from respondents, primarily because of the smaller market when compared to IoT communications, and real-time rendering received the lowest level of interest at 18%. These results do not necessarily gauge the importance of each technology, but are also an indication of relative market size. It is therefore clear that IoT communications, video streaming, and automation in anticipation of Industry 4.0 are among the largest potential markets that would target 5G-capable networks.

In terms of revenue streams, Gartner [15] also provided the findings on the expected 5G use cases that communications service providers (CSPs) expected to drive revenue, enable critical processes, improve operational efficiency, or effect improvements in other CSP sectors. Interestingly, the largest percentage of respondents (38%) expect AR, VR, immersive video or holograms to have the largest revenue stream, followed by smart cities (33%) and high-performance edge analytics (27%). IoT communications received the second-lowest indication of increased revenue, with only 13% of respondents predicting increased revenue through IoT communications, second only to non-critical sensors at 8%. In terms of the Industry 4.0 revolution, 25% of the respondents expected a drive in revenue for control and automation, according to the Gartner [15] survey. In terms of improving operational efficiency, 64% of respondents estimated that non-critical sensors would play a large role, with 58% vouching for fixed wireless access and 50% for IoT communications. A full set of the results is available from Gartner [15].

In terms of MNOs and their strategies towards 5G, the following paragraphs highlight some arguments and strategies by MNOs in South Africa and other countries in the world. South Africa, specifically, is having problems with its business case for 5G and not attracting adequate investor interest.

1.10.1 The South African 5G Milieu

The chief technology information officer of MTN, one of the largest MNOs in South Africa and the largest in Africa, Giovanni Chiarelli, expressed his views on the future of 5G technology in South Africa at the 5G Future Africa Summit. The summit was held in Pretoria, South Africa on 6 May 2019 (with Chiarelli's presentation entitled "*The Impact of 5G on South Africa's Telco Landscape*"). According to Chiarelli, although MTN South Africa is determined to invest heavily in 5G (approximately US $3 billion starting in 2018) towards the commercial deployment of 5G technologies, two major hurdles must still be overcome, these being

- a lack of spectrum and incompetence and lackluster commitment from local regulators to allocate spectrum for 5G, and
- lower than expected investor backing for the 5G business cases presented by MTN South Africa.

Chiarelli also referred to the three fundamental pillars on which 5G is built (eMBB, mMTC and URLLC) and how each of these pillars can benefit usage scenarios where 5G is used as the communications standard. Use case examples included 5G FWA for video and music streaming, smart homes, VR and AR and cloud gaming, as well as private mobile networks for specific industries (such as the mining sector). Also highlighted by Chiarelli were the key enablers to launch 5G services and applications in South Africa commercially, which are relevant to other developing and developed countries as well. These key enablers include

- spectrum readiness,
- a mature and inexpensive end-user device ecosystem (comparing the high retail prices of current-generation 5G-enabled devices),
- access evolution through deployment of 5G NR equipment and massive MIMO, and a
- virtualized evolved packet core and evolving the core and transport networks using network function virtualization and software-defined networking.

Vodacom South Africa, the largest telecommunications MNO in South Africa with a customer base of over 43 million, also expressed its public commitment to commercially available 5G networks. Vodacom South Africa is committed (March 2019) to connect its customers to commercial 5G networks to ensure that the country is not left behind from participation in Industry 4.0, pending spectrum allocations.

Vodacom Lesotho was the foremost corporation to inaugurate 5G commercially on the African mainland in September 2018; however, Vodacom South Africa was stifled from following suit by spectrum allocation delays. Reported by Gilbert [16], Vodacom's chief technology officer, Andries Delport, said in November 2018 that "*5G is going to happen; the question is just how quickly it's going to happen.*" According to Delport, early use cases of 5G are likely to be the deployment of fiber-like 5G fixed wireless connectivity services and Vodacom South Africa had invested US $280 million in 5G infrastructure expansion by September 2018. Delport also acknowledged (similar to Chairelli from MTN South Africa) the skepticism about 5G business cases, and the current price of 5G end-user devices [16], although maintaining that the future of mobile communications lies with 5G and beyond.

1.10.2 The North American 5G Milieu

In the USA, as in many other markets around the world, 5G deployment follows a phased approach, starting with NSA architectures (where 4G and 5G technologies can coexist) before eventually transitioning to SA infrastructure. Policy and regulation developments are also required in the USA to realize the full potential of 5G, as reported by GSMA [13]. These do not deviate much from those in most other countries worldwide, but specifically refer to,

- prioritizing spectrum allocation,
- deploying infrastructure, and
- long-term investments in the economic benefits of 5G (assuming that business cases presented by MNOs align with investors' activities).

In the USA, MNOs such as Verizon have been offering (since April 2019), a US $1 million prize to an entrepreneur that can justify 5G. The competition requires a team to produce the best new product or service using 5G, showing that large MNOs such as Verizon are also struggling to justify 5G as a business case, apart from fast streaming of video services and video games [24]. Verizon admitted in April 2019 that small cell 5G technology is vulnerable to distance and current rollouts are limited to addresses located strategically from 5G small cells. Verizon found that based on the eligibility of addresses that can receive 5G broadband connections (from strategically placed small cells) and the low levels of penetration from users that require the bandwidth and latency offered by 5G, earning an attractive income from the service would remain challenging.

Other MNOs such as AT&T, Sprint and T-Mobile have also contributed to developing 5G infrastructure in the USA. AT&T expects to have nationwide 5G network deployment using the sub-6 GHz spectrum by early 2020 (www.about.att.com). Sprint (www.sprint.com) announced in April 2019 that it was doubling its quarterly investments in 5G to US $1.4 billion. In May 2019, Sprint enabled its 5G service in four cities in the USA on the 2.5 GHz mind-band spectrum used on Sprint's present

4G cellular locations, on condition that a virtually equal imprint for both 2.5 GHz LTE and 5G NR signal coverage.

1.10.3 The 5G Milieu in BRICS Nations (Excluding South Africa)

Brazil's telecommunications regulator, Anatel, reports that the country will enter the 5G domain by 2020, following the government's goal to auction 5G spectrum in March 2020 (The Brazilian Report 2019) The importance of mobile connections for the success of Brazil's broadband programs was a crucial incentive for the expansion of 3G and 4G, but the scenario has changed with 5G. Brazil launched an initiative in 2017, *Projeto 5G Brasil*, to build an ecosystem to take 5G forward in Brazil and through this initiative Brazil is able to participate in international discussions (along with the European Union, the USA, South Korea, Japan and China) on the technology. Furthermore, there is a drive from industry and MNOs to stimulate 5G development, especially from the MNO Claro/Embratel. Claro/Embatel announced in early 2018 that it had started investing in its IP networks to support 5G, with 15% of its existing network upgraded to operate with software-defined networking (SDN) and virtualization in January 2018 [38].

Russia forecasts 5G network coverage to reach a portion of higher than 80% of the country's inhabitants by 2025 [13]. MNOs in Russia are planning their 5G launch for 2020 and lead the Commonwealth of Independent States (CIS) in its 5G efforts. According to GSMA [13], *"The CIS region is one of the most highly penetrated mobile markets in the world, behind only Europe and North America. At the end of 2017, there were 232 million unique mobile subscribers in the region, equivalent to 80% of the population. Russia accounted for more than half (128 million) of this total and is the region's most highly penetrated market at 89 per cent.*" Discussions by the *big four* MNOs in Russia (MTS, Megafon, Beeline and Tele2) are pointing towards creating a single infrastructure MNO for the development of 5G technology. This partnership will reduce the investment to develop 5G in Russia from US $2.4 billion if done separately, to US $1 billion, according to the Ministry of Communications and Mass Media [9] in Russia.

The Indian telecommunication commerce could need a supplementary venture of US $60–70 billion in 5G, according to the Telecom Regulatory Authority of India [53]. The Chinese company Huawei believes that India will become the largest 5G market after China in the next decade [27]. India's government set up the 5G High-level Forum in September 2017 to communicate the visualization for 5G in India and to endorse program inventiveness and action tactics to fulfill this idea. In terms of standards, the Department of Telecommunications and the Telecommunications Standards Development Society in India have successfully implemented "*low-mobility large-cell use cases accepted by the IMT-2020*".

[9] http://government.ru.

China is somewhat of a special case when it comes to BRICS nations investing in 5G. China has made it clear on numerous occasions that it aims to have the largest 5G footprint globally, with the primary goal to pull ahead of the USA in terms of investment and infrastructure development. In 2017, China was set to spend US $411 billion on 5G mobile networks between 2020 and 2030 [41]. In December of 2018, the Ministry of Industry and Information Technology in China awarded 5G spectrum licenses to its three telecom MNOs. This spectrum allocation allows the MNOs to perform trials on the 5G network before the planned commercial roll-out in 2020 [45]. This also shows that China is not wasting any time on allocation of 5G spectrum, unlike most other countries.

1.10.4 The European 5G Milieu

The European Commission is determined to make 5G an actuality for all its people and industries in participant states by 2020. The Europe Action Plan[10] was created in September 2016 to spearhead and monitor the progress of this goal. The Europe Action Plan serves as a plan for both civic and private venture to develop a 5G infrastructure that provides 10 Gbps broadband speeds and latency below 5 ms. To achieve the suggested roadmap, the European Commission proposes the following measures:

- alignment of both roadmaps and priorities for synchronized 5G deployment across all member states towards large-scale commercial access by 2020,
- provisionally making spectrum available for prototyping purposes,
- promoting preliminary deployments along transport paths and in urban areas,
- promoting a 5G business solution through Pan-European multi-stakeholder trials,
- supporting 5G innovation by implementing industry-led venture funding, and
- promoting global standards by uniting leading actors in 5G development.

In 2019, inCITES Consulting published a Europe 5G Readiness Index, assessing how ready each member state in Europe was for 5G deployment. The full set of results is given in Fig. 1.14.

In summary, the highest-scoring member states were, with their respective number of 5G pilots (where a higher number of pilots do not directly translate to a higher rank),

1. Finland (7),
2. Sweden (22),
3. Switzerland (8),
4. The Netherlands (8),
5. Denmark (16), and
6. Norway (12).

[10]Publicly available at http://ec.europa.eu.

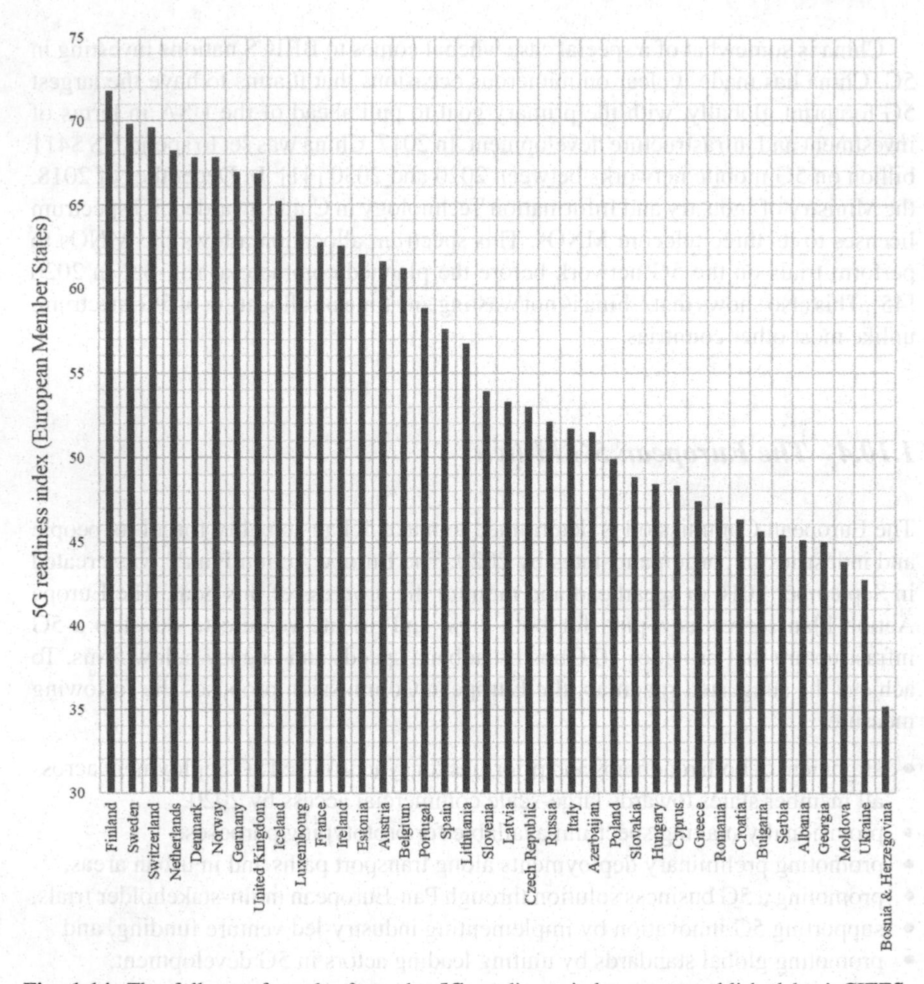

Fig. 1.14 The full set of results from the 5G readiness index report published by inCITES Consulting in 2019. The results are with respect to the total score achieved by each country, a combined score consisting of "*infrastructure and development, regulation and policy, innovation landscape, human capital, country profile and demand*" (for 5G)

A key result from the analysis shows that there is a noteworthy difference amid Western and Eastern Europe in terms of 5G readiness; the western part of Europe is generally closer to introducing widespread commercial 5G. A second key result shows that there is an association between the geographical position of a country and its 5G readiness; primarily concerning countries that are known to be technological pioneers. Gross domestic product and human capital also, perhaps expectedly, play a significant role in European countries' 5G readiness. Most European countries, although not offering widespread commercial 5G connectivity as of mid-2019, are in the phase of launching 5G pilots, developing infrastructure, or at least planning for 5G according to the European Commission roadmap for 5G.

5G could be used to develop effective last mile technologies [30], especially considering the modular approach it allows through network slicing and compatibility for future upgrades. The following section reviews the principles of possibly introducing unique costing approaches of 5G architectures that can benefit emerging markets, which typically lack full financial backing from local and international investors. Further detail on this concept is presented in Chap. 5 of this book.

1.11 Costing Approaches in Unequal Markets

Since data services emerged and became commercially accessible, the communications value chain has evolved along with the development of service features [10]. Each successful generation of mobile network technology has aimed to

- generate new competition drivers within the sector and present unique use cases to take advantage of the benefits over older generations, and
- create new markets for the MNOs to mature into.

Adapted from Frias and Martínez [10], Table 1.3 lists the competition drivers and markets created in transitioning between earlier generations of mobile communications.

As summarized in Table 1.3, there were clear benefits to transitioning between earlier generation mobile communications. However, as we are in the early stages of the transitioning phase between 4G and 5G, the benefits (other than higher speed and lower latency) must still be realized. Note that the *unknown statement in Table 1.3 might be considered somewhat harsh; it essentially only means that we are uncertain of the paradigm shifts [5] for which 5G will be responsible. This section is not aimed at repeating the statements made earlier in this chapter on the business case for 5G, and this will become apparent soon. The reasoning here is aimed at determining how 5G can be applied in both developed and emerging markets, not from the standpoint of the MNO or the industry, but for the individual user. How will 5G benefit users, how much will it cost and how accessible will it be in emerging markets or rural

Table 1.3 Competition drivers and markets created in transitioning between earlier generations of mobile communications, adapted from Frias and Martínez [10]

Transition	Competition drivers	New markets
1G → 2G	Conversion from analogue to digital	Wealthy individuals to mass commercial
2G → 3G	Shift from voice-centric to data-centric	New revenue streams from data (with voice declining)
3G → 4G	Low data rates to broadband mobile	Mobile market revenue streams shared by MNOs and developers
4G → 5G	Evolution to ultra-fast low latency broadband	Technical needs of vertical industries

areas; and is it necessary to be accessible in these areas? It might also be useful to reconsider the challenges and limitations of earlier generation mobile networks and determine if there are similarities. In 2001, Mullany [37] argued that the high levels of investment required in commercializing 3G would lead to 4G being an evolutionary update rather than a revolutionary one. Rapeli [43] presented a picture of the mobile communications market and anticipated development of systems and technologies, as 3G was due for planned commercialization. In Rapeli [43], there were also unanswered questions regarding the annual sales of mobile devices (and their prices), spectrum allocations and the cumulative number of mobile subscriptions from 2001 onwards, a similar question posed in many recent articles regarding 5G. In Hu et al. [22], research questions are asked in terms of the strategic positions of China's MNOs and the government on 3G networks, with reference to the interest from policy makers and the industry. This book will refer to such questions directed at earlier generation mobile networks and determine the effectiveness of certain decisions and how successful policies can be adapted for similar 5G issues. Importantly, emerging markets will benefit most from successful policies, as 5G is not only an upgrade in mobile network communications, but is aligned with Industry 4.0 in terms of timelines and technology.

5G has a uniquely modular architecture, which separates it from prior generations of wireless networks, and these characteristics have already been identified to provide services to users based on the immediate requirements through network slicing (the example of video streaming versus metadata encryption). This capability of 5G should raise the question of whether it is possible to provide users in emerging markets with network slices of adequate bandwidth at reduced prices, or have zero-rating services for specific online services that require 5G bandwidth. 5G networks will in fact permit changeable functions of the network layer to be provided as services [10]. This could therefore mean that *anything* can be provided as a service, at varying cost; perhaps a use case for 5G in emerging markets could lie in this argument. MNOs are, however, currently on the mission-critical path to obtain spectrum and build infrastructure for 5G deployments, and even if the use-cases for users are accepted, emerging markets are still going to struggle to deploy the services to users with similar coverage compared to earlier generations. The supply-and-demand of 5G must also be considered and assessed, in particular the revenue streams from services such as IoT and smart cities compared to those of consumer demand.

In Oughton et al. [40], it is identified that current weakening revenue growth in the technology-evolution space in the telecommunications industry has lowered the craving for infrastructure investment. Oughton et al. [40] research a qualitative framework to support business model adaptation for MNOs that are faced with increasing growth in traffic but with declining revenue owing to the cost of technology and maintenance. Three important research questions were presented by Oughton et al. [40], namely

- how will demographics and a surge in data usage have an impact on the need for mobile infrastructures,

- how do various supply-side alternatives do if verified alongside estimated future-scenarios of data traffic, and
- how can these results be related to the wider mobile telecommunications changes?

Oughton et al. [40] concluded that (in Britain) only 8% of the growth in mobile communications towards 2030 will be due to demographic growth and 92% will be "*from per-user data demand*". This translates that technological advancement takes account for above 90% of the progress in overall data demand. This could mean that for emerging markets, the demand for data will also surge, but the supply, which is extremely technology-dependent, might not be able to keep up if there is insufficient investment.

In terms of supply-side infrastructure changes, Oughton et al. [40] used the following four scenarios:

- Minimum intervention: The study concluded that with this scenario, long-term demands for data would not be met.
- Spectrum performance strategies: The conclusion was that an only spectrum-based methodology is not vigorous and can have mixed results.
- Deployment of (many) small cells: The conclusion was reached that deploying more small cells will aggregate system capacity but will not be economically viable over the long term.
- Hybrid approach (spectrum management and small cells): Similar results were obtained as when adopting the small cells approach, primarily because of the varying demand in rural and urban areas.

The results presented in Oughton et al. [40] were very briefly listed above, only to highlight that they point to the fact that intervention (either economic or technological) is needed to spearhead *and to sustain* 5G. Oughton et al. [39] did a similar survey in the Netherlands with relatively similar results. Schneir et al. [47] estimated and analyzed 5G-related costs for central London over the period 2020–2030 and found that the overall business case for 5G over this period is positive, but dependent on various factors with potential risks closer to the end of the timeline. A sensitivity analysis conducted by Schneir et al. [47] found that a negative return on investment is likely if both traffic and cost (of technology and infrastructure) are significantly higher than the baseline forecasts. Schneir et al. [47] also concluded that network sharing has a positive and sustainable effect on the business case of 5G.

1.12 Conclusion

Applied communication and information theory have evolved significantly from ancient times up to what is perceived as modern communication in 2019. Primarily, the amount of data that we can send between a transmitter and a receiver has increased significantly (bandwidth), associated with a reduction in the time it takes for the information to reach its target device (latency). As bandwidth and latency have steadily improved over the course of history, so have the applications that we invent to take advantage of these improvements.

Currently, Industry 4.0 is evolving, the fourth instalment in the series of industrial revolutions, and it is a data-centric revolution. As a result, the requirement to transfer large quantities of data has become an incentive for future applications. At the forefront of wireless technologies of the modern age is 5G, backed by mm-Wave carrier frequencies to enable bandwidth and latency requirements needed by applications such as VR, AR, autonomous vehicles and the IoT. 5G offers not only speed and latency improvements over earlier generations, but also has significant characteristics that enable a modular approach in its global deployment. The 5G NR, network slicing and aggregation are among these characteristics that are reviewed in this book and introduced in this chapter. These reviews are aimed at not being a duplication of the existing body of knowledge; the focus is placed on relating these technologies to Industry 4.0 and what it could mean for socioeconomic growth in emerging markets and in developing countries.

Though 5G and mm-Wave communication are steadily maturing and becoming more attractive for MNOs to implement, the business case for 5G is still lacking, not only in emerging markets, but also worldwide. MNOs are struggling to present feasible business cases to stakeholders to invest heavily in the global deployment of 5G. Furthermore, lackluster effort from government agencies to release broadband spectrum to MNOs is delaying the process of 5G deployment, with some countries leading in spectrum allocation and already seeing positive results from 5G applications. In emerging markets, for instance in South Africa, there is still a long way to go until agencies recognize the socioeconomic benefits that 5G could present and start prioritizing spectrum allocation to MNOs. An advantage of 5G is its modularity and the prospect of offering unique costing approaches, especially in unequal markets, is another argument for making the technology widely available.

This book researches these aspects from a technological and an economic perspective and provides insights on the implications of specifically mm-Wave technologies for 5G in the era of Industry 4.0.

Appendix 1

3GPP *release 15* offers two frequency ranges (FR) and these are defined as FR1 and FR2, specifically allocated to 5G communications. The frequency ranges of FR1 and FR2 are listed in Table 1.4.

Table 1.4 The 3GPP *release 15* specification frequency bands FR1 and FR2 for 5G communications

Band designation	Frequency range (GHz)	Comment(s)
FR1	0.450–6	Sub-6 GHz
FR2	24.25–52.60	mm-Wave

The 3GPP *release 15* further categorizes the available spectra into frequency division duplex (FDD) and time division duplex (TDD) allocations. In FDD separate frequency bands are used by the transmitter (uplink) and the receiver (downlink) for its send and receive operations respectively. As a result, the sending and receiving operations cannot interfere with each other and are commonly preferred for voice applications in broadband wireless networks as a security measure. FDD therefore requires two radios at each end of the link [17] and is an efficient standard if voice traffic is approximately perpetual and stable in both directions. As spectrum becomes more of a commodity, sending and receiving on different frequencies becomes less cost-efficient and as a result, TDD radios have become more commonplace. TDD is also capable of handling bursts and asymmetric data better when compared to FDD, a characteristic that is crucial in microwave backhaul for cellular traffic [17]. In Goosen [17], a graphical representation of the difference between FDD and TDD is presented and this figure is adapted and presented in Fig. 1.15a, b.

The 5G NR FDD frequency bands are listed in Table 1.5 showing the variations in uplink and downlink frequency, as well as the channel bandwidth for each designation. The NR bands are designated with a prefix "n" for 5G allocations, whereas the "B" prefix was used for LTE designations. The average bandwidth of the FDD bands is also presented in Table 1.5.

The 5G NR TDD frequency bands are listed in Table 1.6 showing the uplink and downlink frequencies, as well as the channel bandwidth for each designation. The average bandwidth of the TDD bands is also presented in Table 1.6, noticeably larger when compared to the average FDD channel bandwidth.

Supplemental downlink (SDL) and supplemental uplink (SUL) bands enable bonding of unpaired spectrum with FDD bands in an effort to enhance network downlink capacity. It was first proposed by 3GPP *release* 9 and the technique allows

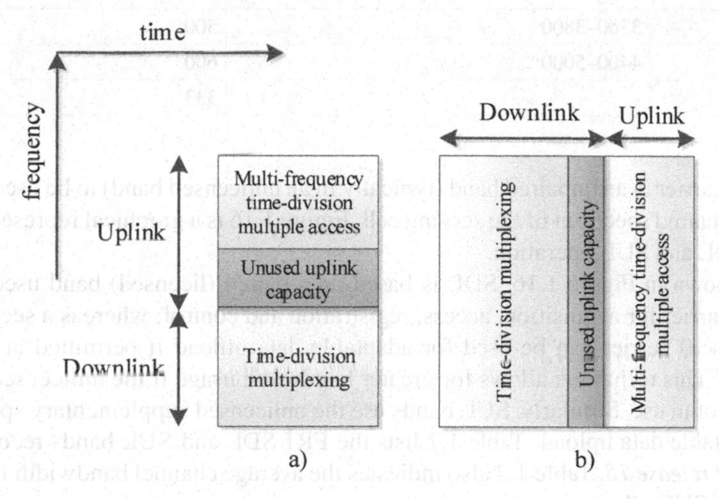

Fig. 1.15 Visual representation of the primary difference between **a** FDD and **b** TDD, adapted from Goosen [17]

Table 1.5 The 3GPP *release 15* FR1 FDD frequency bands for the 5G NR

Designation	Uplink frequency range (MHz)	Downlink frequency range (MHz)	Channel bandwidth (MHz)
n1	1920–1989	2110–2170	60
n2	1850–1910	1930–1990	60
n3	1710–1785	1805–1880	75
n5	824–849	869–894	25
n7	2500–2570	2620–2690	70
n8	880–915	925–960	35
n20	832–862	791–821	30
n28	703–748	758–803	45
n66	1710–1780	2110–2200	90
n70	1695–1710	1995–2020	15/25
n71	663–698	617–652	35
n74	1427–1470	1475–1518	43
Average	–	–	49

Table 1.6 The 3GPP *release 15* FR1 TDD frequency bands for the 5G NR

Designation	Uplink/downlink frequency range (MHz)	Channel bandwidth (MHz)
n38	2570–2620	50
n41	2496–2690	194
n50	1432–1517	85
n51	1427–1432	5
n77	3300–4200	900
n78	3300–3800	500
n79	4400–5000	600
Average	–	333

a single carrier in an unpaired band (typically in an unlicensed band) to be used along with the paired spectrum of the serving cell. Figure 1.16 is a graphical representation of an SDL and SUL operation.

As shown in Figure 1.16, SDL is based on a paired (licensed) band used as an anchor carrier for acquisition, access, registration and control; whereas a secondary (unlicensed) carrier can be used for adaptable data offload if permitted at a time instance. This technique allows for greater bandwidth usage if the unlicensed spectrum is not in use. Similarly, SUL bands use the unlicensed supplementary spectrum for adaptable data upload. Table 1.7 lists the FR1 SDL and SUL bands recognized by 3GPP *release 15*. Table 1.7 also indicates the average channel bandwidth for FR1 SDL and SUL schemas.

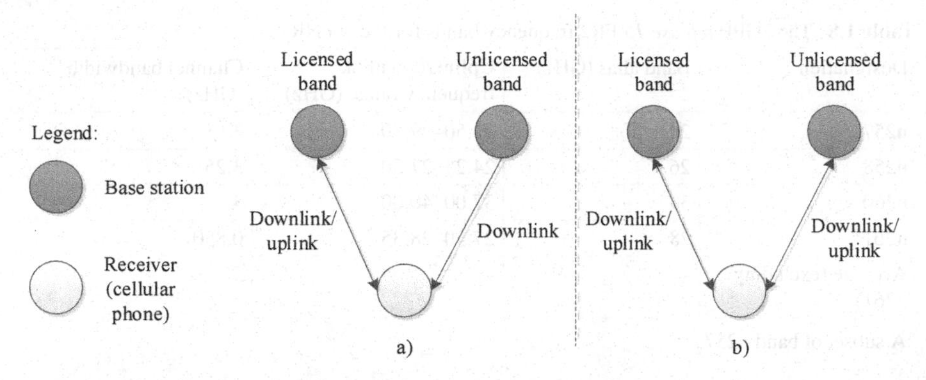

Fig. 1.16 Representation of **a** SDL utilizing a licensed band as the anchor carrier and a secondary unlicensed downlink carrier and **b** SUL using the unlicensed supplementary spectrum for adaptable data uplink and downlink

Table 1.7 The 3GPP *release 15* FR1 SDL and SUL bands for the 5G NR

Designation	Type	Uplink frequency range (MHz)	Downlink frequency range (MHz)	Channel bandwidth (MHz)
n75	SDL	–	1432–1517	85
n76		–	1427–1432	5
n80	SUL	1710–1785	–	75
n81		880–915	–	35
n82		832–862	–	30
n83		703–748	–	45
n84		1920–1980	–	60
Average		–	–	48

For the FR2 band, 3GPP *release 15* does not separate the uplink and downlink bands using SDL or SUL schemas, and each designation uses similar frequency band for both uplink and downlink. Since FR2 spectrum is significantly more than in FR1, efficient channel spacing in abundant bandwidth allows transmitting and sending operations that do not interfere with each other. Table 1.8 lists the FR2 frequency bands for the 5G NR, as well as the average channel bandwidth in this allocation.

Table 1.8 The 3GPP *release 15* FR2 frequency bands for the 5G NR

Designation	Band alias (GHz)	Uplink/downlink frequency range (GHz)	Channel bandwidth (GHz)
n257	28	26.50–29.50	3
n258	26	24.25–27.50	3.25
n260	39	37.00–40.00	3
n261[a]	28	27.50–28.35	0.850
Average (excluding n261)	–	–	3

[a]A subset of band n257

Appendix 2

To calculate[11] the entropy of the arbitrary string, "*The fourth industrial revolution will rely heavily on the availability of broadband, low latency communication*", first determine the characters present in the sentence. These are: [space], [comma], t, a, b, c, d, e, f, h, i, l, m, n, o, r, s, t, u, v, w and y. There are 110 characters in the string, including spaces and punctuation. For each character in the string, determine the number of occurrences through division by the total amount of characters, as summarized in Table 1.9.

Table 1.9 The characters present in the string, "*The fourth industrial revolution will rely heavily on the availability of broadband, low latency communication*", as well as the amount of occurrences and the amount of occurrences divided by the total amount of characters, including spaces and punctuation

Character (x_i)	Occurrences (#)	Probability (p_i)	Character (x_i)	Occurrences (#)	Probability (p_i)
[space]	14	0.127	l	10	0.091
,	1	0.009	m	2	0.018
t	1	0.009	n	7	0.064
a	9	0.082	o	9	0.082
b	3	0.027	r	5	0.045
c	3	0.027	s	1	0.009
d	3	0.027	t	7	0.064
e	6	0.055	u	4	0.036
f	2	0.018	v	3	0.027
h	4	0.036	w	2	0.018
i	10	0.091	y	4	0.036

[11]Various online tools such as http://www.shannonentropy.netmark.pl/ can also be used to calculate Shannon entropy.

The following step is taken to determine the entropy of each (case insensitive) character, by

$$H(x_i) = p_i \log_b p_i \tag{1.4}$$

where the logarithmic base 2 is used, assuming a binary representation of the message. The entropy of the space character would therefore be

$$H(space) = 0.127 \log_2 0.127 \tag{1.5}$$

which is equal to -0.379. Finally, determine the sum of all entropies, through

$$H(X) = -\sum p_i \log_2 p_i \tag{1.6}$$

which results in a Shannon entropy of 4.12461. This indicates that the minimum amount of bits per symbol required to encode the data in binary format (therefore the logarithmic base of 2) is 4.12461, which is typically rounded up, therefore 5. To encode the entire string optimally, in total $110 \times 5 = 550$ bits are required. Furthermore, the metric entropy for this example is $4.12461 \div 110 = 0.0375$.

Appendix 3

Deloitte [6][12] argues, and rightfully so (especially concerning extending fiber to rural and underserved communities), that deep-fiber (deploying fiber closer to the customer) is needed to reveal the full vision of 5G in the USA. Fiber density is often lacking in access networks to make the bandwidth advancements required to expand the pace of innovation and economic growth [6]. 5G offers increases in speed and capacity and will rely on network densification. Mobile network operators (MNOs) will therefore deploy numerous small cells and hotspots within their proposed coverage radius and will need a backhaul that supports the increase in network traffic. Deloitte [6] offers a compelling argument on the necessity for deep fiber to support internet access in the USA, and therefore only the concept of deep fiber is adapted from Deloitte [6] and presented in this appendix. Figure 1.17 shows a comparative view of a traditional fiber deployment versus that of a deep-fiber deployment [6].

The illustrative comparison in Figure 1.17 shows how deep fiber can enable proficient transport of increased wireless traffic from 5G densification. Rather than MNOs and carriers having to purchase additional spectrum to mitigate capacity constraints or erecting macro-towers with mid- or low-band spectrum, reliance is placed on (dense) small cells and hotspots with much smaller coverage radius. Deep fiber therefore serves as the backhaul for these small cells and future-proofs potential

[12]Publicly available from http://www2.deloitte.com.

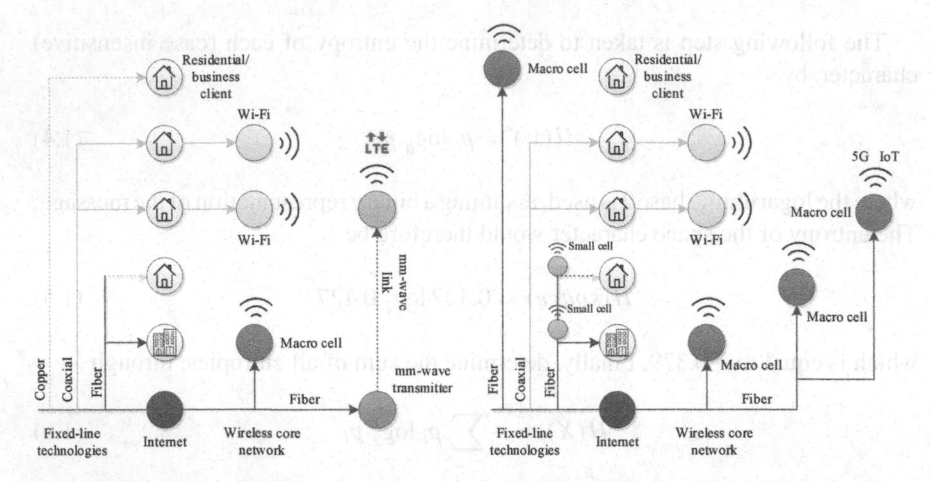

Fig. 1.17 A comparative view of a traditional fiber deployment versus that of a deep-fiber deployment, adapted from Deloitte [50]

backhaul capacity limitations when 5G uptake is high and network traffic increases dramatically.

Appendix 4

The European Commission has a keen interest in the potential of mm-Wave 5G systems. It is preparing the workforce for the fourth industrial revolution accordingly. In May 2019, the European Commission started advertising completely subsidized PhD opportunities in the discipline of mm-Wave antennas, integrated circuits and signal processing. This association is associated with eight leading European research and development laboratories from technology institutes, universities and industry, across The Netherlands (Eindhoven University of Technology), Sweden (Chalmers), Belgium (Keysight Technologies) and Germany (Karlsruhe Institute of Technology and Fraunhofer).

The training network (termed MyWave) acknowledges technologies such as *"distributed massive multiple-input multiple-output (DM-MIMO)"*, innovative antenna concepts on silicon and III–V materials, as well as signal-processing concepts and radio-over-fiber interconnects for future IoT, smart city, self-governing cars and smart industry development. The projects that are offered in the MyWave program are listed in Table 1.10.

From the 5G mm-Wave-based proposed PhD topics listed in Table 1.10, it is clear that Europe aims to increase its level of proficiency in 5G mm-Wave technologies, with specific focus in the fields of DM-MIMO, antenna and signal processing. It shows that there is a shift in interest levels in 5G technologies, but also that there are various issues that need to be resolved relating to the technology and its road to

Table 1.10 The fully funded 5G mm-Wave-based proposed PhD topics offered by the MyWave consortium in 2019 across The Netherlands, Sweden, Belgium and Germany

Host company	Project description	Secondment
Eindhoven University of Technology	System architecture definition and development of radio frequency synchronization concepts	Ericsson
Keysight Technologies	Best practices design guidelines to cope with uncertainty propagation through the design cycle	Karlsruhe Institute of Technology NXP Semiconductors
Chalmers	Power amplifier and antenna co-design strategy for optimized efficiency	NXP Semiconductors
Chalmers	Reconfigurable active-antenna array architectures for mobile users in mm-Wave communications	IMST GmbH Gapwaves
Karlsruhe Institute of Technology	Strategies for energy-efficient high equivalent isotropically radiated power generation in mm-Wave wireless radio links	Ericsson
Karlsruhe Institute of Technology	Energy-efficient and low-cost active front-ends for DM-MIMO	Keysight Technologies NXP Semiconductors
Eindhoven University of Technology	Analogue radio-over-fiber-fed antennas for massive deployment	IMST GmbH Gapwaves
Chalmers	Co-optimized antenna-circuit module integrating contactless interconnects	NXP Semicondutors
Fraunhofer	Highly efficient digital amplifier architecture based on gallium nitride (GaN) technology	United Monolithic Semiconductors
Fraunhofer	Efficient power combining of GaN mm-Wave amplifiers	Ericsson
Eindhoven University of Technology	Channel emulation platform for system testing mm-Wave mobile user scenarios	Ericsson
Eindhoven University of Technology	Multi-physics modeling for improving design-time and energy-efficiency of highly integrated active antenna arrays	Keysight Technologies
Eindhoven University of Technology	Energy efficient signal processing techniques for DM-MIMO systems	Ericsson
Chalmers	Digital array calibration techniques and synchronization for DM-MIMO	NXP Semiconductors
Chalmers	Digital radio-over-fiber for flexible mm-Wave DM-MIMO systems	IMST GmbH

maturity. The stimulation of mm-Wave (below 100 GHz) to back greater data rates and surge the ability of mobile wireless devices and systems to empower the future pass 5G infrastructures is therefore a primary focus point of the MyWave consortium.

References

1. Alter J (2018) The incredible carrier pigeons of the first world war. Retrieved April 25, 2019 from http://www.iwm.org.uk
2. Atoui I, Ahmad A, Medlej M, Makhoul A, Tawbe S, Hijazi A (2016) Tree-based data aggregation approach in wireless sensor network using fitting functions. In: 6th international conference on digital information processing and communications (ICDIPC), pp 146–150
3. Banerjee S, Mishra M, Rai S (2018) Overview: the economics of releasing V-band and E-band spectrum in India. Retrieved May 03, 2019 from http://medium.com
4. Blatner D (2014) Spectrums: our mind-boggling universe from infinitesimal to infinity. Bloomsbury USA. ISBN 987-1620405208
5. Chih-Lin I, Han S, Xu Z, Wang S, Sun Q, Chen Y (2016) New paradigm of 5G wireless internet. IEEE J Sel Areas Commun 34(3):474–482
6. Deloitte (2017) Communications infrastructure upgrade. The need for deep fiber. Retrieved April 29, 2019 from http://www2.deloitte.com
7. Deloitte (2018) 5G: The chance to lead for a decade. Retrieved April 30, 2019 from http://www2.deloitte.com
8. Dent S (2019) Sony built an IoT chip with a 60 mile range. Retrieved May 28, 2019 from http://www.engadget.com
9. Fisher C (2019) Senators ask the FCC to limit 5G auction to protect weather forecasts. Retrieved May 15, 2019 from http://www.engadget.com
10. Frias Z, Martínez JP (2018) 5G networks: will technology and policy collide? Telecommun Policy 43:612–621
11. Friis HT (1946) A note on a simple transmission formula. Proc IRE 24(5):254–256
12. GPPP (2017) View on 5G architecture. Version 2.0 December 2017. Retrieved May 3, 2019 from http://5g-ppp.eu
13. GSMA (2018) Road to 5G: introduction and migration. Retrieved April 27, 2019 from http://www.gsma.com
14. GSMA (2017) An introduction to network slicing. Retrieved April 27, 2019 from http://www.gsma.com
15. Gartner (2018) Market guide for 5G new radio infrastructure. Retrieved April 29, 2019 from http://www.gartner.com
16. Gilbert P (2018) Vodacom sees 5G as use-case-driven. Retrieved April 22, 2019 from http://www.itweb.co.za
17. Goosen M (2006) TDD versus FDD. Elektron J SAIEE 23(1):33
18. Graham B (2018) At last there is a business case for 5G, but maybe we didn't need one? Retrieved April 22, 2019 from http://inform.tmforum.org
19. Grigas G, Juškevičienė A (2018) Letter frequency analysis of languages using Latin alphabet. Int Linguist Res 1(1):18–31
20. Guey J, Liao P, Chen Y, Hsu A, Hwang C, Lin G (2015) On 5G radio access architecture and technology [Industry perspective. IEEE Wirel Commun 22(5):2–5
21. Hauser MD (1996) The evolution of communication. A bradford book. MIT Press. ISBN 0262581558
22. Hu H, Wan X, Lu K, Xu M (2012) Mapping China's 3G market with the strategic network paradigm. Telecommun Policy 36:977–988
23. Jaber M, Imram MA, Tafazolli R, Tukmanov A (2016) 5G Backhaul challenges and emerging research directions: a survey. IEEE Access 4:1743–1766

24. Jasinski N (2019) Verizon is offering $1 million prize to the entrepreneur who can justify 5G. Retrieved April 30, 2019 from http://www.verizonwireless.com
25. Karisan Y, Caglayan C, Trchopoulos GC, Sertel K (2016) Lumped-element equivalent-circuit modeling of millimeter-wave HEMT parasitics through full-wave electromagnetic analysis. IEEE Trans Microw Theory Tech 64(5):1419–1430
26. Kaur S, Gangwar RC (2016) A study of tree based data aggregation techniques for WSNs. Int J Database Theory and Appl 9(1):109–118
27. Khan D (2019) India can become 2nd largest 5G market in 10 years: Huawei. Retrieved April 29, 2019 from http://economictimes.indiatimes.com
28. King BJ (2013) When did human speech evolve? Retrieved April 11, 2019 from http://www.npr.org
29. Lambrechts JW, Sinha S (2016) Microsensing networks for sustainable cities. Springer International Publishing Switzerland. ISBN 978-3-319-28358-6
30. Lambrechts JW, Sinha S (2019) Last mile internet access for emerging economies. Springer International Publishing Switzerland. ISBN 978-3-030-20956-8
31. Larsson C (2018) 5G networks: planning, design and optimization. Academic Press. ISBN 0128127082
32. Liang C, Razavi B (2009) Systematic transistor and inductor modeling for millimeter-wave design. IEEE J Solid-State Circ 44(2):450–457
33. Lieberman P, Fecteau S, Théoret H, Garcia RR, Aboitiz F, MacLarnon A, Melrose R, Riede T, Tattersall I (2007) The evolution of human speech: Its anatomical and neural bases. Curr Anthropol 48(1):39–66
34. Marchant J (2016) A journey to the oldest cave paintings in the world. Retrieved May 4, 2019 from http://www.smithsonianmag.com
35. Marr B (2018) How much data do we create every day? The mind-blowing stats everyone should read. Retrieved May 21, 2019 from http://wwww.forbes.com
36. Mellor D H (1990) Ways of communicating. Cambridge University Press. ISBN 0521370744
37. Mullany FJ (2001) High-speed downlink access in 3G systems: a portent for the evolution of 4G systems? Wirel Pers Commun 17:225–235
38. Nes C (2018) State of 5G in Brazil. Retrieved April 29, 2019 from http://techinbrazil.com
39. Oughton EJ, Frias Z, van der Gaast S, van der Berg R (2019) Assessing the capacity, coverage and cost of 5G infrastructure strategies. Analysis of the Netherlands. Telematics Inform 37:50–69
40. Oughton EJ, Frias Z, Russel T, Sicker D, Cleevely DD (2018) Towards 5G: Scenario-based assessment of the future supply and demand for mobile telecommunications infrastructure. Technol Forecast Soc Chang 133:141–155
41. Perez B (2017) Why China is set to spend US$411 billion on 5G mobile networks. Retrieved April 29, 2019 from http://www.scmp.com
42. Pourghebleh B, Navimipour NJ (2017) Data aggregation mechanisms in the internet of things: a systematic review of the literature and recommendations for future research. J Network Comput Appl 97:23–34
43. Rapeli J (2001) Future direction for mobile communications business, technology and research. Wireless Pers Commun 17:155–173
44. Razavi B (2008) A millimeter-wave circuit technique. IEEE J Solid-State Circuits 43(9):2090–2098
45. Ren S (2019) China's 5G riches are a blocked number for investors. Retrieved April 29, 2019 from http://www.bloomberg.com
46. Sahoo BPS, Chou C, Weng C, Wei H (2019) Enabling millimeter-wave 5G networks for massive IoT applications. A closer look at the issues impacting millimeter-waves in consumer devices under the 5G framework. IEEE Consumer Electronics Magaz 8(1):49–54
47. Schneir JR, Ajibulu A, Konstantinou K, Bradford J, Zimmermann G, Droste H, Canto R (2019) A business case for 5G mobile broadband in a dense urban area. Telecommun Policy. https://doi.org/10.1016/j.telpol.2019.03.002

48. Schumm L (2018) Who created the first alphabet? Retrieved April 25, 2019 from http://www.history.com
49. Series M (2015) IMT vision: framework and overall objectives of the future development of IMT for 2020 and beyond. Recommendation ITU-R, Rep. M.2083-0
50. Shannon CE (1948) A mathematical theory of communication. Bell Syst Tech J 27(3):379–423
51. Shannon CE (1949) Communication theory of secrecy systems. Bell Syst Tech J 28(4):656–715
52. Sørlie IE (2019) 5G & Industry 4.0 at Hannover Messe 2019. Retrieved April 5, 2019 from http://www.ericsson.com
53. TRAI (2019) Enabling 5G in India: a white paper. Retrieved April 29, 2019 from http://main.trai.gov.in
54. Verdú S (1998) Fifty years of Shannon theory. IEEE Trans Inf Theory 44(6):2057–2078

Chapter 2
5G and Millimeter-Wave Key Technologies for Emerging Markets to Participate in the Fourth Industrial Revolution

Abstract 5G is fundamentally different from its predecessors on certain levels of its design and implementation. These differences primarily lead to higher capacity and speed, as well as lower latency, and reduced energy consumption (compared to earlier generation networks). However, the demand for 5G and mm-Wave networks varies significantly between developed countries and emerging markets. This chapter is therefore focused on identifying the technologies and capabilities of 5G and mm-Wave networks that specifically benefit emerging markets' needs, which are fundamentally different from the needs of the developed world. This book also reviews techniques to prepare the emerging market for the fourth industrial revolution (Industry 4.0) and this chapter is central in identifying and reviewing the characteristics of next-generation communications technology that would enable these markets to participate in Industry 4.0, specifically from a telecommunication (transfer of information) perspective.

2.1 Introduction

There has been a significant departure from the way in which telecommunications operators operated and offered internet-related services on previous generation mobile networks. Fifth-generation (5G) and millimeter-wave (mm-Wave) have opened up new avenues to explore technology, revenue streams, and service delivery to a wider gamut of subscribers. However, as with any new technology, service providers that make use of it are now faced with three elements they need to define in order to succeed in the next generation of service delivery, namely

- the services to be offered on their mobile broadband networks and use cases based on consumers, primarily the differences among developing countries, emerging markets, urban, rural, and suburban markets,
- optimal techniques to monetize these networks and create value for consumers, especially in unserved markets, and

W. Lambrechts and S. Sinha, *Millimeter-wave Integrated Technologies in the Era of the Fourth Industrial Revolution*, Lecture Notes in Electrical Engineering 679,
https://doi.org/10.1007/978-3-030-50472-4_2

- the technical, economic, and socioeconomic capabilities to be adapted, incorporated, or introduced to succeed in the market.

The shift towards 5G and mm-Wave networks and the accompanying changes in the way of doing business are directly related to their technical capabilities and most importantly, the differences between 5G and earlier generation networks. 5G and mm-Wave networks are deemed a paradigm shift from earlier generation networks, not only owing to their increased capacity, speed, and latency, but also owing to key enabling technologies that set them apart. According to Akyildiz et al. [1], ten key enabling technologies of 5G have been identified, namely

1. *"wireless software-defining networks (SDNs),*
2. *network function (NF) virtualization (NFV),*
3. *the mm-Wave spectrum (reviewed in Chap.* 1 *of this book),*
4. *massive multiple-input multiple-output (mMIMO),*
5. *network ultra-densification,*
6. *(big data and) mobile cloud computing,*
7. *scalable internet of things (IoT),*
8. *device-to-device (D2D) connectivity with high mobility,*
9. *green communications, and*
10. *new radio (NR) access techniques*".

This chapter aims to review these enabling technologies (and services) and relate how these key technologies can benefit a new generation of connectivity in previously unserved markets. From a technical perspective, this review is relevant in defining strategies to price internet connectivity in unequal markets, presented in Chap. 6 of this book. Emerging markets and particularly rural areas have historically been a part of the digital divide for several reasons. One major reason, as identified in Lambrechts and Sinha [8], is the costs involved in last mile connectivity in these areas (as a function of geographical distance) as well as the challenges of recovering the costs due to a low, and typically poor, subscriber base. Wired networks such as digital subscriber lines and fiber are costly to distribute to rural areas as the backhaul, and in terms of mobile broadband, the coverage areas of operators typically do not extend to these areas, or if so, are expensive in terms of data consumption. Large base stations (BSs) (macro-cells) are needed to cover wide areas and capital investment for these tends to be large. Consumers in these areas therefore need to settle for below-average broadband connectivity at high prices, limiting the socioeconomic impact that the internet can have in these areas.

In this chapter, the primary theme of this book, the use of 5G and mm-Wave, is investigated as an alternative to earlier generation technologies, particularly to bridge the digital divide, introduce broadband to unserved areas in emerging markets, and prepare these markets for Industry 4.0. Mobile broadband is not new, and considering the lack of coverage in many areas, the question has been asked what makes 5G different.

Firstly, the key enabling technologies listed in Akyildiz et al. [1] will be reviewed in this chapter in respect of their implication for emerging markets. PwC [15] also

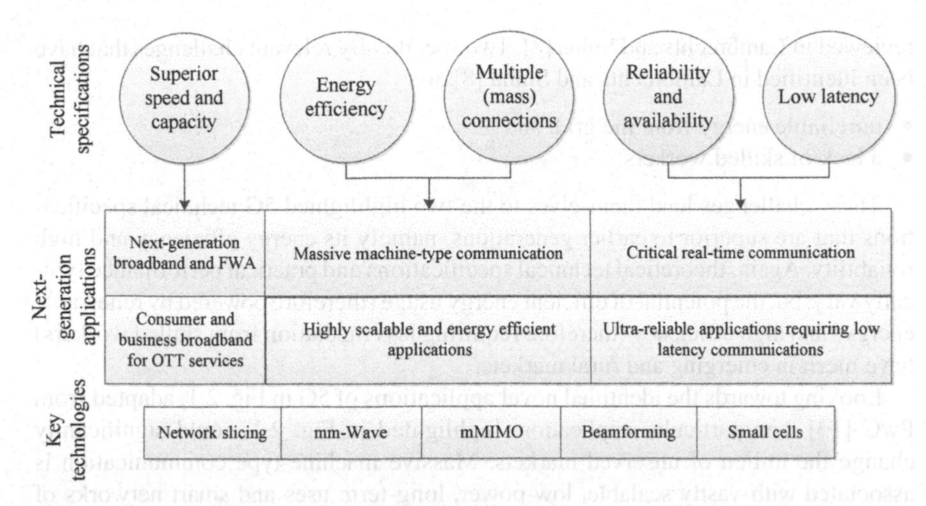

Fig. 2.1 The key capabilities, use cases, and technologies of 5G as identified by PwC [13]

identified key capabilities of 5G along with its view on the key technologies, which in many ways overlaps with the opinions of Akyildiz et al. [1]. In Fig. 2.1, the capabilities, use cases and key technologies of 5G, adapted from PwC [15], are presented, and the characteristics that are relevant to expanding the technology to emerging markets and rural areas are highlighted and reviewed.

Figure 2.1 is a summary of what sets 5G (and indirectly mm-Wave networking) apart from earlier generation networks; however, not all of these differences are relevant in emerging markets and rural areas. Therefore, in Fig. 2.1, the specifications, applications, and technologies that could potentially have a large impact in emerging markets are highlighted. PwC [15] identifies five 5G technical specifications that set it apart from earlier generations, these being its sheer increase in speed compared to 4G, its energy efficiency usage, handling of mass connections, high reliability, availability, and low latency. These are technical specifications and in most practical scenarios, actual performance will vary significantly. For example, the speed, efficiency, latency, and reliability in Singapore will be vastly different from those in a rural African village, primarily owing to the investment in backhaul, support, upgrades, and maintenance. However, the potential (theoretical maximums) in both scenarios are equal if, hypothetically, similar hardware is implemented. Therefore, from an emerging market perspective, the potential of 5G and mm-Wave networking is high. Investment in such a network should have long-term benefits in economic and socioeconomic growth. Two technical specifications of 5G are highlighted in Fig. 2.1 and earmarked to have an impact in emerging markets, namely

- energy efficiency, and
- high reliability.

In emerging markets and in rural areas, challenges exist that might be unbeknown to (or at least underestimated by) the developed world, many of these identified and reviewed in Lambrechts and Sinha [8]. Two specifically relevant challenges that have been identified in Lambrechts and Sinha [8] are

- unreliable energy from the grid, and
- a lack of skilled workers.

These challenges lend themselves to the two highlighted 5G technical specifications that are superior to earlier generations, namely its energy efficiency and high reliability. Again, theoretical technical specifications and practical performance typically vary, but the potential of efficient energy usage (therefore powered by renewable energy) and high efficiency (therefore requiring less mediation from skilled workers) have merit in emerging and rural markets.

Looking towards the identified novel applications of 5G in Fig. 2.1, adapted from PwC [13], one particular application, highlighted in Fig. 2.1, could significantly change the milieu of unserved markets. Massive machine-type communication is associated with vastly scalable, low-power, long-term uses and smart networks of electricity grids. The term *scalable* is important in the context of this book, as this is another potential key characteristic that can benefit emerging markets and rural areas. Capital investment in the telecommunications industry tends to be large and natural monopolies are characteristic in this industry, as reviewed in Chaps. 5 and 6 of this book. However, 5G and mm-Wave networks are highly scalable, therefore allowing new and existing operators to enter markets on a small scale and expand over time, as consumer demand increases. One key 5G technology that, in our opinion, is considered to spearhead the scalability of 5G, is network slicing, also highlighted in Fig. 2.1. Combined with the bandwidth advantages of mm-Wave, mMIMO, beamforming, and small cells, 5G network slicing allows

- vast levels of scaling,
- relatively low capital investment in areas where only *basic* internet is supplied at first,
- innovative pricing schemes that could benefit (poor) consumers, and
- new techniques of monetizing the network and recover costs.

In view of these benefits and the aim to bridge the *digital divide in emerging markets and rural areas*, this chapter is dedicated to reviewing key enabling technologies of 5G and mm-Wave from a technical perspective (followed by an economic perspective in Chaps. 5 and 6 of this book), starting with 5G network slicing in the following section.

2.2 5G Network Slicing

In Chap. 1 of this book, a brief introduction on 5G network slicing was given. The importance of network slicing is, however, significant when considering its ability to tailor services and therefore the cost to the consumer. In emerging markets, poor areas, and many rural areas, internet connectivity is limited or non-existent, and

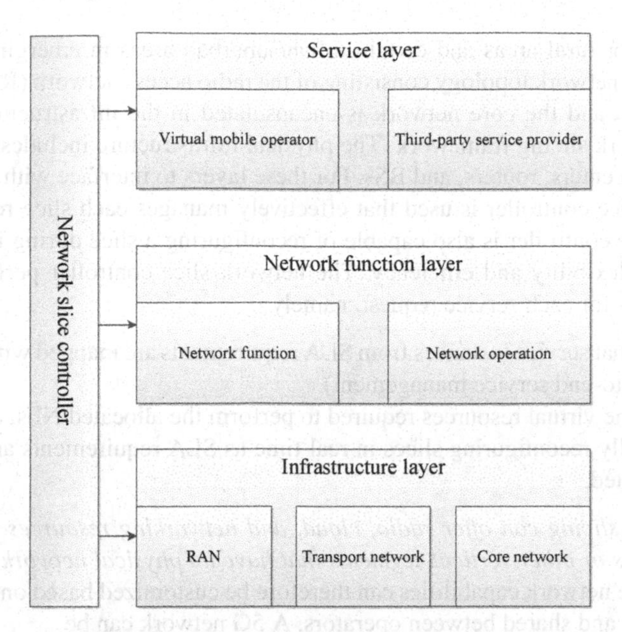

Fig. 2.2 A simplified framework of 5G network slicing to identify the primary layers. Most of the proposed 5G network slicing frameworks are based on this architecture

infrastructure development requirements in these areas are typically large and the subscriber base too small to recover the costs. Network slicing allows operators to provide broadband internet services scaled to the demands of the local population. Infrastructure developments therefore also scale in these areas and operators can invest significantly less and recover costs more quickly. These techniques are reviewed in Chap. 6 of this book. In this section, an overview of the concepts of 5G network slicing is reviewed. In Fig. 2.2, a simplified framework of 5G network slicing is presented and the different layers of the generalized architecture are identified. This framework is adapted from various resources and represents a common and well-known framework.

As shown in Fig. 2.2, within the simplified framework of 5G network slicing, a service layer is defined to interface directly with business entities such as virtual mobile operators or third-party service providers. These entities effectively share the underlying physical network and take advantage of the service requirements. Each service (service instance) embeds all the network characteristics in service-level agreements (SLA) that are fulfilled by a network slice. Below the service layer is the NF layer that creates a network slice based on a request from the service instance. A set of network operations configures the NF and manages the slice during its lifecycle. NFs can be simultaneously shared by different slices to increase resource efficiency, at the cost of reduced complexity of operations management. For relatively simple service instances, this efficiency can be taken advantage of by the business entities to provide lower quality of service (QoS) to a larger subscriber base, ideal for rural areas and certain urban/suburban areas in emerging markets. The physical network topology consisting of the radio access network (RAN), transport network, and the core network is encapsulated in the infrastructure layer of the 5G network slicing framework. The physical infrastructure includes equipment such as data centers, routers, and BSs. For these layers to interface with each other, a network slice controller is used that effectively manages each slice request. The network slice controller is also capable of reconfiguring a slice during its lifecycle to improve flexibility and efficiency. The network slice controller performs three primary tasks for each service request, namely

- ensuring that service instances from SLA requirements are mapped with sufficient NFs (end-to-end service management),
- defining the virtual resources required to perform the allocated NFs, and
- dynamically reconfiguring slices in real-time as SLA requirements are modified or completed.

"*Network slicing can offer radio, cloud, and networking resources to application providers or other vertical segments that have no physical network infrastructure*" [3]. The network capabilities can therefore be customized based on the service requirements and shared between operators. A 5G network can be

- dynamically automated and optimized as the resource requirements vary,
- scaled to guarantee performance for each user, and
- isolated and secured (slice security and slice privacy) for each individual user.

Through network slicing, service providers are given flexibility to up- and downscale amenities to handle shifting customer needs, decrease their capital spending, as well as their operational expenditure, and reduce their time to market. Operators and third-party service providers can rely on cloud computing to offer on-demand application, platforms, and infrastructures.

According to Barakabitze et al. [3], 5G network slicing is an exchange-to-exchange logical network or cloud service operating on a shared core physical or virtual structure. These entities are mutually sequestered and have self-governing control and administration that can be produced as the need arises. "*Each network slice can comprise cross-domain components from separate domains in the same or in different administrations*". The slices also share modules relevant to the access network (communication links), core network (interconnected routing devices), and edge network (applications and hosts) [3]. Essentially, 5G network slicing includes slicing the 5G RAN, the core network, and even end-user devices [9]. Barakabitze et al. [3] define eight 5G network slicing enabling technologies, namely

- "*SDN,*
- *NFV,*
- *communications protocol in an SDN environment to enable the SDN controller to interact directly with the forwarding plane of network devices,*
- *cloud computing through software as a service, platform as a service, and infrastructure as a service,*

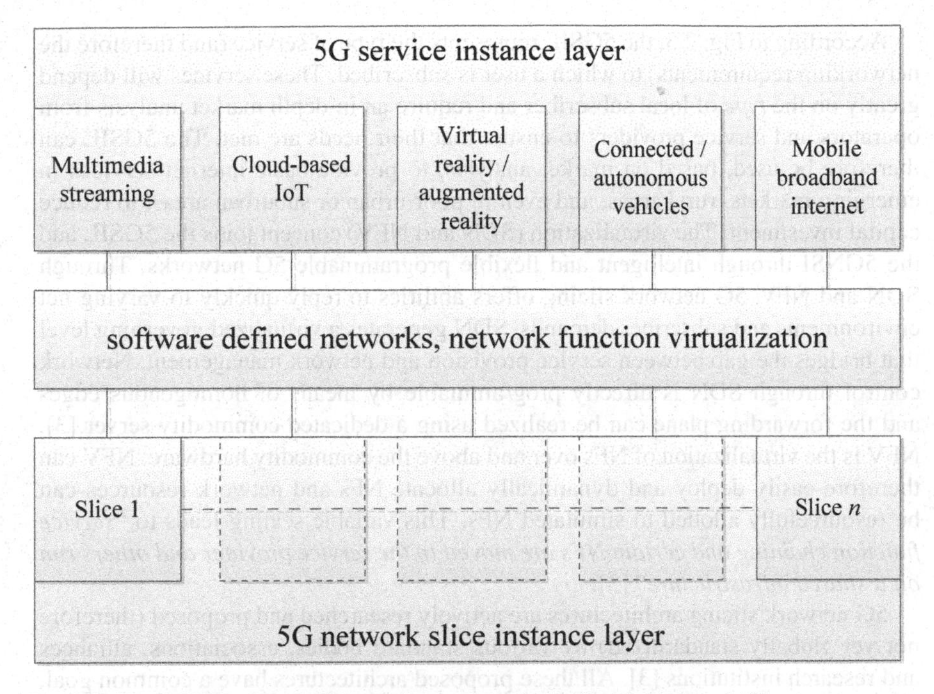

Fig. 2.3 The NGMN network slicing notion as described in Barakabitze et al. [3]

- *multi-access computing that allows application and content providers cloud computing capabilities, as well as an internet technology service environment at the edge of the mobile network,*
- *network hypervisors to isolate network slices logically, and*
- *virtual machines (or containers)*".

SDN and NFV can arrange the virtual network resources in a flexible and real-time environment. Importantly, network slices are consequently self-contained, mutually isolated (important for security provisions), separately manageable, and programmable (software-defined). The Next Generation Mobile Networks (NGMN[1]) defines three layers to depict the 5G slice capabilities, namely the

- 5G Service Instance Layer (5GSIL),
- 5G Network Slice Instance (5GNSI), and
- 5G Resource Layer (5GRL).

Importantly, the 5GSIL is where diverse services that need to be maintained are defined and provided by operators. The 5GNSI offers the system characteristics that are needed by the 5GSIL, and the 5GRL are the physical and logical resources dedicated to an NF or pooled between a group of NFs. The NGMN network slicing notion is represented in Fig. 2.3 [3].

[1] http://www.ngmn.org.

According to Fig. 2.3, the 5GSIL represents the type of service (and therefore the networking requirements) to which a user is subscribed. These services will depend greatly on the *type* of local subscriber and require an in-depth market analysis from operators and service providers to ensure that their needs are met. The 5GSIL can therefore be used, based on market analysis, to provide basic internet services in emerging markets, rural areas, and even in poor urban or suburban areas, to reduce capital investment. The virtualization (SDN and NFV) concept joins the 5GSIL and the 5GNSI through intelligent and flexible programmable 5G networks. Through SDN and NFV, 5G network slicing offers abilities to reply quickly to varying net environments and subscriber demands. SDN generates a virtualized governing level that bridges the gap between service provision and network management. Network control through SDN is directly programmable by means of homogenous edges and the forwarding plane can be realized using a dedicated commodity server [3]. NFV is the virtualization of NFs over and above the commodity hardware. NFV can therefore easily deploy and dynamically allocate NFs and network resources can be resourcefully allotted to simulated NFs. This variable scaling leads to "*service function chaining and certain NFs are moved to the service provider and others run on a shared infrastructure*" [3].

5G network slicing architectures are actively researched and proposed (therefore not yet globally standardized) by various standard bodies, associations, alliances and research institutions [3]. All these proposed architectures have a common goal, satisfying the needs of the end user while meeting the requirements of the different verticals. Barakabitze et al. [3] have identified several cooperative 5G network slicing studies and provides conceptual definitions of each of these. The identified 5G network slicing studies include

- "*5G Exchange (5GEx).*
- *MATILDA,*
- *SliceNet,*
- *5GTANGO,*
- *5G NORMA,*
- *SONATA,*
- *5G-MoNArch,*
- *5G-Transformer,*
- *5G-Crosshaul,*
- *5G!PAGODA, and*
- *NECOS*".

It is evident from the work published by Barakabitze et al. [3] that 5G network slicing is currently still under research and development and will possibly only be deployed as a standard in 5G and mm-Wave networks in the near future. Barakabitze et al. [3] also review the listed 5G network slicing schemes, compare each of them to the generalized architecture and highlight the major differences among them. Finalizing 5G network slicing protocols is crucial to emerging markets, considering lower capital investment from providing scaled services. In developed countries, 5G and mm-Wave deployments are already being rolled out, and network slicing will only be implemented at a later stage. Since there are fewer financial constraints in developed countries than in emerging markets, such a strategy can be followed.

The following section, the 5G NR mMIMO technology, is reviewed from the perspective of being an enabler for 5G and mm-Wave networks in emerging markets.

2.3 5G New Radio Massive MIMO

MIMO systems are founded on a amalgamation of antenna expansions and complex algorithms and have been used in wireless communications for a while. MIMO algorithms control how data maps into antennas and where to focus energy in space. There needs to be tight coordination between network devices and mobile devices to make MIMO work. Followed by the design of the 5G NR, MIMO has also evolved and has become crucial for 5G NR deployments. This principle of traditional MIMO is briefly reviewed before the enhancements related to 5G NR massive MIMO are presented. Fundamentally, MIMO in traditional wireless and radio frequency (RF) communications aims to improve on the signal-to-noise ratio (SNR) of the transmitted signal. Through diversifying the signal, the receiver will be able to use several varieties of the equivalent transmitted signal to decode the information. By design, the signals are prepared to be influenced in dissimilar means by the path (medium or channel) and therefore the likelihood that the signals will be influenced at the identical interval becomes unlikely. Several types of signal diversifying exist, including

- time diversity,
- frequency diversity, and
- space diversity.

Through time diversity, the signals are able to simply be transferred at altered times, for example by means of varied timeslots and channel coding. By using different frequencies or modulation schemes, frequency diversity can be introduced, and space diversity, the basis of MIMO, "*uses antennas located at different positions to take advantage of the different signal paths in an environment and improve on the overall SNR*" [6]. The simplified channel capacity [6] and therefore spectral efficiency of a MIMO link can be described by

$$C \approx m.\log(1 + SNR) \tag{2.1}$$

where m is the smallest amount of antennas in either the transmitter or receiver. MIMO therefore offers spectral efficiency that increases linearly with the (smallest) amount of antenna elements in the transmitter or receiver. In Fig. 2.4, the simplified architecture of a MIMO configuration is presented.

As shown in Fig. 2.4, the signal from each transmitting antenna element follows a different path towards each of the receiving antenna elements. The interference and degradation of each of these paths differ, and through MIMO, several techniques can be employed to recover the original information. As described by Idowu-Bismark

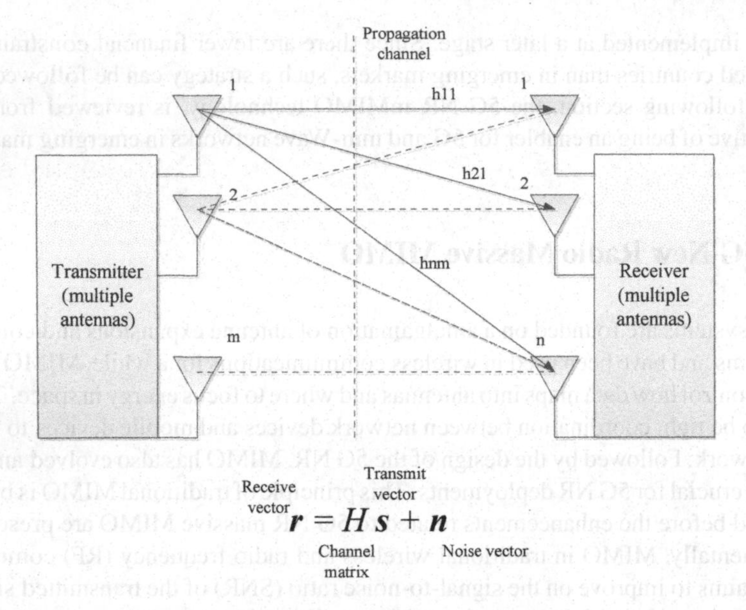

Fig. 2.4 The simplified architecture of a wireless MIMO configuration

et al. [6], all the transmit signals pass through a channel matrix or paths, consisting of *n* transmit and *n* receive paths, equal to the amount of transmitting and receiving antennas. The multiple receiving antennas decode the information to its original representation. To model this system, the following set of equations is defined:

$$\boldsymbol{r} = H\boldsymbol{s} + \boldsymbol{n} \tag{2.2}$$

where $\boldsymbol{r}$ is the receive vector, $\boldsymbol{s}$ is the transmit vector, $\boldsymbol{n}$ is the noise vector, and H is the channel matrix defining the state information of each signal path. The information in the channel matrix includes a distortion coefficient acting on the "*phase and amplitude of the transmitted signal*" in the time domain. Importantly, H is constructed by the receiver, since the transmitter assumes a perfect channel. It is therefore an open loop system and the transmitter transmits at performance characteristics regardless of the state information of each channel, although a closed loop system with feedback is also possible. To recover the transmitted information, three primary techniques are used by MIMO systems:

- spatial diversity—a more defined term than space diversity, spatial diversity denotes the transmit and receive variety and the propagation characteristics of each individual path,
- spatial multiplexing, and
- beamforming (reviewed separately in the following section).

Forming part of spatial diversity (sending the same signal over different paths), spatial multiplexing uses this concept but sends multiple messages simultaneously without these messages interfering, since they are separated in space. Through spatial multiplexing, additional data capacity can be utilized to increase the data throughput of the channel, carrying additional traffic.

Building on MIMO, multi-user MIMO (MU-MIMO) permits numerous operators to share equal network resources simultaneously. MU-MIMO permits numerous independent radio stations admission to a structure, improving the communication abilities of respective terminals. MU-MIMO therefore entails spatial sharing of wireless channels between multiple users, achieved by additional hardware resources, but not at the cost of added bandwidth. By using additional antenna elements (and digital signal processing), the interference between user streams can be mitigated. Furthermore, and reviewed in the following section, while a transmitter is transmitting to *all* clients simultaneously, beamforming can be employed to enhance the signal path for each client. Consequently, in crowded parts with a large amount of active users, the efficacy of information transport in MU-MIMO allows each user's QoS to be maintained. Furthermore, the network can dynamically switch between serving a single user or multiple users, adjusting the directionality and power of the beam(s).

Furthermore, mMIMO builds on MU-MIMO by introducing a number of antennas that exceed the number of users, an ability that can be taken advantage of in 5G and mm-Wave networks, since antenna dimensions shrink significantly as a function of the operating frequency. The 5G NR can therefore take advantage of nMIMO together with the large quantities of available bandwidth to serve dense areas (of subscribers) and scale dynamically based on real-time demand. The primary benefits of nMIMO include

- enlarged network capability by permitting 5G NR distribution at mm-Wave frequencies and by employing MU-MIMO to serve multiple consumers simultaneously using similar frequency resources,
- improved coverage through more uniform and dynamics transmission characteristics and adjusting the transmission path(s) in areas where users have weak coverage, and
- improved user experience resulting from the above benefits.

In the context of the 5G NR, nMIMO can radically change the way that mobile devices in urban and rural areas receive wireless content. Importantly, users' experience will improve drastically and not at the expense of power consumption (from the BS) or available bandwidth, which are important figures of merit in emerging markets with limited resources. In terms of the overall theme of this book, supplying mm-Wave broadband to emerging markets and rural areas, 5G NR mMIMO technology is a fundamental enabler to deploy future-generation internet networks in unserved areas, taking advantage of the technical benefits and inevitably realizing economic benefits, including

- lower cost in infrastructure where the subscriber base is low (or subscribers cannot afford high-cost bandwidth) will incentivize providers to expand to these markets and address the digital divide in terms of Industry 4.0,

- lower initial capital investment from operators and stakeholders, but with the ability to scale and expand on infrastructure as the user needs increase,
- guaranteed QoS for users, and
- lower power consumption of BSs, which will result in more opportunities to employ renewable energies to keep BSs powered and active reliably, in areas where power from utilities is often unreliable.

Since nMIMO essentially transmits to multiple users simultaneously, using similar frequency resources, a form of directionality is required, and a technique to achieve this is beamforming, as discussed in the following section.

2.4 Beamforming

As 5G can take advantage of the mm-Wave spectrum, it follows that at these higher frequencies the physical dimensions of the antennas are significantly reduced. However, at these frequencies and as discussed in Chap. 1 of this book, higher propagation loss is experienced. Beamforming, focusing a wireless signal as opposed to it radiating in all directions, has therefore become an important characteristic of 5G and mm-Wave networks. Beamforming can be applied to both the transmitting and receiving side of a wireless network. Because of the principle of reciprocity, transmit and receive beamforming are identical. A common technique to achieve beamforming is having multiple antennas in close proximity, all broadcasting at the same frequency but at a slightly offset time delay. Overlapping waves can be designed to produce constructive or destructive interference and lead to more focused waves towards a receiver. The beam response ξ is a linear combination of the outputs of each element, and can be described by

$$\xi(\theta) = \sum_{k=0}^{k=N} a_k(\theta).x_k \tag{2.3}$$

where N is the amount of antenna elements, a_k is the complex coefficient of each element, x_k is the voltage response from each element, and θ is the angle of the main lobe of the beam. The intricacy of the beam response emanates from determining the complex coefficients of each element to enhance the gain of the antenna in a specific direction. An analysis on the derivations and calculations of beamforming techniques is presented in Dietrich [6]; it falls outside the scope of this book. It is, however, relevant to mention that beamforming can essentially be accomplished by three primary methods [14], these being

- "*analog beamforming,*
- *digital beamforming, and*
- *hybrid analog-digital beamforming*".

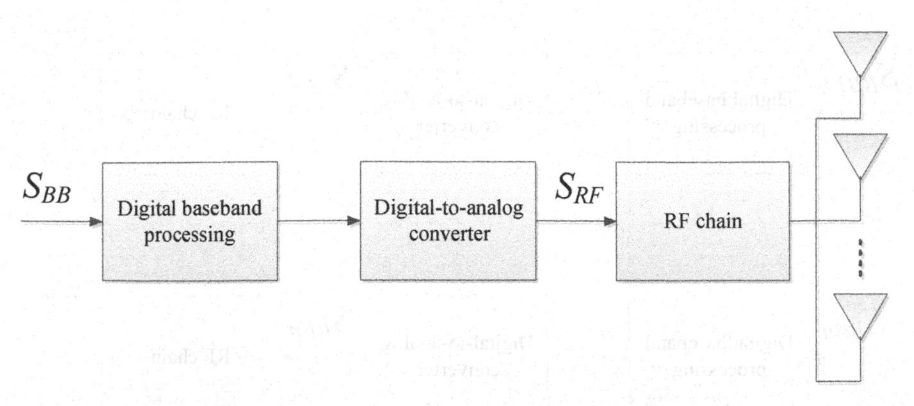

Fig. 2.5 Analog beamforming architecture, adapted from Rohde and Schwarz [14]

Analog beamforming has been used since the early 1960s [14] and consists of a single "*RF chain and multiple phase shifters that feed an antenna array*". The analog beamforming architecture is represented in Fig. 2.5.

According to Fig. 2.5, the steering of the wireless signal is carried out by "*a selective RF switch and fixed phase shifters*" [19]. Published works such as that of Zhang et al. [19] argue (rightly so) that traditional analog beamforming used costly, large, and high-power analog phase shifters and needed to be replaced with alternative methods. Early passive architectures steered the beam to virtually any angle using active beamforming antennas; this could be achieved in the RF or intermediate frequency domain. According to Rohde and Schwarz [14], analogue beamforming in mm-Wave is less expensive (and less complex) compared to fully digital or hybrid implementations and is used in radar and short-range communications. However, analog beamforming has a frequency-dependent phase shift component that can lead to beam squint [4].

Since traditional analog beamforming is typically limited to a single RF chain, digital beamforming supports a number of RF chains equivalent to the amount of antenna elements. Furthermore, with efficient precoding on the digital baseband signal, even higher flexibility in transmitting and receiving signals can be achieved. Adapted from Rohde and Schwarz [14], Fig. 2.6 represents the simplified architecture of a fully digital beamforming system.

Digital beamforming as shown in Fig. 2.6 has advantages in terms of additional degrees of freedom that are ideally leveraged by multi-beam MIMO systems and mitigating issues such as beam squint in analog beamforming. However, in 5G and mm-Wave networks, digital beamforming can become complex and expensive, and present high-energy consumption from the computing requirements on each RF chain. As a result, digital beamforming is more suitable for BSs as opposed to mobile devices, and also preferred in urban areas rather than in areas where energy is scarce.

A potential solution to take advantage of the resolution and lower complexity of analog beamforming and the flexibility of digital beamforming is the implementation of a hybrid approach, as shown in Fig. 2.7.

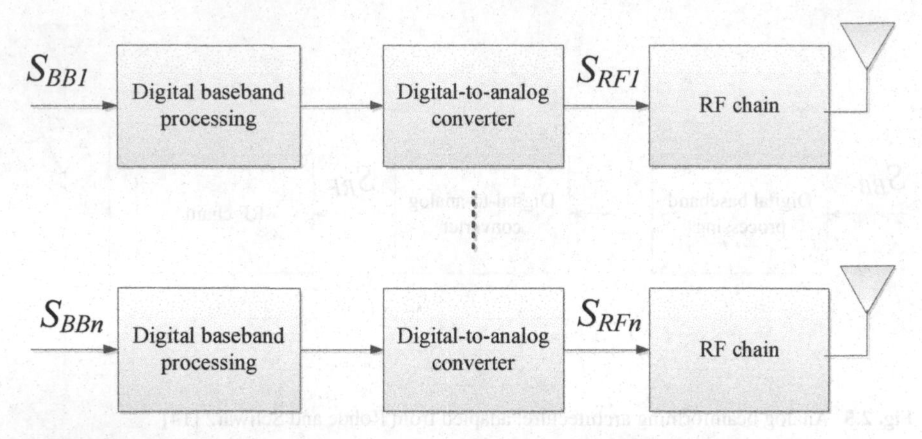

Fig. 2.6 Digital beamforming architecture, adapted from Rohde and Schwarz [14]

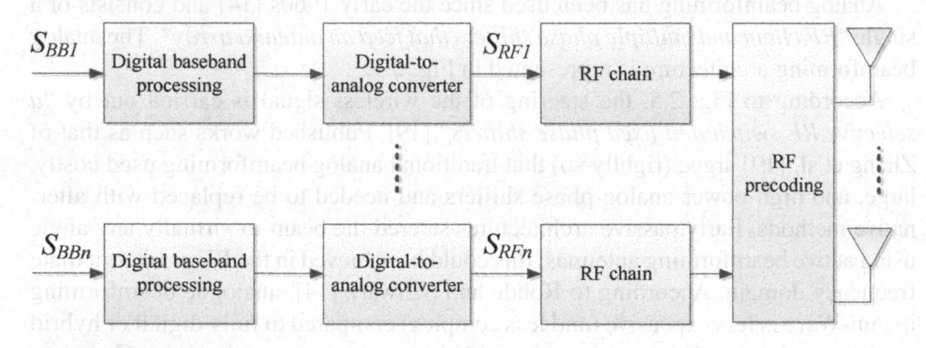

Fig. 2.7 Hybrid beamforming architecture, adapted from Rohde and Schwarz [14]

The hybrid beamforming architecture has significantly fewer digital-to-analog converters than the number of antennas, leading to lower complexity and power consumption, but significantly higher flexibility compared to the analog beamforming architecture.

Since antenna dimensions are significantly reduced at mm-Wave frequencies, it presents an opportunity to incorporate a great amount of antennas in a relatively small space to improve signal propagation for 5G and mm-Wave signals. Beamforming has several advantages, including

- relatively low overall power consumption,
- increased directionality for lower interference,
- higher security (potential for signal interception), and
- a higher SNR that produces fewer errors in the transmission.

However, beamforming also has some disadvantages, which include

- complexity in antenna design,

- a requirement for resource-intensive computing (dedicated digital baseband and RF chain per antenna to control both amplitude and phase of the transmitted signal across all antennas on large arrays can negate the power consumption benefits and the cost can be prohibitive owing to the complexity), and
- a requirement for directionality between the transmitter and receiver.

In terms of 5G and mm-Wave networks in rural areas, energy efficiency becomes a crucial factor, since in many of these areas, and typically, in emerging markets, energy supplied from utilities is not reliable. Using renewable energy to power 5G and mm-Wave BSs may be more sustainable. Karlsson et al. [6], performed a study specifically on the effect of beamforming as well as cell discontinuous transmission (DTX) in 5G networks, operating specifically at 28 GHz with a bandwidth of 100 MHz. Karlsson et al. [6] proposed a "*rural area model that captures a non-uniform distribution of users (of approximately 100 inhabitants/km^2) and uses the generalized Lloyd algorithm to determine the deployment of the BSs to satisfy a stated QoS*". Karlsson et al. [6] performed simulations and found that beamforming considerably decreases the amount of necessary BSs by improving the signal power and subduing the interfering signals. For rural networks, according to Karlsson et al. [6], beamforming is a fundamental enabler of power conserving in 5G and mm-Wave systems. Karlsson et al. [6] also highlighted that the simulation profiles of their proposed models can be improved to yield scenarios that are more realistic; however, the initial results do point towards beamforming being an important factor in reducing energy consumption in 5G and mm-Wave networks. It also reduces the cost of implementing broadband in rural areas, where the subscriber base is typically low, as reviewed in Chap. 6 of this book. In terms of DTX, Karlsson et al. [6] concluded that only minor benefits were observed, as beamforming leads to relatively high cell loads, since each BS serves many subscribers. Vodafone Germany,[2] for example, launched beamforming in numerous rural areas across Germany to improve mobile broadband coverage and effectively reduce reliance on outdated wired infrastructure in these areas. At the time of writing, little detail was available on the type of beamforming implemented or the impact on the broadband deployments. For an in-depth analysis on beamforming at mm-Wave frequencies, specifically at 60 GHz, it is recommended to consult Yu et al. [18].

Small cell BSs are another key characteristic of 5G and mm-Wave networks and again a result of the high operating frequencies. Small cell BSs, specifically the number of small cells used in a deployment, have an impact on the power consumption, capacity, and throughput of a 5G mm-Wave network, again with advantageous consequences for emerging markets and rural areas, as reviewed in the following section.

[2]http://www.vodafone.de.

2.5 Small Cell Base Stations

Small cells (which include femtocells, picocells, and micro-/metrocells as reviewed in Chap. 1 of this book) are essentially traditional radio communications equipment shrunk down. There are numerous benefits to reducing the size of transceivers, especially the power amplifier (PA) as reviewed in Chap. 4 of this book, but importantly, the technology must allow this. 5G and mm-Wave networks operate at higher frequencies compared to traditional cellular technology, as summarized in Chap. 1 of this book, and therefore, through the inverse proportionality between frequency and the physical dimensions of (most) electronic components, miniaturization can occur. Small cells and therefore small cell BSs are therefore a result of miniaturization. It has numerous advantages, including lower cost and ubiquity; however, it has also been met with much criticism. Sbeglia [17] argues, and rightly so, that numerous small cell deployments across an urban area, which also holds true in rural and suburban areas, are at the least aesthetically displeasing, or in the worst case, a health concern. The entire premise of small cells, especially in 5G and mm-Wave networks where it is a necessity, is deploying a large amount of BSs to increase the total subscribers the system can support (and coverage), increase ubiquity, increase capacity, and decrease latency. Furthermore, again out of necessity, mm-Wave frequencies have very poor penetration through objects such as walls and require numerous small cells to mitigate this.

Lemieux and Zhao [10] describe how small cells fit into the 5G ecosystem, primarily emphasizing their ability to increase network capacity. Large cell towers (macro-cells) have large coverage areas that serve thousands of subscribers [10] over large areas, with minimal obstructions. To extend the coverage area, distributive antenna systems (DASs) can be used in conjunction with the macro-cells. However, these DASs do not increase capacity, only coverage, whereas increasing the number of BSs is a technique to increase network capacity. Macro-cells are large, costly, and require high capital investment and environmental impact studies to deploy. Small cells can be used to complement these macro-cell BSs and offload some of the data capacity from the macro-cells [10] and improve the robustness of the network. Although small cells do not have the same coverage as macro-cells, through ubiquity the coverage area can be increased. However, the drawback is the sheer number of small cell BSs needed (notwithstanding the disadvantages, as argued by Sbeglia [17]) and the time to deploy each cell. PwC [13] reported that in the USA, the Federal Communications Commission estimated that at least 800,000 small cells are needed to make 5G a reality. According to PwC [13], the International Data Corporation estimates this number at closer to 2 million. However, according to PwC [13], it could take up to two years to deploy a single small cell, as well as significant cost (apart from permit and rental costs), unless there is a shift in industry collaboration to transform small cell permitting and deployment strategies. Chatchaikarn [5] listed additional challenges for the small cell infrastructure market. These challenges include heat generation from each small cell's PA, which will lower component lifetime and considering the number of small cells needed in urban areas, could potentially lead to urban heat islands [7]. Careful consideration is therefore crucial in designing and implementing the small cell PAs in terms of their

- efficiency,
- operating cost,
- reliability,
- gain,
- coverage, and
- compatibility (standardization).

Furthermore, according to Chatchaikarn [5], powering small cells remains a challenge. Although power over Ethernet (PoE) has been identified as a viable method to build a backhaul and power small cells, there is a limit on the amount of power that can be offered over PoE. Chatchaikarn [5] points out the relationship (inverse proportionality) between the amount of power available to a small cell and the number of bands from different carriers (through carrier aggregation) it can support. Another common goal of combining small cells and macro-cells and different technologies, is to create a heterogeneous network (HetNet) of 5G and mm-Wave connectivity and earlier generation networks (2G, 3G, 4G). In summary, small cell BSs are crucial in realizing true 5G and mm-Wave networking. There are, however, advantages and disadvantages to small cells and these need to be considered during the planning phases of 5G and mm-Wave rollout.

In the following section, several additional key enabling technologies are reviewed, although their significance in emerging markets and rural areas is less prevalent. These technologies are still worth mentioning, since they have an impact on the ubiquity and cost of deploying 5G and mm-Wave networks, and this overall cost does have an impact on emerging markets' decision to roll out these technologies.

2.6 Other Key Enabling Technologies of 5G and mm-Wave

The remaining key enabling technologies identified by Akyildiz et al. [1], namely

- "*network ultra-densification,*
- *(big data and) mobile cloud computing,*
- *scalable IoT,*
- *D2D connectivity with high mobility, and*
- *green communications, and*
- *NR access techniques*"

are briefly reviewed in this section. These technologies could in some ways specifically benefit emerging markets and rural communities in preparing for Industry 4.0 and are therefore worth reviewing.

2.6.1 Network Ultra-densification

There are numerous ways to add capacity to a network, primarily by adding spectrum or by using it more efficiently. Network ultra-densification [or ultra-dense cellular networks (UDNs)] is another method to add capacity to an existing network, if additional spectrum is not available. According to Wei and Hwang [17], "*UDNs have evolved from the traditional macro-cell-only homogeneous network to a multiband HetNet where macro-cells operating at lower frequency bands (and high power) underlie small cells operating in high (ideally mm-Wave) frequency bands*". UDN is essentially an evolution of HetNets through additional densification of small cells, especially in 5G and mm-Wave networks, considering the advantages of these networks as reviewed in this chapter. Densification is adding more (small) cells to a site to increase the amount of available capacity, as reviewed in the small cell section of this chapter. Urban areas and big public locations are typical contenders for network ultra-densification, and for many emerging markets, this is a relatively cost-effective means to increase network capacity, especially considering low capital investment (compared to 4G) for initial 5G and mm-Wave deployments. Network ultra-densification therefore benefits the areas where there is a high concentration of mobile users, and public free-internet areas in emerging markets/rural areas, which are also reviewed in Chap. 5 of this book. Wei and Hwang [17] determine the optimal cell size in UDNs and have identified several challenges to this technique that should be addressed and overcome to realize ubiquitous 5G and mm-Wave networks in urban areas. A number of these challenges are also applicable to rural areas. According to Wei and Hwang [17], these challenges include:

- network planning and deployment (architecture, resource management, interference management, managing the complexity of the network, proper optimization, and scheduled maintenance are all key for a successful deployment),
- identifying and considering the fundamental limits of network densification, therefore not *overdesigning* the network and effectively introducing interference between small cells,
- managing the tradeoff between the cost of the network and the QoS to the users, essentially limiting the number of small cells and guaranteeing a pre-specified QoS, and
- taking into account that at mm-Wave, adequate knowledge and considerations for line-of-sight (LOS) and non-LOS communications are needed and must be a key consideration during the planning of network densification. Furthermore, in dense urban areas, the tradeoffs between low frequencies (high penetration but higher interference from adjacent cells) and high frequencies (more bandwidth, less interference, but high losses) must be considered during planning.

Network densification comprises densification over space, therefore the physical placement of small cells, as well as densification over frequency, exploiting greater slices of the radio spectrum in assorted bands. Large-scale and low-cost spatial densification is enabled by self-organizing networks and interference managing

(SDR and NFV). The complete potential of network densification can also be harnessed by combining these techniques with backhaul densification and intelligent receivers, able to dynamically cancel interference. Additionally, and as identified and researched by Andreev et al. [2], modern urban areas (especially megacities) rarely rely only on static or semi-static mobile broadband topologies, and the internet is *moving* in these areas. Architecture planning therefore also needs to account for moving access networks (MANs), and in the case of 5G networks, mm-Wave MANs.

Consider a scenario in a dense urban area where access networks are provided by drones, from mobile hotspots in autonomous vehicles, and D2D connectivity with high mobility. An MAN therefore, and particularly if operating in the mm-Wave spectrum, requires additionally complex planning and knowledge of radio propagation models and channel modeling. The technique to undertake such planning and massive characterization of moving access points (MAPs) is cumbersome, complex and resource-intensive, and furthermore, specific to each area (typically generic modeling does not apply). In emerging markets, adequate skills training and investments in higher education are needed to ensure that local communities can socially and economically develop towards planning and maintaining these systems locally.

In summary, network ultra-densification and UDNs are feasible techniques to add capacity to dense urban areas and in many circumstances, in rural areas as well (with the advantage of lower complexity compared to dense urban areas). However, initial planning, cost, considering MAN and MAP, and extending these networks require a significant effort to effect successful and efficient implementation. Andreev et al. [2], for example, propose techniques to optimize UDNs, although these techniques are specific to the area considered in that work; there are generic methods and considerations that can be applied to other scenarios. Another characteristic of Industry 4.0 that is spearheaded by enabling technologies is cloud computing. In emerging markets, especially in rural areas, mobile cloud computing is an important and valuable characteristic that allows geographically removed communities to take part in Industry 4.0.

2.6.2 Mobile Cloud Computing

According to Wang et al. [20], "*mobile computing, wireless networks and cloud computing are three technologies that are rapidly converging into*" the umbrella of mobile cloud computing (MCC). The benefits of 5G and mm-Wave in terms of capacity, speed, and latency have further spearheaded development of mobile services, driven by the demands in resource management, data storage and mobile sensing. MCC is a solution for mobile services to improve the computational abilities of resource-constrained mobile devices and exploit elastic resources of cloud computing such as virtual servers, services or application platforms. Pricing strategies of pay-as-you-use or fixed rental are typically applied to monetize MCC, and with 5G and mm-Wave networks, these services can excel beyond the bottlenecks of internet connectivity. In emerging markets and rural areas, the benefits of MCC through 5G and mm-Wave are twofold: firstly, a population in an emerging market or rural area could attract investors and external stakeholders to build infrastructure that hosts the services and applications in these areas. Socioeconomic growth in these areas can be increased through upskilling the local population and creating new jobs to expand, maintain or use this infrastructure. Secondly, providing an area has access to (scalable) 5G mm-Wave networks, MCC allows the local population to access computer-intensive services on the cloud and innovate and build custom services and applications applicable to the local communities. 5G and mm-Wave will also allow scalable and ubiquitous internet access, and in many rural communities (especially in agriculture), the IoT can benefit local communities, as briefly reviewed in the following section.

2.6.3 Scalable Internet-of-Things

IoT is not defined in this section, since it is assumed that the reader is acquainted with the concept (or consider consulting [11]), 5G and mm-Wave, primarily through network slicing, ultra-reliable low latency communication, and non-public networks/industrial IoT, offers new levels of scalability to the IoT. Scalability in the context of the IoT can occur in two primary *directions*, namely:

- vertical scaling (or scaling up) that escalates the capacity of current hardware components or software implementations through the addition of additional resources to it, or
- horizontal scaling that escalates the capacity by joining multiple hardware components or software resources to operate as a sole component.

Essentially, the goal of scaling of IoT is to enhance the network in terms of its

- capacity,
- portability (and heterogeneity),
- adaptability (a form of standardization),
- usability (ease of deployment),
- efficiency,
- safety, and
- security.

Additionally, IoT platforms must be highly available, maintainable, and most importantly, relevant to the (local) consumers. Complementary to the reviews on how 5G and mm-Wave, through network scaling, shift the potential of next generation networks, the IoT can also benefit from its capabilities in many of these segments. 5G has improved efficiency, safety, security, and scalability, and can be adopted and integrated in IoT architectures to yield mm-Wave massive IoT systems, as reviewed in Chap. 1 of this book. Lambrechts and Sinha [7] also provide a review on how the IoT and WSNs can be implemented to achieve sustainable and smart cities. The following section reviews D2D connectivity, specifically its high-mobility aspect and what this means when operating in the mm-Wave spectrum.

2.6.4 Device-to-Device Connectivity with High Mobility

In providing users with mobile broadband services, user mobility is a key requirement for mobile networks, as it affects network performance and perceived quality of service [15]. It is essential to define mobility-aware network merits such as "*handoff rate, handoff probability, sojourn time, direction switch rate, throughput, and coverage for efficient network dimensioning and optimization*" [15]. According to Tabssum et al. [15], as a result of key technologies of 5G such as

- "*spatial randomness of network tessellation,*
- *heterogeneity of BSs,*
- *the ultra-dense nature of network deployment, and*
- *diversified mobility patterns of users, devices, and network nodes*"

the modeling and parameterization of mobility in these networks are crucial (and complex) in optimizing the design and performance of these networks. Tabassum et al. [15] highlight that mobility in 5G networks has an impact on

- resource management in terms of channel allocation schemes, multiple access tools, estimating network capability, call blocking levels, traffic capacity per single cell, QoS, signing, and traffic load approximation,
- radio propagation characteristics like disparity in signal power, interference, call dropping intervals, and handoff processes centered on signal power, and
- location administration characteristics like position scheduling, data location policies, and database demand.

In heterogeneous and ultra-dense random mobile networks such as 5G, mobility characterization becomes particularly challenging. Tabassum et al. [15], for example, define various mobility models such as

- "*purely random models (random walk, random way point, random direction),*
- *spatially correlated models (pursue mobility, column mobility), and*
- *temporally correlated models (Gauss-Markov, Levy flight)*".

The differences in these models are primarily due to their statistical properties, and Tabassum et al. [15] deliver an in-depth analysis of the advantages and disadvantages of each of these models. In general, a mobility-aware 5G network not only improves QoS for users, but also optimization of resources. This is again an important factor in emerging markets and rural areas, and initial planning (and modeling) of 5G networks should not be overlooked or underestimated prior to deployment. The mobility modeling approach can be complex and numerous models exist; however, in Tabassum et al. [15], a summary of and in-depth analysis of these models are

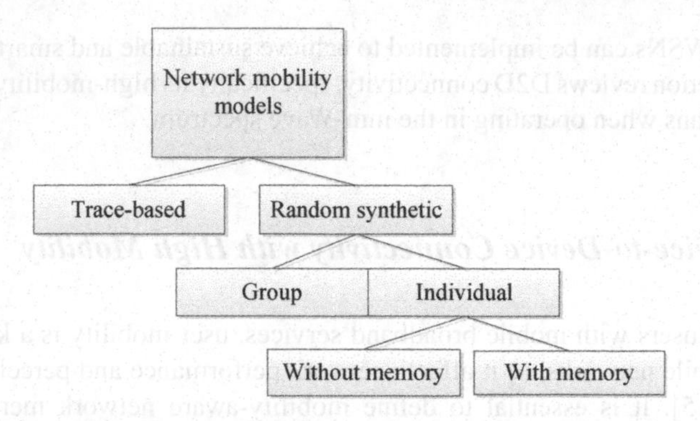

Fig. 2.8 The classification of network mobility models to model, analyze and optimize wireless networks, adapted from Tabassum et al. [15]

presented. Adapted from Tabassum et al. [15], the various mobility models are outlined in Fig. 2.8.

According to the mobility model classification presented in Fig. 2.8, the two primary categories are trace-based and random synthetic models. Trace-based representations are acquired by measurement of installed structures and are accurate based on the movement and topology in specific areas. Traces are used in assessing performance as well as for the optimization of handoff protocols. According to Tabassum et al. [15], a disadvantage of trace-based mobility models is that they are area-specific (from measurements) and can often not be universal for an assortment of circumstances. Arbitrary artificial representations are mathematical models (such as Bayesian, Brownian, and Levy patterns) that characterize the movement of networked devices in an area. Generally, these models are relatively simple and analytically traceable; however, the models may not reflect mobility patterns. According to Tabassum et al. [15], random synthetic models are subcategorized as individual (with and without memory) and group models. Individual mobility models without memory simulate the movement of a mobile subscriber that is not dependent on additional nodes. The location, speed and movement bearing of one node are not influenced by nodes in the surrounding area or by any previous activity. Although these models are mathematically traceable, they may not be representative of real-world activity, and individual models with memory are typically preferred. These models are also independent of other nodes; however, the next location of the node is a function of the previous location and velocity of movement. A further extension on individual models are group models. These models are based on mathematical functions to simulate the behavior of a group of nodes and therefore have spatial dependencies. Mathematically, the simplified representation of mobility models is defined as a stochastic sequence, S:

$$S = \left(L, \Theta, T, T_p\right) \tag{2.4}$$

where L is the distance that a mobile subscriber has covered (flight length), Θ is the direction change from the previous flight, T is the flight time, and T_p is a pause interval. Using this stochastic process, users would choose each variable before taking another flight permitted to their corresponding distributions. The mobile velocity is conveyed by $V = L/T$ and the beginning of each transition is called a way point. The trajectory is conveyed as a group of waypoints and the lines produced by S [15]. Simulating and analyzing the mobility models are not within the scope of this chapter and it is recommended that readers consult Tabassum et al. [15] for this. The next characteristic of 5G and mm-Wave that is reviewed, again important in areas where energy is limited or unreliable, is its energy efficiency.

2.6.5 Green Communications (Energy Efficiency)

A review on small cells is presented in this chapter, as well as the characteristics of 5G and mm-Wave that make it energy-efficient (through small cell deployments and inherent characteristics of the technologies). Mowla et al. [12] highlight that the power consumption of a HetNet should be considered from the perspective of both the access network and the backhaul. Although small cell networks can increase the amount of available bandwidth in a 5G HetNet, the increasing number of cells that become uncoordinated or lightly loaded contribute to the total power consumption of the access network. Mowla et al. [12] compare such a system with a smart grid utility that has various means of generating energy (coal, diesel, or renewable energy) and needs to distribute these methods as the demand varies during the day. In a 5G Hetnet, keeping all small cells operating at full capacity would ensure maximum bandwidth at all times; however, as demand changes during off-peak times, it will lead to an oversupply, increased operational costs, and high power consumption. Mowla et al. [12] therefore propose an energy-efficient communications model for 5G HetNets. This analytical model calculates the optimal number of required active small cells during various hours of the day. From the results, the model can save energy by putting redundant cells into sleep mode while maintaining QoS. Mowla et al. [12] investigate and introduce two energy-efficient backhaul solutions to complement the proposed 5G access network model. Vyas [16] investigates the potential of using renewable energy for 5G small cell deployments without degrading the performance of the network. Building on works such as those of Mowla et al. [12] and Vyas [16] will be beneficial in designing and deploying energy-efficient 5G and mm-Wave networks. Furthermore, upskilling local communities on energy efficiency and communications technology will have long-term socioeconomic benefits, as reviewed in Lambrechts and Sinha [8].

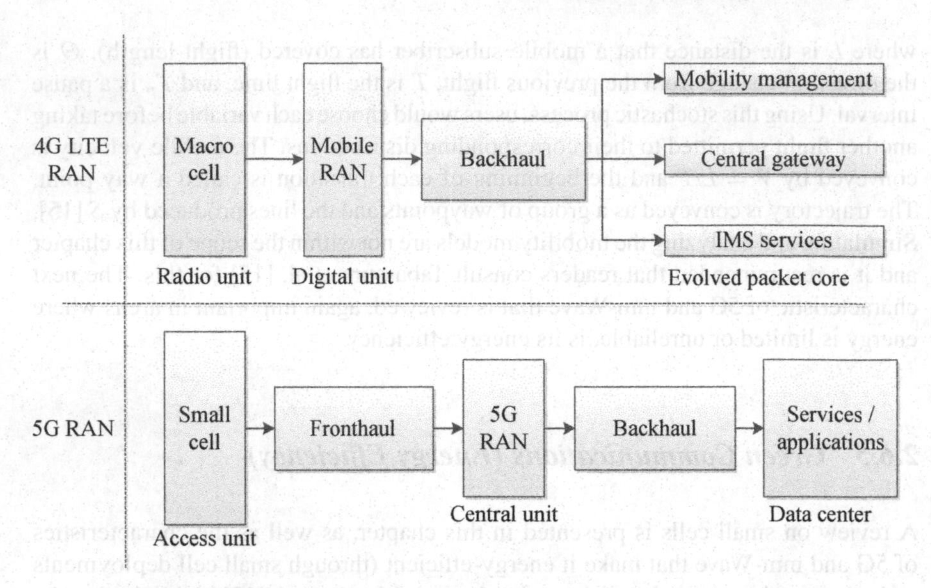

Fig. 2.9 The key differences between 4G RAN and 5G RAN as supplied by Samsung

2.6.6 New Radio Access Techniques

5G radio access deployments are characterized by their dense, throughput-focused, and software-driven nature. Also, the logical separation of the 5G RAN components differs from that of earlier generation networking technologies such as 4G. The 5G RAN is not constrained by BS proximity or complex infrastructure. It operates as disaggregated, flexible, and virtual entities with new interfaces that can add additional access points in a network. Samsung[3] provides a summary of the key differences between 4G RAN and 5G RAN, adapted and presented in Fig. 2.9.

The extent of these differences shown in Figure 2.9 is not within the scope of this chapter; however, it is vital to be informed that the 5G RAN real-time functions contain

- network scheduling,
- link adaptation,
- power control,
- interference coordination,
- retransmission,
- modulation, and
- coding.

[3]http://www.samsung.com.

These tasks entail high real-time merits and substantial calculating load and distribution sites should contain devoted hardware with large accelerator processing conditions and merit, while being in adjacent vicinity to services. RAN non-real-time functions comprise

- cell handover,
- cell selection and reselection,
- user-plane encryption, and
- multiple connection convergence.

These functions are not dependent on high real-time performance characteristics and latency requirements and a common processor can be used to service these requests. From this brief summary of the 5G RAN, it is evident that the provision of next-generation mobile communication services is becoming increasingly complex. The rollout of efficient, successful, and sustainable 5G and mm-Wave networks is more challenging than in previous generations and requires knowledge, skills, and deployment of radio, edge, transport, core, and cloud infrastructure. In terms of the complexity consequences for emerging markets and rural areas, Chaps. 5 and 6 of this book propose strategies to finance deployment, and innovate pricing strategies to ensure availability of these networks in these communities, while focusing on upskilling the local communities for long-term socioeconomic benefits [8]. 5G RAN architecture is not reviewed in detail in this book, as it falls outside the general scope, although it is worth mentioning to grasp the complexity of these systems.

2.7 Conclusion

This chapter is dedicated to identifying and reviewing the key technologies of 5G and mm-Wave networks that could have a significant impact on its deployment and ubiquity in emerging markets, especially in rural areas. Emerging markets often have financial constraints in many sectors, including the telecommunications industry, and alternatives that are less costly are typically implemented. In 5G and mm-Wave networks, certain technologies allow for less costly implementation, without compromising on the equipment and infrastructure. These networks are modular and in essence allow for scalable implementation. Furthermore, as the subscriber base increases and the demand for network resources increases, 5G and mm-Wave networks can be upgraded and services can be added, while ensuring that the technology (in terms of capacity, speed, and latency) remains future-proof.

Numerous references to these enabling technologies are made in Chaps. 5 and 6 of this book, and the reader is urged to refer back to this chapter for a brief review of each technology. These technological benefits are therefore a baseline for economic and socioeconomic growth in emerging markets.

References

1. Akyildiz F, Nie S, Lin S, Chandrasekaran M (2016) 5G roadmap: 10 key enabling technologies. Comput Netw 106:17–48
2. Andreev S, Petrov V, Dohler M, Yanikomeroglu H (2019) Future of ultra-dense networks beyond 5G: harnessing heterogeneous moving cells. IEEE Commun Magaz 57(6)
3. Barakabitze AA, Ahmad A, Mijumbi R, Hines A (2020) 5G network slicing using SDN and NFV: a survey of taxonomy, architectures and future challenges. Comput Netw 167:106984
4. Cai M, Laneman JN, Hochwald B (2017) Beamforming codebook compensation for beam squint with channel capacity constraint. IEEE International Symposium on Information Theory (ISIT). IEEE, 25 June 2017, pp 76–80
5. Chatchaikarn A (2017) Top 10 small cell infrastructure questions. Understanding the trends and challenges for today's small cells and how power-efficient power amplifiers can help overcome them. Retrieved December 20, 2019 from http://www.mouser.cn
6. Dietrich CB (2000) Adaptive arrays and diversity antenna configurations for handheld wireless communication terminals. PhD Dissertation. Virginia Tech
7. Idowu-Bismark OB, Ibhaze AE, Atayero AA (2017) Mimo optimization techniques and their application in maximizing throughput for 3GPP HSPA+. J Wireless Networking Commun 7(1):1–8
8. Karlsson A, Al-Saadeh O, Gusarov A, Challa RVR, Tombaz, S, Sung KW (2016) Energy-efficient 5G deployment in rural areas. In: IEEE 12th international conference on wireless and mobile computing, networking and communications (WiMob). IEEE,, 17 Oct 2016, pp 1–7
9. Lambrechts JW, Sinha S (2019) Last mile internet access for emerging economies. Springer International Publishing Switzerland. ISBN 978-3-030-20956-8
10. Lambrechts JW, Sinha S (2016) Microsensing networks for sustainable cities. Springer International Publishing Switzerland. ISBN 978-3-319-28358-6
11. Lemieux N, Zhao M (2019) Small cells, big impact: Designing power solutions for 5G applications. Retrieved December 18, 2019 from http://www.ti.com
12. Li X, Samaka M, Chan HA, Bhamare D, Gupta L, Guo C, Jain R (2017) Network slicing for 5G: challenges and opportunities. IEEE Internet Comput 21(5):20–27
13. Mavromoustakis C, Mastorakis G, Batalla JM (eds) (2016) Internet of things in 5G mobile technologies. Springer International Publishing Switzerland. ISBN 978-3-319-30911-8
14. Mowla M, Ahmad I, Habibi D, Phung QV (2017) A green communications model for 5G systems. IEEE Trans Green Commun Networking 1(3):264–280
15. PwC (2018) Why 5G can't succeed without a small cell revolution. Retrieved January 5, 2020 from http://www.pwc.com
16. Rohde and Schwarz (2016) Millimeter-wave beamforming: antenna array design choices & characterization. White Paper, pp 1–26
17. Sbeglia C (2019) Opposition to 5G small cell deployment spreads across US. Retrieved December 15, 2019 from http://www.rcrwireless.com
18. Tabassum H, Salehi M, Hossain E (2018) Mobility-aware analysis of 5G and B5G cellular networks: A tutorial. arXiv preprint arXiv:1805.02719
19. Vyas A (2019) 5G green communications. a review report. Adani Institute of Infrastructure Engineering. https://ssrn.com/abstract=3442579 or http://dx.doi.org/10.2139/ssrn.3442579
20. Wang X, Han G, Du X, Rodrigues JJPC (2015) Mobile cloud computing in 5G: emerging trends, issues, and challenges. IEEE Netw 29(2):4–5
21. Wei Y, Hwang S (2018) Optimization of cell size in ultra-dense networks with multiattribute user types and different frequency bands. Wireless Commun Mobile Comput 2018:1–10
22. Yu Y, Baltus PGM, Van Roermund AHM (2011) Integrated 60 GHz RF beamforming in CMOS. Springer Science and Business Media. Springer, Dordrecht. Print ISBN 978-94-007-0661-3
23. Zhang S, Guo C, Wang T, Zhang W (2018) ON-OFF analog beamforming for massive MIMO. IEEE Trans Veh Technol 67(5):4113–4123

Chapter 3
Transceivers for the Fourth Industrial Revolution. Millimeter-Wave Frequency Mixers and Oscillators

Abstract The fourth industrial revolution (Industry 4.0) demands ubiquitous, low-power and high-speed communications between a vast number of devices for applications in association with the internet of things and wireless sensor networks. Traditional receiver architectures such as the super-heterodyne transmitter and receiver are widely used and have proven to be adequate for a new generation of communication technologies, but not without considerations to adapt to millimeter wave (mm-Wave) and teraherz (THz) frequency operation. On system level, next-generation communication protocols such as fifth-generation (5G) are pushing the limits of the hardware. On subsystem level, improvements are being researched to adapt traditional architectures and topologies to cope with a significant increase in operating frequency. On component level, integrated circuit designers and researchers are adapting and improving the layout, geometry and physical characteristics of active and passive components and taking advantage of physics to mitigate limitations incurred by mm-Wave and THz circuits. In this chapter, a review on system level of traditional transceivers is presented as a precursor to the challenges and limitations of mm-Wave 5G-capable transceivers. This is followed by an in-depth critical review at subsystem level of the frequency mixer. Circuit operation and performance metrics of the frequency mixer are presented and followed by a discussion on mm-Wave 5G-capable circuits presented in literature, highlighting the shortcomings and/or innovations to adapt these circuits for a new revolution in communication systems.

3.1 Introduction

Heightened awareness of next-generation radio frequency (RF) wireless technologies like fifth-generation (5G) is matched by a surge in the development of high-performance, low-cost and low-power transceivers [19]. 5G allows for wireless communications below 6 GHz (sub-6 GHz band); however, this spectrum is already becoming crowded. To its advantage and as a result of the modularity of the 5G

W. Lambrechts and S. Sinha, *Millimeter-wave Integrated Technologies in the Era of the Fourth Industrial Revolution*, Lecture Notes in Electrical Engineering 679,
https://doi.org/10.1007/978-3-030-50472-4_3

new radio, millimeter-wave (mm-Wave) frequencies are also allocated to 5G wireless communication. Bands such as the 28–33 GHz spectrum are popular because of minimal atmospheric attenuation. Transceivers in this band can take advantage of this through transmitting at lower power, hence achieving lower overall power consumption (which is important for IoT applications reliant on renewable energy). There are, however, additional practical implications and hardware-constrained limitations when implementing mm-Wave communication systems compared to sub-6 GHz realizations [45].

Microelectronic engineering of electronic circuits at mm-Wave frequencies is not an easy task, with numerous second- and third-order effects from lossy interconnects and parasitic effects of active components such as transistors to be dealt with. Package integration of mm-Wave circuits generates additional cost, requiring high-quality materials and additional simulation and design considerations that are crucial to ensure accurate operation. A typical transceiver of a communication system is a combination of various subsystems or building blocks, each of these blocks designed to operate for its specific application and to be able to interconnect with other subsystems.

Since many of the subsystems of a transceiver are required to operate at mm-Wave frequencies and are interconnected, challenges and limitations (along with the advantages of high-speed communications) must be considered during the design, development and deployment phases. These challenges are not limited to subsystems, but also apply to individual active (transistors and diodes) and passive (resistors, capacitors and inductors) components used to realize each circuit. Lambrechts and Sinha [24] reviewed the effects of mm-Wave frequencies on active and passive components, as well as techniques to mitigate some of these unwanted effects. Božanić and Sinha [6] reviewed packaging considerations for mm-Wave subsystems to mitigate external factors and interference. In this chapter, the effects and necessary considerations of mm-Wave signal processing of subsystems in a typical 5G communication system are considered, with reference to sources detailing effects from individual active and passive components, and also packaging. According to Yang et al. [45], the most common hardware-constrained characteristics of mm-Wave systems are

- phase noise in oscillators,
- non-linear power amplifiers (PAs) and low-noise amplifiers (LNAs),
- in-phase or quadrature (I/Q) imbalance in mixers,
- active phase shifters in large-scale phased antenna arrays versus lens topologies, and
- data processing challenges for digital equipment.

This chapter and the subsequent chapter examine these limitations in detail based on a critical evaluation of the subsystems where the constraints are typically found (oscillators, PAs, LNAs, and mixers). Furthermore, since modern transceivers are typically realized primarily in semiconductor processes, the challenges and limitations of the subsystems are reviewed with specific reference to integrated circuits (ICs) in complementary metal-oxide semiconductor (CMOS) or bipolar

CMOS (BiCMOS) technologies, which cover a wider array of topics, such as "*wave propagation, antenna design and communication capacity and limits*" [18].

The following section reviews the fundamental principles of transmitter and receiver systems and provides an overview of the challenges and limitations of specific subsystems when operated in the mm-Wave spectrum. Subsequent sections critically review the major considerations, challenges, issues and limitations of subsystems operated in mm-Wave with reference to effects generated by individual active components, passive components, packaging, and interconnects.

3.2 Traditional Transceiver Architecture Considerations

RF transmitters and receivers (transceivers) have been used extensively to transmit radio signals between modules and transfer information between two points. The traditional architectures and fundamental principles of transceivers are well known and are used in any electronic device capable of wirelessly transmitting data. However, more recently, adapting wireless transceivers to operate at high frequencies, especially in the mm-Wave bands, have received more attention owing to the inherent advantages of transmitting information on higher carrier signals [45]. Although the fundamental principles of these components and systems remain unchanged from traditional systems, there are numerous challenges accompanying communication at very high frequencies, with specific focus on mm-Wave frequencies used in 5G communication systems. Challenges occur at component level, subsystem level, at interconnects as well as at integration (packaging) and affect many stages of the traditional transceiver.

In this section, a brief review of traditional transceivers is presented, followed by reviews on the challenges in each subsystem of the receiver and transmitter. Researchers, designers and practitioners that develop and research technologies that are complementary to Industry 4.0 should have a common understanding of the fundamental principles of wireless transceivers. Firstly, *receiver* architecture and mm-Wave challenges are presented, followed by a similar approach to *transmitters*.

3.2.1 Traditional Receivers

In its simplest form, a traditional receiver is represented by a direct-conversion (also called homodyne or zero-intermediate frequency (IF) receiver) receiver architecture [2], where an arriving RF signal is transformed in a single-stage solution. These architectures are typically low-cost, as they require few components, which is an advantage in terms of integrated technologies owing to cost savings from smaller chip area requirements, but they do suffer from performance limitations. Figure 3.1 is a representation of a direct-conversion receiver.

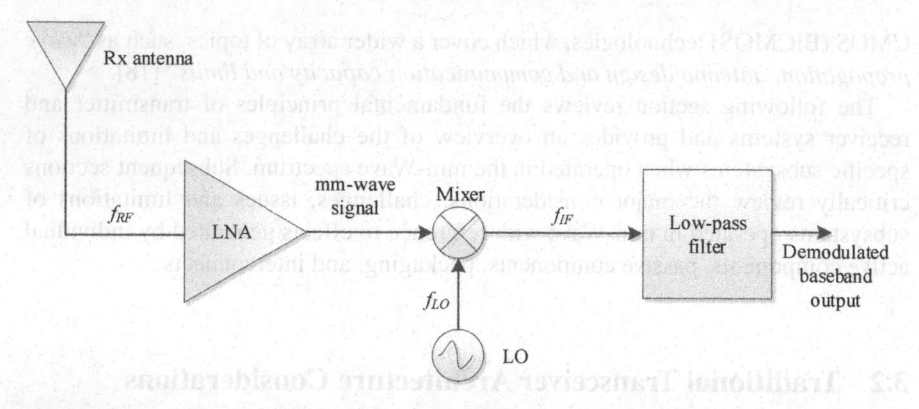

Fig. 3.1 Simplified representation of a direct-conversion receiver architecture

The direct-conversion receiver architecture in Fig. 3.1 receives an incoming wireless RF signal at the receiving antenna and this (weak) signal is at first enlarged by an LNA. The RF signal is then fed to a three-port device, the mixer, along with a local oscillator (LO) signal (the second input to the mixer) with a frequency ideally identical to the carrier frequency. The mixer multiplies the two input signals and at the output of the mixer, the multiplied signal is at 0 Hz (direct-current), because RF and LO have the same frequency, and is filtered to remove unwanted frequency components resulting from the multiplication of sine waves. The filtered signal is enlarged and demodulated and the baseband information sent by the transmitter is extracted from the signal. In some circumstances direct-conversion receivers are implemented using two stages of mixers to create an I/Q demodulator. A single LO is used to drive both mixers, but each with a 90° phase shift at the two mixer LO inputs.

In higher frequency communications and as an improvement on the direct-conversion receiver, another common architecture translates an RF input signal into a lower-frequency signal, referred to as the IF, and is referred to as a super-heterodyne receiver [3]. Figure 3.2 is a simplified representation of the super-heterodyne receiver architecture.

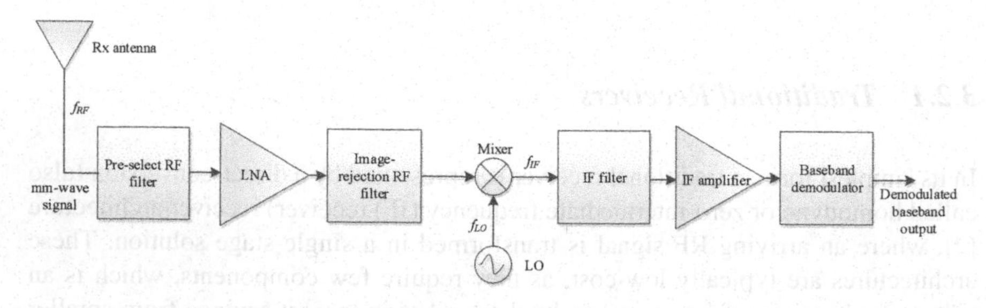

Fig. 3.2 Simplified representation of a super-heterodyne receiver architecture

From Fig. 3.2, in the super-heterodyne receiver, the incoming RF signal (from the receiving antenna that is matched to the LNA) is immediately fed through a bandpass filter (called the pre-select filter) to reject unwanted and *out-of-band* frequency components and fundamentally limit the bandwidth of the system. The LNA amplifies the RF signal and the noise figure (NF) of the LNA essentially determines the NF of the receiver. This amplified (in magnitude) representation of the RF signal is again filtered by an image-reject filter to remove any undesirable images from the signal and further limit its bandwidth as it is fed to the mixer subsystem. The mixer, driven by a dual input, the incoming RF signal plus the LO, down-converts the RF signal to an IF signal, essentially through multiplication of sine waves (the RF carrier and the generated LO signal). The IF signal at the output of the mixer is equal to the difference (and sum) of the RF and LO frequencies. Another bandpass filter is used to remove frequency artifacts from the multiplication process. The IF signal is then amplified to provide a significant amount of gain to the signal, used by the demodulator to extract the originally transmitted information. Super-heterodyne receivers are commonly implemented in two separate frequency-conversion stages, especially in higher frequency applications ($\gg$1 GHz). One crucial advantage of the super heterodyne receiver is that much of the signal processing can occur at the lower (IF) frequency, reducing the cost and complexity of the on-chip integrated components. Another advantage of the super heterodyne receiver is its superior sensitivity; the super heterodyne architecture filters out unwanted and spurious frequency components in the IF band, as opposed to the RF band, which is typically easier to achieve, and enhances the overall sensitivity of the receiver.

The following section reviews the fundamental principles of traditional transmitters in a similar approach as that adopted for the traditional receivers.

3.2.2 *Traditional Transmitters*

The traditional transmitter follows a similar strategy as the receiver, where the direct-conversion transmitter [1] is commonly used for lower frequency operation and has advantages in terms of lower cost and complexity. The direct-conversion transmitter also contains digital signal processing components at its initial phases to prepare the digital information for the RF stages. A basic illustration of the direct-conversion transmitter design is presented in Fig. 3.3.

Shown in Fig. 3.3, the direct-conversion transmitter design uses a digital signal processor (DSP) to produce the I/Q signals of the information contained in the baseband signal. The I and Q signals are then both fed to their respective digital-to-analogue converters (DACs), after which the signals are filtered (low-pass) to remove unwanted frequency components. This also limits the bandwidth used by the circuit. Following the filtering phase, the signals are able to be passed through the two mixers, both driven at their secondary inputs by one LO but with a 90° phase shift for the I and Q signals. The RF output signals are then combined and a PA amplifies the

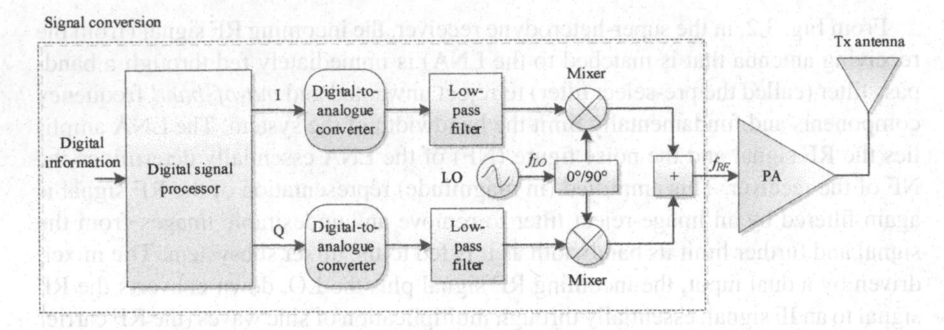

Fig. 3.3 Simplified representation of a direct-conversion transmitter design

subsequent signal. This representation of the signal is transmitted by the transmitting antenna as an RF carrier carrying the baseband information.

For higher frequencies, the super-heterodyne transmitter architecture is preferred, which has higher complexity and costs more, but allows some signal processing to be done at lower IF bands. The simplified super-heterodyne transmitter design is presented in Fig. 3.4.

The super-heterodyne transmitter design in Fig. 3.4 uses a similar DSP technique compared to the direct-conversion transmitter (highlighted in the dashed rectangle in Fig. 3.4); the primary difference is seen only after the process of combining the I and Q signals. In the super-heterodyne transmitter, the combined I and Q signals are in the IF band and pass through an IF filter (bandpass). Following this stage, the signal needs to be stabilized by an automatic gain control (AGC) subsystem to maintain consistent signal amplitudes regardless of any variations in the input (IF) signal. The signal can then be up-converted from the IF band towards RF using a single LO, albeit the second occurence, and the RF signal again passes through a filter to remove artifacts resulting from the multiplication process in the mixer. A PA operating in

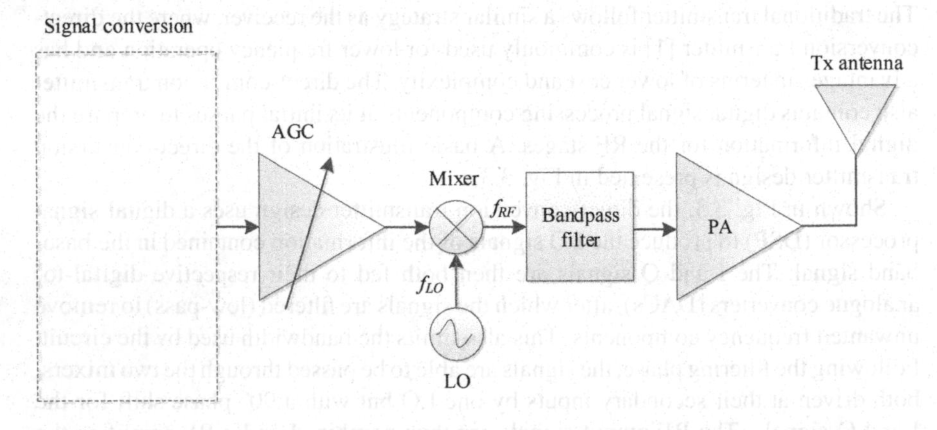

Fig. 3.4 Simplified representation of a super-heterodyne transmitter design

the RF domain (another relatively complex realization in mm-Wave 5G transmitters that require high-frequency and high-power active components) amplifies the signal significantly as it is fed to the (matched) transmitting antenna, alternatively first passing through another band-select filter.

In the super-heterodyne transmitter quite a few correction and equalization stages are needed after the information has been modulated. By using a single IF in the super-heterodyne transmitter, a single type of consecutive stage can be used, which minimizes the complexity of each individual stage and increases the reliability of the architecture. Furthermore, and especially important considering communication protocols such as 5G, the output frequency of the super-heterodyne transmitter can be changed dynamically, and only the RF output stages need to be retuned, as opposed to all intermediate stages in the traditional direct-conversion architecture. Lastly, an advantage of the super-heterodyne transmitter that lends itself to be used in 5G systems is the possibility of using high-speed DACs to generate the IF signals directly from a DSP or microprocessor, permitting more advanced modulating methods and enabling the use of software-defined radios as opposed to costlier and more error-susceptible hardware modulators.

The following section takes into account the fundamental principles of traditional transceivers and summarizes the challenges and limitations of the systems as a whole (either from the perspective of the receiver or the transmitter) before reviewing the challenges of subsystems. Reference to current implementations and research works are made to highlight the challenges experienced in realizing mm-Wave transceivers.

3.3 Transceiver Architecture Considerations in mm-Wave

Fully integrated mm-Wave transceivers can enable multi-Gbps point-to-point wireless transmission. However, traditional approaches with digital baseband processors and high-speed analogue-to-digital converters increase the overall system complexity significantly. In addition, power consumption of these digital components increases considerably when required to process real-time information at mm-Wave speeds [21]. Works such as that of [21] realize full integration of mm-Wave systems with up to 5 Gbps data rates and lower power consumption by integrating direct modulation and mixed-signal demodulation on-chip (realized in standard CMOS), therefore eliminating certain high-speed DSP requirements. The scaling of CMOS (reduction in transistor gate length) has become a promising and low-cost enabler for mm-Wave and THz [38] analogue circuits [12] with various integrated mm-Wave transceivers being proposed [14, 22, 28]. There are, however, still some fundamental challenges of mm-Wave circuits in CMOS [34], such as limited link bandwidth in cascaded subsystems to realize a complete transceiver [12]. Another inherent challenge of having all mm-Wave circuits on chip is generating the LO signal, which will be especially challenging if quadrature LOs are needed. Chi et al. [12] also highlight the difficulties of controlling transceiver performance variation as a function of variations in the CMOS process, as well as external and internal voltage and temperature variations. Most of the effects of such variations are more pronounced at mm-Wave frequencies when compared to lower RF implementations and techniques exist to mitigate or eliminate the effects, but at the expense of higher price and intricacy.

In terms of standard CMOS processes, the performance limitations are dependent on the maximum operational speed of the active modules and the merit of the passive modules. The $f_{\max}$ of a transistor is a good figure of merit (FOM) of its high-frequency performance. This frequency defines where the power gain of the components reaches unity—therefore up to which frequency it will not degrade the switching signal. Intrinsic design parameters such as the transistor gate resistance R_g can be manipulated through layout techniques [24], but there are fundamental limits on the maximum frequency of the metal-oxide semiconductor field effect transistor (MOSFET), derived from the approximation that

$$f_{\max} \approx \sqrt{\frac{f_T}{8\pi R_g C_{gd}}} \tag{3.1}$$

where f_T is the transition or cut-off frequency where the current gain of the transistor is zero and C_{gd} is the gate-drain capacitance of the transistor. For bipolar transistors, the approximation of $f_{\max}$ is given by [24]

$$f_{\max} \approx \sqrt{\frac{f_T}{8\pi R_b C_\mu}} \tag{3.2}$$

where R_b is the base resistance of the transistor and C_μ is the collector-base capacitance [24]. The full derivation of $f_{\max}$ in terms of the transistor geometry and f_T is given in Appendix 1 of this chapter. At this point it is of importance to take note of the relationship as well as the intrinsic limitations that $f_{\max}$ places on standard CMOS processes. Noticeable from both approximations (MOSFET and bipolar), $f_{\max}$ can be increased by lowering the base or gate resistance of the transistor as well as by decreasing the collector-base or gate-drain capacitance of the transistor, which are typically process-specific parameters but also adjustable to an extent through layout techniques [25] and extracting f_T and $f_{\max}$ [41]. The chapter will explore how these limitations affect mm-Wave communications on subsystem level (and how some of the limitations can be overcome to realize fully integrated mm-Wave standard CMOS transceivers). Typically, when integrating mm-Wave circuits in a semiconductor process, important performance variables that should be considered vary from circuits operating at lower frequencies, or at dc. Figure 3.5 is a summary of the performance variables that could influence the operation of a circuit if not considered, typically adding to the complexity and cost of mm-Wave integrated circuits.

As shown in Fig. 3.5, mm-Wave IC performance variables can be divided into three primary performance-related categories, namely

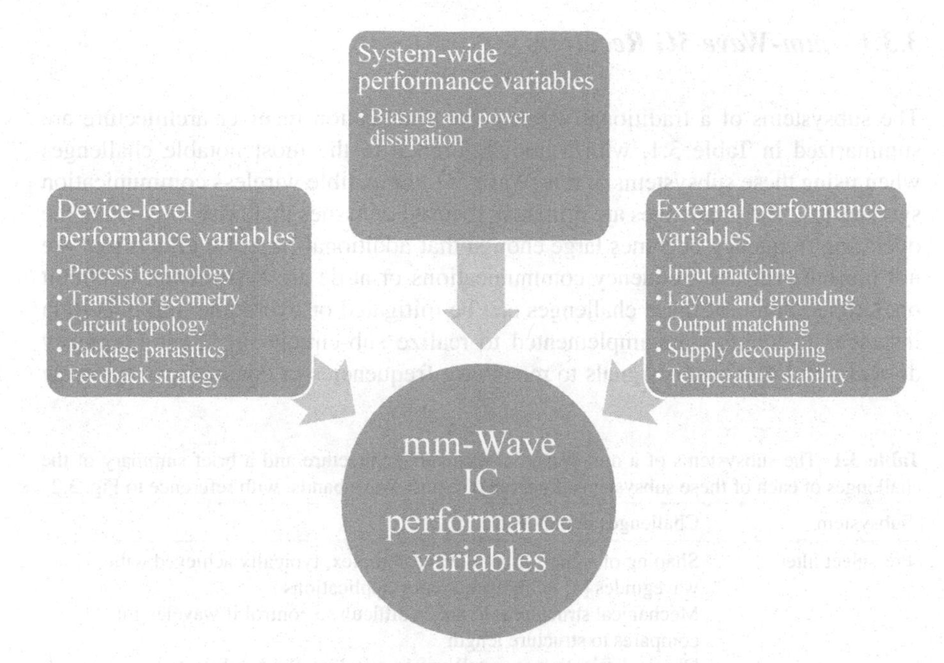

Fig. 3.5 Performance variables for high-frequency ICs required to realize efficient and effective mm-Wave 5G-compatible communication systems

- device-level,
- system-wide, and
- external.

Device-level performance relates to the physical characteristics of the components (usually CMOS/BiCMOS integrated active and passive components) that require attention and have unique limitations based on the operating frequency of the system. In mm-Wave 5G transceivers, for example, device-level performance characteristics require accurate simulations of electronic components (and full circuits) to ensure that real-world performance is comparable to the design requirements. System-wide performance is related to the biasing and power dissipation of the circuits as a system and is also influenced by component-level biasing, depending on the operating voltage of the semiconductor process. Thirdly, external performance variables such as matching, grounding, decoupling and temperature stability are needed to ensure consistent operation as the outside environment changes.

The typical subsystems that are used in an mm-Wave 5G receiver are summarized in the following section in terms of the challenges and limitations each subsystem poses at these high frequencies. The summary of the subsystems in the following section is expanded upon in subsequent sections of this chapter, highlighting the fundamental considerations when implementing mm-Wave wireless communications in proposed 5G systems.

3.3.1 mm-Wave 5G Receivers

The subsystems of a traditional wireless communication receiver architecture are summarized in Table 3.1, with a brief reference to the most notable challenges when using these subsystems in mm-Wave 5G-compatible wireless communication systems [7]. The challenges are primarily focused on issues that arise only when the operating frequency becomes large enough that additional parasitic effects that are not present in lower frequency communications or at dc are detrimental to circuit operations. Some of these challenges can be mitigated or overcome and in certain instances techniques are implemented to realize sub-circuits in lower frequency domains and up-convert signals to mm-Wave frequencies in consecutive, typically

Table 3.1 The subsystems of a mm-Wave 5G receiver architecture and a brief summary of the challenges of each of these subsystems if operated in mm-Wave bands, with reference to Fig. 3.2

Subsystem	Challenges in mm-Wave
Pre-select filter	Shaping of signals in mm-Wave complex, typically achieved with waveguides [4] in high-frequency applications Mechanical structure tolerances difficult to control if wavelength compares to structure length Ohmic dissipation at mm-Wave large (skin effect, substrate losses, metal losses)
LNA	Maintaining low NF while switching active components at mm-Wave frequencies Significant trade-off between high-gain and low-noise amplifier design; cascading typically needed, increasing complexity and cost Linearity (1-dB compression) a stringent requirement at increasing frequency operation High quality factor (Q-factor) passive components a limitation in microelectronic circuits
Image-reject filter	Similar challenges as pre-select filter; needs to reject unwanted components at very high frequencies
Sub-harmonic mixer	Mixers inherently sensitive to amplitude and phase imbalances; more difficult to mitigate at mm-Wave Signal isolation from LO port more difficult to achieve at mm-Wave (leaked signals at transistor-level)
LO	I/Q imbalances more prevalent at high-frequency and high-bandwidth signals. I/Q imbalance compensators increase cost and complexity Phase accuracy influenced by numerous factors such as phase jitter, thermal noise and signal interference Spurious tones and phase noise more difficult to optimize at mm-Wave frequencies, transferred to RF signal as additional noise
IF bandpass filter	Precise tuning required since this filter determines the selectivity of the receiver IF filter typically independent of mm-Wave RF signal (advantage)
IF amplifier	Maintaining low NF and high gain while switching active components at mm-Wave frequencies

cascaded, stages. The detailed techniques for each subsystem are studied in extra detail in Chap. 4 of this volume.

Noticeable from Table 3.1 is that subsystems such as filters typically suffer from high losses at mm-Wave frequencies as a function of the material of the passive components. Passive components are mechanical structures and physical tolerances and material losses are intensified as the wavelength of the operating frequency becomes equal to the dimensions of these components. The complexity and cost involved to realize accurate and efficient filters in mm-Wave are among the challenges in high-frequency receivers. Active subsystems such as the sub-harmonic mixer and local oscillator typically suffer not only from losses and tolerances of passive components, but also from limitations of the active components (transistors). Amplitude and phase imbalances, weak signal isolation (leaking between stages) and removal of spurious tones are among the major challenges in mm-Wave receivers. Additive noise and linearity from active switching in subsystems such as the LNA have a detrimental effect on the NF of the entire receiver and should ideally be minimized.

A similar correlation between the subsystems and challenges in mm-Wave transmitter architectures can be made. Each subsystem typically requires additional design and simulation when compared to lower frequency operation, and these challenges are presented in the following section.

3.3.2 mm-Wave 5G Transmitters

At the transmitter, IF signals need to be up-converted to a mm-Wave carrier in order to transmit the signal at these frequencies. As a result, the up-conversion and power amplification of the signals require high-frequency and high-power integrated components, where these two characteristics are typically trade-offs. Table 3.2 summarizes the principal subsystems required in a traditional transmitter and gives a brief summary of the challenges of each subsystem if operated within mm-Wave bands. Essentially, a similar trend is observed when compared to a mm-Wave receiver, where active and passive subsystems are fundamentally limited or challenged on component level.

The LO and sub-harmonic mixer of a mm-Wave transmitter have similar challenges when compared to those of the mm-Wave receiver. In terms of the RF bandpass filter, mostly similar limitations are incurred as with the receiver, with physical limitations of passive components a prominent consideration in transmitter design. Apart from physical limitations and difficulty in achieving high-bandwidth capabilities, high selectivity, low insertion loss and good temperature stability, the advantages and disadvantages of RF filter practical placement (at the antenna, behind the PA or on the high-frequency side of the mixers) should be considered as a function of the specific application.

For a multi-Gbps wireless link to exceed multi-kilometer transmission, a watt-level PA is required. However, such output power requirements at mm-Wave on fully integrated transmitters are challenging, especially in traditional CMOS

Table 3.2 The subsystems of a mm-Wave 5G transmitter and a brief summary of the challenges of each of these subsystems if operated in mm-Wave bands with reference to Fig. 3.4

Subsystem	Challenges in mm-Wave
IF bandpass filter	IF filter is typically independent of mm-Wave RF signal (advantage)
AGC	Automatic gain control DSP algorithms for effective power management in transmitters can become complex even in relatively lower IF
	Temperature stability of AGC a crucial design parameter for accurate transmit power variation compensation
Sub-harmonic mixer	Mixers are inherently sensitive to amplitude and phase imbalances; more difficult to mitigate at mm-Wave
	Signal isolation from LO port is more difficult to achieve at mm-Wave (leaked signals at transistor-level)
LO	I/Q imbalances are more prevalent at high-frequency and high-bandwidth signals. I/Q imbalance compensators increase cost and complexity
	Phase accuracy is influenced by numerous factors such as phase jitter, thermal noise and signal interference
	Spurious tones and phase noise are more difficult to optimize at mm-Wave frequencies, transferred to RF signal as additional noise
RF bandpass filter	Insertion loss, temperature stability, system bandwidth and high selectivity of mm-Wave filters are difficult to control
	Physical size and packaging of phased arrays (half-wavelength spacing) lead to very small allowable tolerances
	Several alternatives to filter placement in circuit—considerations are based on system application
	Shaping of signals in mm-Wave is complex, typically achieved with waveguides [4] in high-frequency applications
	Mechanical structure tolerances are difficult to control if wavelength compares to structure length
	Ohmic dissipation at mm-Wave is large (skin effect, substrate losses, metal losses)
PA	High-power and high-frequency operation are needed; choice of process technology influence cost and complexity
	Integration into a single application-specific integrated circuit is ideal but depends on process technology(ies) and PA architecture
	High linearity, power-added efficiency and high bandwidth requirements are difficult to achieve in mm-Wave frequencies
	On-chip integrated area varies significantly with architecture (number of inductors required) and can become large (expensive)

(Li et al. [27]). Increasing the f_{max} of traditional CMOS is promising but the low breakdown voltage limits the output power of mm-Wave CMOS PAs. Additional limitations on packaging and external losses increase the complexity and cost of interconnecting high-power PA outputs to external components. Semiconductor processing in materials with high breakdown voltage and high f_T/f_{max} "*such as gallium-nitride and indium-phosphide (InP)*" as proposed in [24] can increase the output power capabilities of integrated PAs, but increases complexity if combined with traditional silicon-based CMOS processes. Materials such as indium-gallium-arsenide (InGaAs) present a compromise in terms of higher breakdown voltage and higher f_t than traditional silicon CMOS, while being a relatively low-cost and attainable processing technology (Li et al. [27]).

In this chapter, the subsystems of mm-Wave 5G transceivers are critically reviewed in terms of their high-frequency operation and the challenges and limitations associated with these frequencies. Such a review is critical to determine the available techniques to mitigate some of these challenges and ensure efficient and consistent operation of these transceivers. The first subsystem that is critically reviewed is the frequency mixer, which is required to down-convert the RF signal at the receiver and up-convert the signal to RF at the transmitter. Traditional mixers are reviewed as a baseline for identifying, at component level, the limitations of mm-Wave 5G mixers.

3.4 Frequency Mixers

Frequency mixers are predominantly implemented for frequency translation and phase comparison and frequency translation is typically performed with either time-varying or non-linear circuits (in its most basic form it uses diodes [32]). A frequency mixer is essentially a "*three-port electronic device (two inputs and one output) that mixes (multiplies) the two input signals such that the output signal f_{out} is either the sum of or the difference between the two input signals*" [32], f_{in_1} and f_{in_2}. Mathematically, this relationship can simply be shown as

$$f_{out} = f_{in\_1} \pm f_{in\_2} \tag{3.3}$$

or in a more convenient form, as a function of the radial frequency

$$\omega_{out} = \omega_{in\_1} \pm \omega_{in\_2} \tag{3.4}$$

where

$$\omega = 2\pi f \tag{3.5}$$

and is specified in radians per second. The ports of a mixer are typically designated the LO port, RF port and the IF port. A sinusoidal continuous wave (CW) signal or a square wave signal usually drives the LO. This is depending on the application and generated by either a fixed oscillator or a voltage-controlled oscillator (VCO). The LO is always designated as an input to the frequency mixer where the RF and IF signals can be interchanged, depending on where the mixer is used (transmitter or receiver). In the transmitter, the IF signal is typically the second input and RF designated at the output, whereas in a receiver, "*RF is the second input and IF the output*" [32]. The input signals to a mixer are ideally (theoretically) sinusoidal waves (in practice the transfer function of the input signals will represent a non-ideal variation of a sinusoidal signal) and these signals are multiplied by the mixer.

Therefore, it is important to consider the trigonometric identity of

$$\cos(A + B) = \cos A \cos B - \sin A \sin B \tag{3.6}$$

where A and B represent the amplitudes of two arbitrary signals when analyzing the output of a mixer. The significance of this trigonometric identity will become apparent in the following sections where the transmitter- and receiver-side frequency mixers are reviewed. The following section reviews the transmitter-side mixer fundamentals as a precursor to the challenges and limitations of this subsystem in mm-Wave 5G-compatible transmitters.

3.5 Transmitter-Side (Up-Conversion) Mixer Fundamentals

In any transmitting frontend, the frequency mixer has two inputs that are designated IF and LO, and the output is a multiplied version of these signals, designated RF. The ideal transmitting mixer input and output strategy is shown in Fig. 3.6.

The frequency mixer at the transmitter, as shown in Fig. 3.6, modulates the information on the IF (baseband) signal onto an RF carrier produced by the LO whereas the output is therefore a high-frequency representation of the original information. This process is called *up-conversion* and the output of the mixer is designated the RF. In mm-Wave 5G implementations, the RF signal is typically within the mm-Wave 5G band. This higher frequency representation of the signal essentially means that the information can be sent *more quickly* from one point to another through a channel, since the higher the frequency, the higher the rate of communication. This is one distinct advantage of 5G communication. Mathematically, the ideal transmitting-side mixer/multiplier is embodied by considering for example the input signals IF and LO, where the IF (v_{IF}) and LO (v_{LO}) signals are represented by (adapted from Niknejad [33])

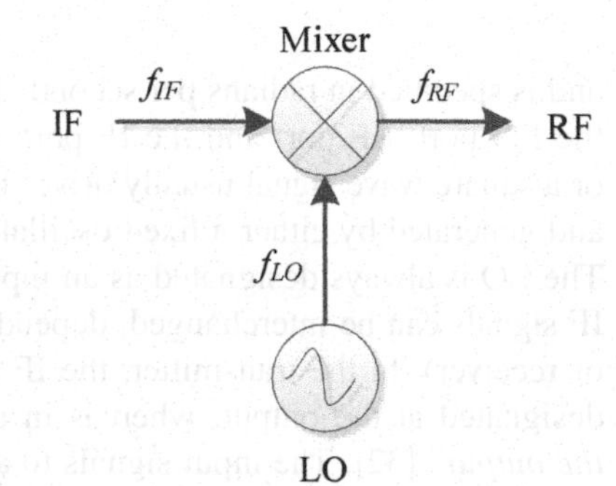

Fig. 3.6 An ideal representation of a frequency mixer at the transmitter, drawn typically with a multiplier signal. The multiplication of IF and LO is a process called up-conversion to an output signal at an RF (mm-Wave) frequency

$$v_{IF} = A_{IF}\cos(\omega_{IF}t) \tag{3.7}$$

where A_{LO} is the time-varying amplitude of the IF signal and $\omega_{IF}t$ is the time-varying frequency of the IF signal. v_{LO} is represented by

$$v_{LO} = A_{LO}\cos(\omega_{LO}t) \tag{3.8}$$

where A_{LO} is the amplitude of the LO and ω_{LO} is the frequency of the LO signal. The mixer output v_{out} would therefore be a multiplied version of the IF and LO signals, denoted by

$$v_{out} = v_{IF} \times v_{LO} \tag{3.9}$$

where the output signal will also be a time-varying signal dependent on the two inputs. If applying the trigonometric identity (3.6), the output signal of the mixer can be determined such that

$$v_{out} = \frac{A_{IF}A_{LO}}{2}[\cos(\omega_{LO} - \omega_{IF})t - \cos(\omega_{LO} + \omega_{IF})t] \tag{3.10}$$

showing that there are in fact two translated frequencies, simplified as LO − IF and LO + IF. The process of multiplying and translating the input frequencies of the transmitting-side mixer and the resulting output frequencies, in the frequency-domain, are shown in Fig. 3.7.

Even in the ideal multiplier, as shown in Fig. 3.7, there are two RF output frequencies, *RF1* and *RF2*, whose second-order product registers the same offset from IF. These two results are equally valid, although one signal is generally referred to as the image and is undesired. As a result of the additional frequency in the translated

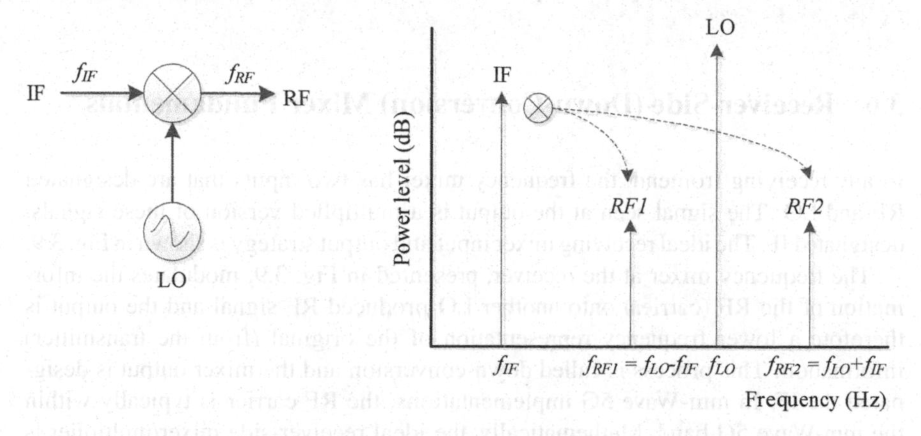

Fig. 3.7 Simplified representation of transmitter-side frequency translation performed by an ideal RF mixer, showing the RF, IF and LO relative frequencies in the frequency domain

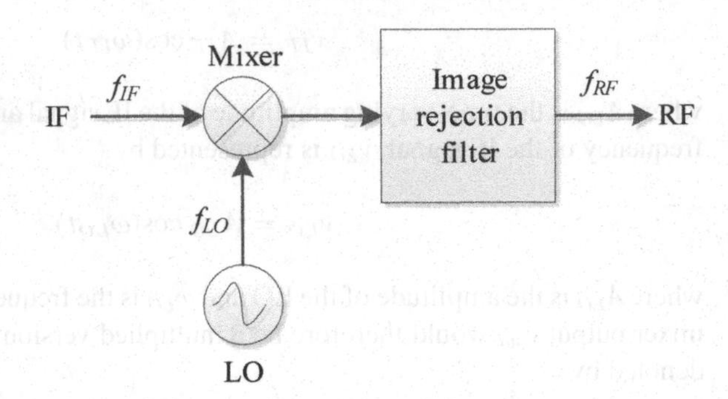

Fig. 3.8 Placement of the high- or low-pass image rejection filter to remove the (undesired) RF upper or lower sideband, depending on the application

output, the upper or lower sideband (additional) frequency must be filtered out by an RF filter. The filter, called the image rejection circuit, is placed at the mixer output, represented by Fig. 3.8.

The RF filter, as shown in Fig. 3.8, is typically placed before the power amplifier to limit the bandwidth requirements of the PA by removing any unwanted frequency components (and not transmitting them). This up-conversion process (heterodyning) in transmitters therefore prepares the baseband signal to be transmitted on an RF carrier, and in mm-Wave 5G communication systems the carrier will be in the mm-Wave spectrum.

The receiver-side mixer operates similarly, the primary difference being that the input signal to the mixer (apart from the LO) is at mm-Wave RF and must be down-converted back to baseband and the information has to be extracted from this signal. The incoming RF signal has therefore been subjected to noise from the environment as it passed through the channel (typically air), and the mixer has stringent performance requirements in terms of not adding additional noise and further degrading the integrity of the information. The following section reviews the fundamentals of the receiver-side mixer, again as a precursor to mm-Wave 5G mixers.

3.6 Receiver-Side (Down-Conversion) Mixer Fundamentals

In any receiving frontend, the frequency mixer has two inputs that are designated RF and LO. The signal seen at the output is a multiplied version of these signals, designated IF. The ideal receiving mixer input and output strategy is shown in Fig. 3.9.

The frequency mixer at the receiver, presented in Fig. 3.9, modulates the information of the RF (carrier) onto another LO-produced RF signal and the output is therefore a lower frequency representation of the original (from the transmitter) information. This process is called down-conversion and the mixer output is designated the IF. In mm-Wave 5G implementations, the RF carrier is typically within the mm-Wave 5G band. Mathematically, the ideal receiver-side mixer/multiplier is embodied by considering for example the input signals RF and LO, where the RF (v_{RF}) and LO (v_{LO}) signals are represented by

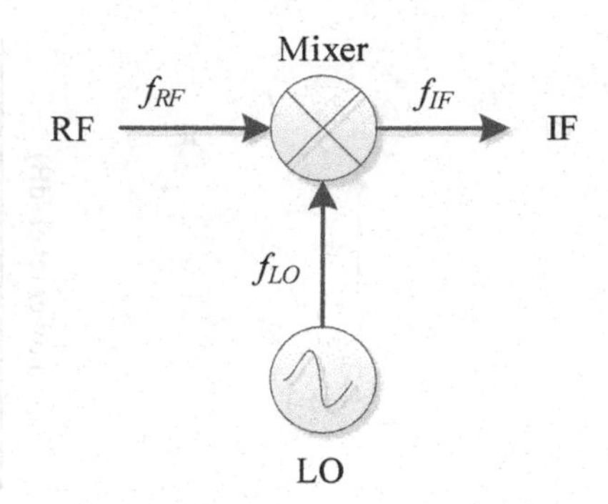

Fig. 3.9 An ideal representation of a frequency mixer at the receiver, drawn typically with a multiplier signal. The multiplication of RF and LO is a process called down-conversion to an output signal at an IF (typically a few MHz) frequency

$$v_{RF} = A_{RF} \cos(\omega_{RF}t + \phi(t)) \tag{3.11}$$

where A_{RF} is the time-varying amplitude of the RF signal and $\omega_{RF}t$ is the time-varying frequency of the RF signal. The term $\phi(t)$ represents a shift in phase in the RF carrier when compared to the LO signal. v_{LO} is therefore represented by

$$v_{LO} = A_{LO} \cos(\omega_{LO}t) \tag{3.12}$$

where A_{LO} is the amplitude of the LO signal and ω_{LO} is the frequency of the LO signal. The output signal from the mixer v_{out} would therefore be a multiplied version of the RF and LO signals, denoted by

$$v_{out} = v_{RF} \times v_{LO} \tag{3.13}$$

where the output signal will also be a time-varying signal dependent on the phase and amplitude of the two inputs. If applying the trigonometric identity (3.6), the output of the mixer, given as

$$v_{out} = \frac{A_{RF}A_{LO}}{2}[\cos((\omega_{LO} + \omega_{RF})t + \phi(t)) + \cos((\omega_{LO} - \omega_{RF})t + \phi(t))] \tag{3.14}$$

showing that there are in fact two translated frequencies, simplified as LO + RF and LO − RF. The process of multiplying and translating the input frequencies of the receiver-side mixer and the resulting output frequencies, in the frequency domain, are shown in Fig. 3.10.

Note that in Fig. 3.10, the LO frequency could potentially be lower than the RF signal (lower-side injection) or higher than the RF signal (high-side injection). As a result of the additional frequency in the translated output, the upper or lower sideband frequency should be filtered out by a baseband bandpass filter. The filter is placed at the mixer output, presented by Fig. 3.11.

The baseband bandpass filter is necessary, since the second frequency that also down-converts to the same IF adds noise and interference in the receiver. This filter is

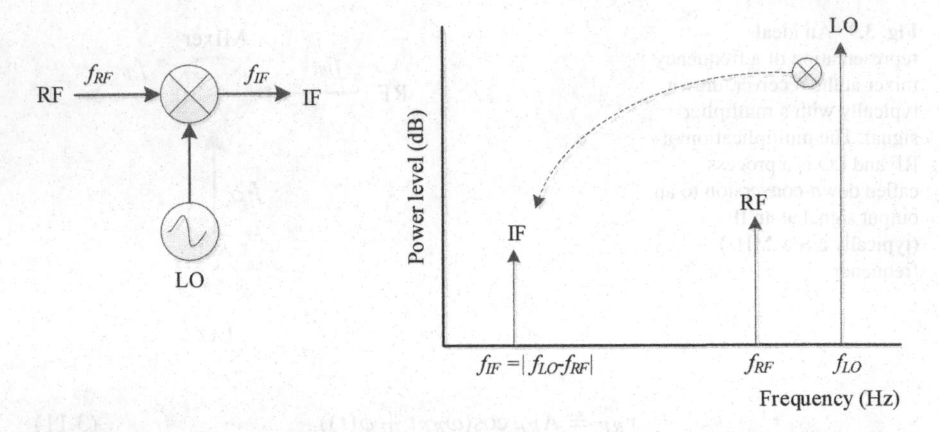

Fig. 3.10 Simplified representation of receiver-side frequency translation performed by an ideal RF mixer, showing the RF, IF and LO relative frequencies in the frequency domain

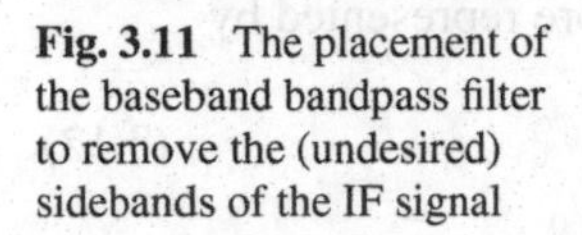

Fig. 3.11 The placement of the baseband bandpass filter to remove the (undesired) sidebands of the IF signal

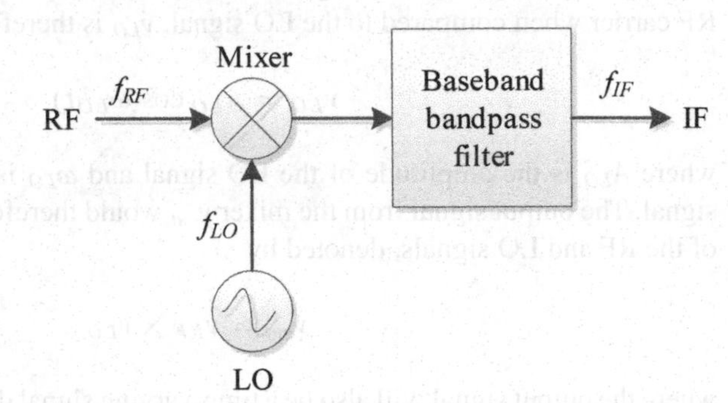

typically placed after the mixer, as opposed to filtering the incoming RF signal and its accompanying images, since such an architecture places a restriction on the selection of the IF frequency as a direct result of the capabilities of the image rejection filter. As the difference between RF and IF increases (with RF ≫ IF), the filter bandwidth (BW) should decrease and the *Q*-factor of the filter increases, through

$$Q = \frac{\omega_{RF}}{BW} \tag{3.15}$$

which could lead to impractical requirements on the filter *Q*-factor, especially when integrating the filter on chip where passive component *Q*-factors are typically limited. The linearity of the down-conversion mixer is an important FOM and typically high-gain mixers have increasingly non-linear output signals. For this reason, to improve the linearity of the receiver, the IF signal generated by the down-conversion mixer is amplified by the IF amplifier before the signal is demodulated.

It is also essential to recognize the performance metrics of the mixer to determine where the challenges and limitations will be experienced. By reviewing the performance metrics, it is possible to identify which frequency-dependent variables on component level and at system level might impose limitations on the design of the frequency mixer. The following section reviews the most commonly used performance metrics of frequency mixers with respect to their frequency dependency.

3.7 Mixer Performance Metrics

Microwave frequency mixers are responsible for translating frequencies and are a vital component in any transceiver. Because of its high-frequency operation, the requirements of the mixer (especially considering low-noise/high-gain functionality) are stringent as well as difficult to realize as the operating frequency increases. At mm-Wave frequencies, these requirements become even more difficult to realize and understanding the performance metrics sheds light on where the frequency limitations lie. This section reviews these performance metrics [29], which include

- conversion loss,
- isolation,
- 1-dB compression,
- multi-tone intermodulation distortion,
- NF, and
- additional performance metrics such as power consumption and chip area.

The first, and arguably the most important, performance metric is the conversion loss of the frequency mixer, basically the change in power level between input and output port.

3.7.1 Conversion Loss

One of the most important FOMs of mixers is the conversion loss, essentially described by the variation in power levels concerning RF input and the IF output (for a down-conversion mixer), simply defined as

$$\text{Conversion loss} = 10\log\frac{A_{RF}(\text{mW})}{A_{IF}(\text{mW})} \tag{3.16}$$

or

$$\text{Conversion loss (dB)} = P_{RF} - P_{IF} \tag{3.17}$$

where P_{RF} and P_{IF} are expressed in dBm. In an up-conversion mixer, the RF and IF terms will be swapped. The conversion loss should ideally be kept at a minimum, therefore as little power loss experienced when translating the RF signal to IF. The

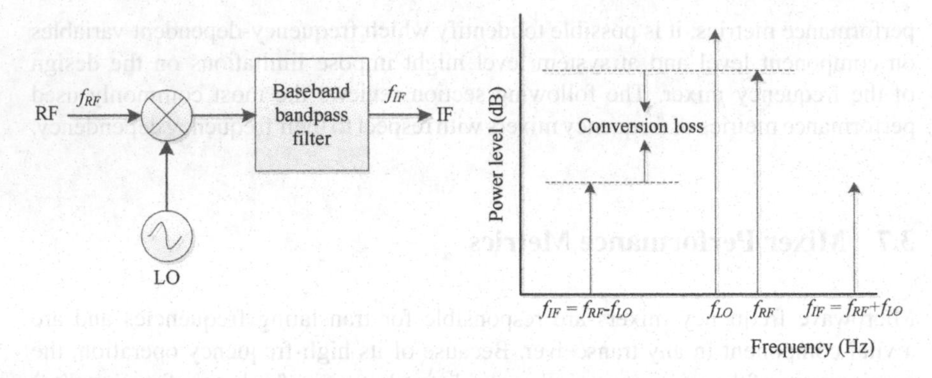

Fig. 3.12 Frequency-domain representation of the conversion loss of a receiver-side down-conversion mixer

conversion loss is a function of the mixer architecture and the active and passive components used to realize the circuit. Additional losses are incurred from transmission line losses, balun mismatches and the imbalance of the mixer. Higher bandwidth mixers are also typically associated with higher conversion losses, as it is difficult to maintain low, and constant, conversion loss over the entire bandwidth. Performance metrics such as isolation and 1-dB compression are in essence dependent on the conversion loss and therefore this is one of the primary quality FOMs to control and minimize in any mixer architecture. Figure 3.12 is a visual representation of conversion loss of a down-conversion mixer in the frequency domain.

Active mixers implement conversion gain through a built-in internal amplifier to ensure that the power of the output exceeds the input power (or at the very least the same). Another conversion FOM is conversion compression. This is the extent of the extreme RF power that will still result in a linear response. Typically, as an RF input signal increases in power, the IF output power will not be constant across the entire bandwidth of the mixer and the conversion compression will be lower. The choice of LO power level also influences the overall conversion compression of the mixer and should ideally not be too high and allow for a relatively high input RF power level.

Another important performance metric of a frequency mixer is its isolation between ports. At high frequencies such as mm-Wave, the isolation can become more complex to achieve owing to parasitic effects on component level. Active devices such as transistors suffer from the frequency-dependent Miller effect, a virtual increased input capacitance as function of gain, leading to low resistance between transistor input and output. Quantifying the isolation of ports for a high-frequency mixer is therefore essential and is reviewed in the following paragraph.

3.7.2 Isolation

Mixer isolation is the extent of power that escapes between ports and port isolation is achieved by mixer balancing and using hybrid junctions. Three common representations of mixer isolation are used as a FOM, namely

- "*L-R isolation between the LO and RF port,*
- *L-I isolation between the LO and IF port, and*
- *R-I isolation between the RF and IF port*" [29].

Graphically, these isolation types are represented in Fig. 3.13.

Mathematically, isolation in the difference between power levels of the input and escaped (leaked) output power, expressed as

$$P_{ISO(L-R)}(\text{dB}) = P_{LO} - P_{RF} \tag{3.18}$$

$$P_{ISO(L-I)}(\text{dB}) = P_{LO} - P_{IF} \tag{3.19}$$

$$P_{ISO(R-I)}(\text{dB}) = P_{RF} - P_{IF} \tag{3.20}$$

and normally only a single measurement (input-to-output or output-to-input) is required to determine the mixer isolation for each case (therefore approximately valid in both directions). L-R isolation is a critical FOM in down-conversion mixers, since LO power can escape (leak) towards the RF electric circuit and interfere with the integrity of the RF signal. In up-conversion mixers, if the LO and RF are close to each other in terms of frequency, interference is also caused between the RF and LO signals that are virtually impossible to filter out owing to its minimal frequency separation. This effect is maximized in mm-Wave frequencies where high-frequency

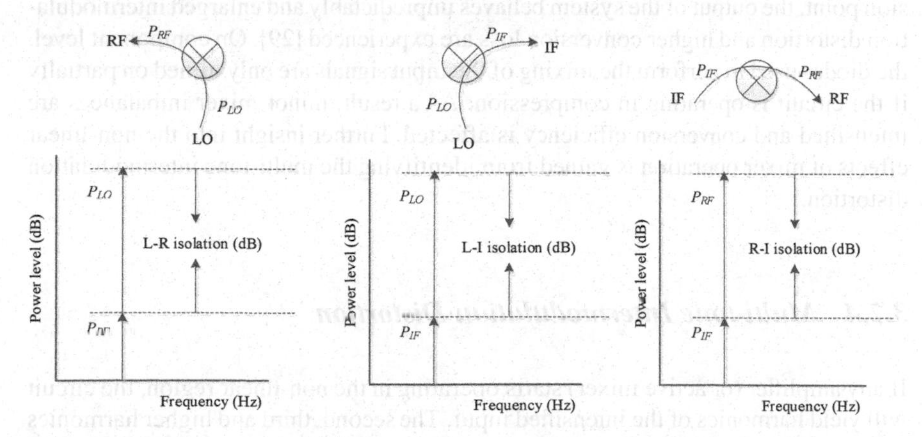

Fig. 3.13 Graphical representation in the frequency domain of **a** L-R, **b** L-I and **c** R-I mixer isolation

filters with extremely narrow bandwidth are difficult to achieve. L-I isolation must limit the outflow between the LO and IF signals when the RF port is terminated with 50 Ω and the largest effect of L-I leaking is experienced when the LO and IF frequencies are adjacent to each other. In this case, the LO signal can interfere with the IF circuitry and if the power level of the LO signal is high, it could saturate the IF amplifier. Finally, R-I isolation between the RF and IF signals is typically less of an issue, since the power levels of the two signals are typically orders of magnitude apart. R-I isolation is, however, "*a diagnostic metric for the total conversion efficiency of the mixer and with high R-I isolation, a mixer is well balanced and conversion loss tends to be low*" [29].

Ensuring that the frequency mixer is operated in its linear region, the 1-dB compression point should be identified. The following review on the 1-dB compression point as a performance metric of frequency mixers identifies the importance of linearity in mixers.

3.7.3 1-dB Compression

The 1-dB compression point of a mixer is fundamentally a measure of its linearity. During linear operation, if the RF signal increases by a set amount of power, the power of the IF signal will increase proportionally. As the RF power level intensifies past a specific level, the proportional change in the IF signal will change and non-linear operation is experienced; this is where the 1-dB compression point is, this point is represented by Fig. 3.14.

As shown in Fig. 3.14, at low RF input power, the slope with respect to IF output power is linear (therefore constant). Though, as the RF input power grows, the slope diverges from the ideal (constant) state and IF output power varies non-linearly with respect to the RF input power. As the mixer operation approaches its 1-dB compression point, the output of the system behaves unpredictably and enlarged intermodulation distortion and higher conversion loss are experienced [29]. On component level, the diodes used to perform the mixing of the input signals are only turned on partially if the circuit is operating in compression. As a result, minor mixer imbalances are intensified and conversion efficiency is affected. Further insight into the non-linear effects of mixer operation is gained from identifying the multi-tone intermodulation distortion.

3.7.4 Multi-tone Intermodulation Distortion

If any amplifier (or active mixer) starts operating in the non-linear region, the circuit will yield harmonics of the intensified input. The second, third and higher harmonics are typically out-of-bounds with the circuit bandwidth and relatively easy to filter out. Conversely, non-linearity will likewise yield a mixing consequence between

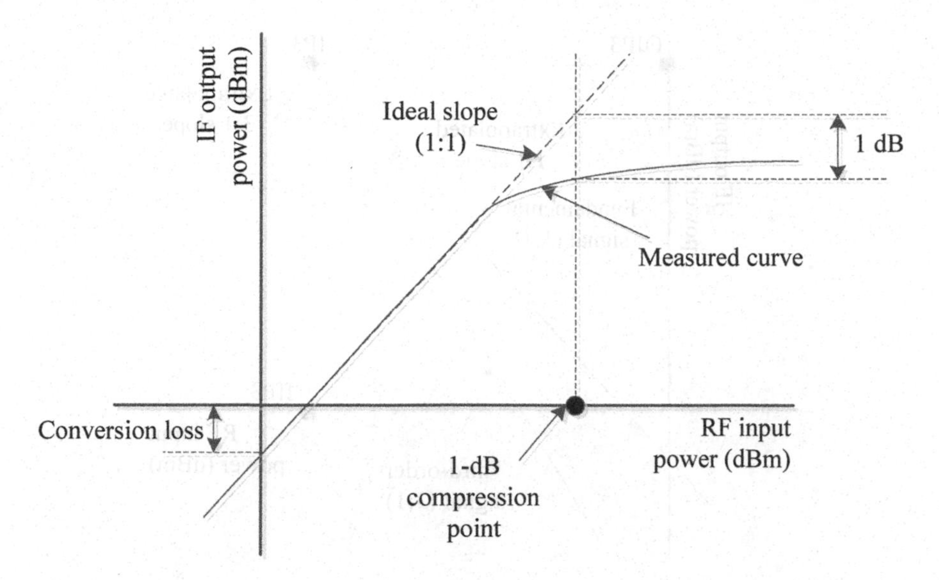

Fig. 3.14 Depiction of the 1-dB compression point of a mixer. Adapted from Marki and Marki [29]

a minimum of two signals. If these signals are adjacent in terms of their respective frequencies, the inter-modulated components cannot be filtered out and become interference in the circuit. As a result, it is important to control and maintain linearity in such circuits as far as possible.

Multi-tone intermodulation distortion (IMD) infers that more than one tone enters the mixer in a single port and intermodulation occurs inside the mixer. These tones also typically mix non-linearly, lead to distortion and interference of the original (wanted) signal and place a hypothetical higher limitation on the dynamic range of the receiving subsystem. The balancing of the mixer and the non-linear characteristics of the active circuit components affect the IMD of a mixer. The most noticeable passband intermodulation initiates as a result of the third-order non-linearity. A general and widely accepted FOM is the "*two-tone third order input intercept point (IP3) to predict the non-linear behavior of the mixer as the RF input power increases*" [15]. The IP3 is determined by providing the circuit with "*a two-tone signal and plotting on a log-log scale the fundamental output power versus the third-order intermodulation distortion products power (IMD3 or IM3) as a function of input power*" [15]. The IP3 therefore describes the capability of the mixer to suppress two-tone, third-order IMD and is the hypothetical position on the IF output power against RF input power arc where the wanted output (both tones) and separate third-order products converge. By plotting the output power versus the input power, the 1-dB compression point can be determined, and such a plot is called the *first-order signal plot*. If the signal power of the third-order products is also plotted on the same figure, the IP3 point can be determined. Figure 3.15 is a graphical representation of the IP3.

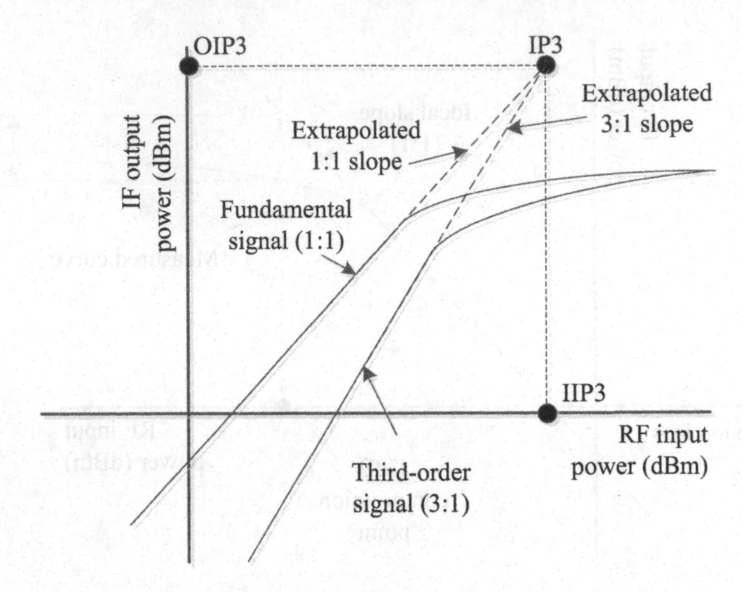

Fig. 3.15 Derivation of the IP3 of a mixer

In Fig. 3.15, the third-order harmonic power level is included in the original 1-dB compression point plot, and this product increases at three times the "*power of the first-order product*" [15] (RF versus IF fundamental components). This three-time increase is denoted as a 3:1 gain rate and is evident from the mathematical derivation of the mixer output. By extending the linear portions of both the fundamental signal power and the third-order signal power, the lines will meet at a point where the third- and first-order signal amplitudes are equal. Although this point cannot be practically achieved in any amplifier or mixer, it is a useful theoretical measure of the linearity of the circuit. The point where the lines meet is therefore called the IP3 and can be interpreted pertaining to the input power or the power at the output. If the point is interpreted pertaining to the input power (therefore the x-axis in Fig. 3.15), the point is called the IIP3, and if it is interpreted pertaining to the output power (y-axis in Fig. 3.15), it is called the OIP3. A higher output at the convergence point relays to improved linearity. This will also then relay to a lower IMD. Thus, the IP3 value gives an idea of how large an input signal a circuit can process before IMD occurs. Furthermore, the two-tone input signals can either be equal in power and frequency or have unequal characteristics. Figure 3.16 presents the frequency and power of the two input tones.

The relevance of Fig. 3.16 is realized when deriving the values of OIP3 and the third-order intermodulation power level IM3. If the two input tones have unequal power and frequency, the input signal is defined as

$$V_{in} = v_{low} \sin(2\pi f_{low} t) + v_{hi} \sin\left(2\pi f_{high} t\right) \tag{3.21}$$

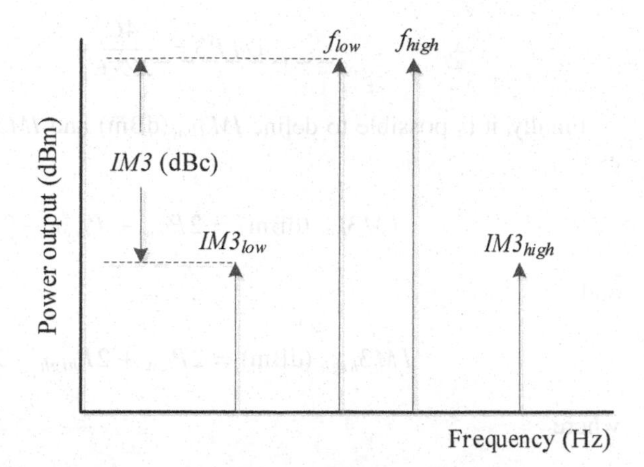

Fig. 3.16 The frequency and power relationship between the two input tones to determine the IP3, plotted in the frequency domain

where v_{low} and v_{high} are the voltage amplitudes of the two tones, respectively, and f_{low} and f_{high} are the respective frequencies. The output signal can be defined as

$$V_{out} = C_0 + C_1 V_{in} + C_2 V_{in}^2 + C_3 V_{in}^3 + \cdots \tag{3.22}$$

where C_0 is the dc output voltage, C_1 is the small-signal voltage gain, C_2 is the second-order distortion coefficient and C_3 is the third-order distortion coefficient. By combining (3.21) and (3.22) and rearranging the equations as a summation of sinusoidal expressions, the small-signal two-tone transfer is obtained as

$$\begin{aligned} V_{out} &= C_1 v_{low} \sin(2\pi f_{low} t) + C_1 v_{low} \sin(2\pi f_{low} t) \\ &+ \frac{3}{4} C_3 v_{low}^2 v_{high} \sin\left[2\pi \left(2 f_{low} - f_{high}\right) t\right] \\ &+ \frac{3}{4} C_3 v_{high}^2 v_{low} \sin\left[2\pi \left(2 f_{high} - f_{low}\right) t\right] \\ &+ \cdots + \cdots \end{aligned} \tag{3.23}$$

where the first dual expressions are the fundamental voltage intensified by C_1 and the next dual expressions are the IM3 components. Further expressions (not presented) represent higher harmonics that are typically negligibly insignificant or drop out of band. If the two input tones have equal power, therefore $v_{low} = v_{high} = v$, OIP3 is simplified to

$$OIP3 = \frac{(C_1 v)^2}{Z_L} = \frac{\left(\frac{3}{4} C_3 v^3\right)^2}{Z_L} \tag{3.24}$$

where Z_L is the load impedance. Solving for OIP3, the following equation holds:

$$OIP3 = \frac{4C_1^3}{3Z_L C_3}. \tag{3.25}$$

Finally, it is possible to define $IM3_{low}$(dBm) and $IM3_{high}$(dBm) mathematically as

$$IM3_{low}\,(\text{dBm}) = 2P_{low} + P_{high} - 2OIP3 \tag{3.26}$$

and

$$IM3_{high}\,(\text{dBm}) = 2P_{low} + 2P_{high} - 2OIP3 \tag{3.27}$$

where

$$P_{low} = \frac{(C_1 v_{low})^2}{Z_L} \tag{3.28}$$

and

$$P_{high} = \frac{\left(C_1 v_{high}\right)^2}{Z_L}. \tag{3.29}$$

Essentially, if two signals having varying frequencies are used in a non-linear structure, the output demonstrates certain mechanisms that are not classified as harmonics of the frequencies at the input, and these are the inter-modulation products.

3.7.5 *Noise Figure*

The NF of a mixer is an extent of the noise introduced by the mixer itself and is an important FOM to minimize, since this noise will also be converted to the IF output. The basic definition of the noise factor (*F*) in a two-port system comes from Friis [16] and is the signal-to-noise ratio (SNR) of the output power compared to that of the input power, expressed by

$$F = \frac{SNR_{in}}{SNR_{out}} \tag{3.30}$$

where the SNR quantities are power ratios and when expressed in dB is termed NF, where

$$NF = 10\log_{10} F \tag{3.31}$$

and conveniently expressed as

$$NF = SNR_{in,\text{dB}} - SNR_{out,\text{dB}} \tag{3.32}$$

which is effective when the "*input termination is at standard noise temperature of 290 K*" (Friis [16]). The output SNR is a function of the gain of the system and its output-referred noise, such that

$$SNR_{out,\text{dB}} = \frac{S_i G}{N_a + N_i G} \tag{3.33}$$

where S_i is the input signal to the system, G is the gain, N_i is the input noise and N_a is the input-referred noise. The F of any component is associated with its *noise temperature* such that

$$F = 1 + \frac{T_e}{T_0} \tag{3.34}$$

where T_e is the noise temperature and T_0 is the normal *noise temperature* of 290 K as stated above. The general definition of F is therefore given by

$$F = 1 + \frac{N_a}{N_i G} \tag{3.35}$$

and the NF of a network is the degrading of the SNR as the signal moves in the network of devices. A low NF is an indication of a good amplifier and means that very little noise is added to the system during the active processing of the signal. Although NF is typically related to two-port networks, mixers are treated similarly when discussing NF considering the LO is connected to the third port. A mixer, however, down-converts both the desired IF signal and its image and this means that the overall noise at IF is dependent on the noise

- at the wanted RF band down-converted to IF,
- at the image RF band down-converted to IF, and
- added by the non-ideal mixer circuit itself.

Noise in mixers is generated by the active components and by "*thermal sources due to resistive losses*" [40], especially at mm-Wave frequencies. The NF of a mixer is governed by determining if its input is a

- single-sideband (SSB) or a
- double-sideband (DSB).

signal. The definition for SSB NF assumes that there is no signal at the IF apart from the noise source, which is often the case when the image is suppressed by an image filter in advance of being fed to the mixer input port. The SNR of the SSB mixer output can be derived as

$$SNR_{SSB,out} = \frac{S_d}{N_d}\left(\frac{1}{2+\frac{2N_{mixer}}{N_d}}\right) \tag{3.36}$$

where S_d is the desired signal, N_d is the noise in the desired band and N_{mixer} is the noise generated by the mixer. The noise factor is then simplified to

$$F_{SSB} = 2 + \frac{2N_{mixer}}{N_d} \tag{3.37}$$

and for an ideal (noiseless) mixer F_{SSB} ($N_{mixer} = 0$) is equal to 2 (and logarithmically equal to 3 dB due to noise folding). For a practical mixer (non-ideal), the noise from multiple sidebands can therefore fold into the IF signal and degrade the overall NF of the mixer. The DSB NF definition assumes that the image signal contains both noise and an image signal identical to the desired signal and its noise. This definition is useful in direct-conversion receivers where the image is in fact the desired signal. For DSB, derivation shows that

$$F_{SSB} = 1 + \frac{2N_{mixer}}{N_d} \tag{3.38}$$

and for an ideal (noiseless) mixer F_{DSB} ($N_{mixer} = 0$) is equal to 1 (and logarithmically equal to 0 dB). Therefore, the relationship between F_{SSB} and F_{DSB} is

$$F_{DSB} = \frac{F_{SSB} + 1}{2}. \tag{3.39}$$

Note that in a cascaded system [37], which is especially useful in mm-Wave systems where down-conversion is often done in stages, the total noise at the output is calculated by taking into account the gain and NF at each stage. Figure 3.17 shows a system with three gain blocks cascaded.

As shown in Fig. 3.17, each gain stage has its own gain and NF, and the overall noise at the output is calculated by

$$F_{out} = F_1 + \frac{F_2 - 1}{G_1} + \frac{F_3 - 1}{G_1 G_2} \tag{3.40}$$

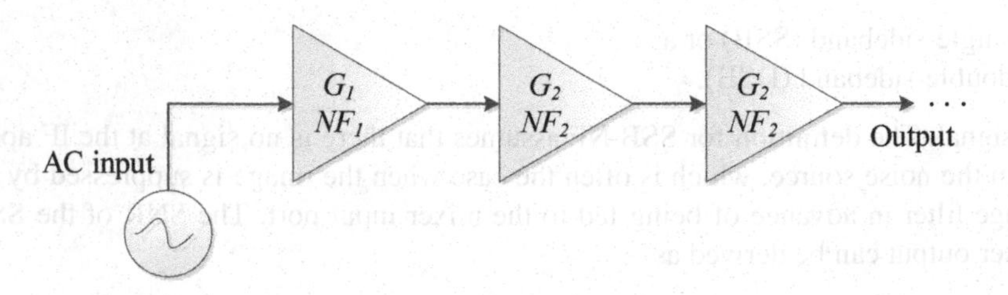

Fig. 3.17 Three cascaded gain stages to derive the total noise at the output

and extending this equation to further stages is relatively trivial. Additional performance metrics of mixer circuits should also be taken into account, as with any other integrated electronic circuit. Such performance metrics include for example the power consumption of the circuit, as well as the physical geometry, since ICs are typically limited by the available area on chip.

3.7.6 Additional Performance Metrics

As with any electronic circuit, power consumption (energy efficiency) in mW (or dBm) is an important FOM and depends on the architecture of the circuit, gain, active components and intrinsic losses. Power consumption in modern electronics has also become more crucial for various reasons, including a shift to using renewable energy, the increasing price of coal energy (especially in emerging markets), technologies such as the IoT requiring numerous systems to work together and the necessity of mobility for modern devices. In terms of mobility, energy efficiency is crucial, but this places additional requirements and limitations on the physical size of systems when realized as SoC using integrated technology. Chip area in mm^2 is therefore another FOM that most ICs regard as crucial, and in mixer design, the same philosophy holds.

3.8 CMOS/BiCMOS Mixers for 5G Communication

3.8.1 Down-Conversion Mixers

In [13] an mm-Wave (60 GHz) down-converter mixer in CMOS, with an LO buffer, that includes output matching as well "*as a noise and distortion canceling active balun*", is presented. The mm-Wave mixer in [13] is proposed for 5G communication systems to enable multi-Gbps wireless data transfer. Using a CMOS process implies low biasing voltages and this is typically associated with a loss in linearity across the system bandwidth, as reviewed in this chapter. In order to linearize the mixer with a low-voltage supply in addition to operating at mm-Wave frequencies, Choi et al. [13] propose an "*on-chip transformer-based topology*". Furthermore, for SSB and IF (12 GHz) output matching, an "*active balun is used with common-source and common-drain configurations to cancel IP3*". Adapted from Choi et al. [13], a simplified representation of the suggested mm-Wave mixer and the variations from the traditional architecture is presented in Fig. 3.18.

As shown in Fig. 3.18, the proposed down-conversion mixer incorporates an LO buffer (also in CMOS) as well as an active balun at the mixer output. The G_m gain stage, as well as the down-conversion mixer, is realized by one single MOSFET architecture (reducing the number of active components that require biasing) to ensure high linearity at the "*low-supply voltage*", in this case 1 V. The input impedance of

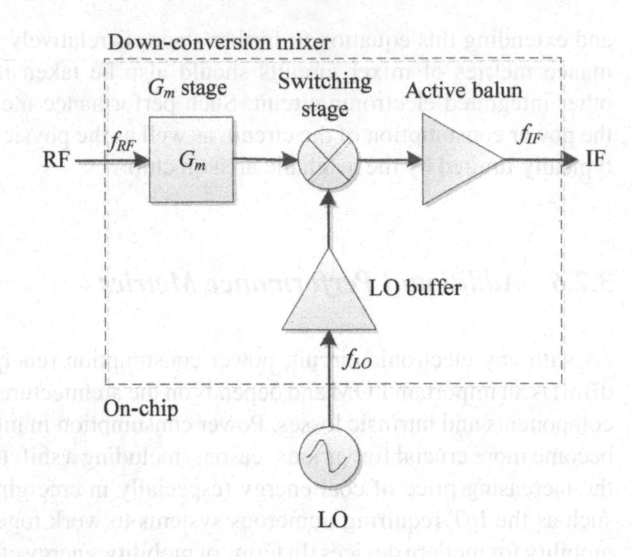

Fig. 3.18 Variation of mm-Wave 5G (60 GHz) down-converter CMOS mixer proposed in the work of Choi et al. [13]

the gain stage is approximated in [13] as

$$Z_{in} = \frac{g_{m1} L_S}{C_{gs}} + j\omega(L_G + L_S) + \frac{1}{j\omega C_{gs}} \tag{3.40}$$

where g_{m1} and C_{gs} are the transconductance and the gate-source capacitance of the single transistor, and L_G and L_S are the on-chip transformer components of the gate and source of the transistor. This strategy effectively removes the imaginary terms of the input impedance Z_{in} at the operating frequency and allows for impedance matching by designing the real term to be 50 Ω. The active balun also serves to improve the IIP3 through a common-drain path, as described in [13]. Table 3.3 gives the performance metrics for the proposed down-conversion mixer listed in [13].

Table 3.3 Performance metrics reviewed in this chapter and obtained by the down-conversion mixer presented in Choi et al. [13]

Performance metric	Specified unit	Value obtained
Operating frequency	GHz	57–66
Gain	dB	>5.6
OIP3	dBm	>12.4
NF	dB	<11 (SSB)
1-dB compression	dBm	−7
L-R isolation	dB	>35
Power consumption	mW	18
LO power	dBm	−10
Area	mm^2	0.22

The performance metrics listed in Table 3.3 give an indication of *good* values obtainable by a mm-Wave 5G-capable down-conversion SSB mixer, as presented in [13]. The proposed mixer is specifically aimed at 5G mobile phone usage where prototypes displayed good performance with low power consumption and at relatively low cost, as realized in a regular CMOS node.

In the conference proceedings presented by Singh and Kumar [39], a wideband mm-Wave 5G down-conversion mixer is presented, with claimed improved NF and IIP3. The mixer proposed by Singh and Kumar [39] also implemented an active balun (load) in the bias network to increase the linear operation of the system, as well as the conversion gain. Additionally, a transimpedance amplifier is used as an IF buffer to increase linearity further. The suggested mixer is based on the popular Gilbert mixer design [42], realized in a standard CMOS technology. The work does not give performance metrics on all parameters as reviewed in this chapter, but states an NF of 5.9 ± 0.4 dB and an IIP3 of 11.78 ± 1.75 dBm.

3.8.2 Up-Conversion Mixers

Chen et al. [9] proposed an *"mm-Wave (27.5–43.5 GHz) up-conversion standard CMOS mixer for 5G communication systems"*, with some variations to the traditional (Gilbert) approach. In [9], the mixer linearity is improved by using coupled resonators and a linearized transconductance (gain) stage that typically dominates the linearity of the mixer as well as a double-balanced Gilbert cell. Furthermore, the two input and output on-chip baluns implement wideband inductive impedance matching to linearize the mixer over a large frequency range. Table 3.4 lists the performance metrics obtained by Chen et al. [9].

The work in [9] does not specify the OIP3, NF or LO power, but the authors still claim that the linearized up-conversion mm-Wave 5G-capable mixer performs

Table 3.4 Performance metrics reviewed in this chapter and obtained by the up-conversion mixer presented in Chen et al. [9]

Performance metric	Specified unit	Value obtained
Operating frequency	GHz	27.5–43.5
Gain	dB	−5
OIP3	dBm	–
NF	dB	–
1-dB compression	dBm	0.42
L-R isolation	dB	>40
L-I isolation	dB	>43
Power consumption	mW	14
LO power	dBm	–
Area	mm^2	0.686

Table 3.5 Performance metrics reviewed in this chapter and obtained by the up-conversion mixer presented in Qayyum et al. [35]

Performance metric	Specified unit	Value obtained (measured)
Operating frequency	GHz	50–67
Gain	dB	11.38
OIP3	dBm	–
NF	dB	–
1-dB compression	dBm	−1.4
L-R isolation	dB	–
Power consumption	mW	52
LO power	dBm	3
Area	mm^2	0.146

adequately for wideband 5G implementations, with low power consumption and operation in "*both the 28 GHz and 38 GHz 5G bands*".

Another mm-Wave (60 GHz) up-conversion mixer is offered in [35] and is realized in SiGe BiCMOS technology. SiGe BiCMOS has distinct advantages in the mm-Wave and THz domains [11] when compared to standard CMOS processes [24], and Quayym et al. [35] take advantage of these characteristics. Apart from the high f_T and $f_{\max}$ (300/400 GHz) of SiGe BiCMOS transistors, the technology also allows for low-noise circuit operation, both of which are crucial performance metrics in up-conversion mixers. The mixer proposed in [35] is based on the double-balanced Gilbert mixer with two on-chip (symmetrical to mitigate imbalance of the mixer) transformer baluns at the LO and the output (RF) port. A differential IF input port for monolithic integration is used. Most of the design complexity of the proposed up-conversion mixer in [35] is focused on the symmetry of the on-chip transformer baluns. Attention was placed on the geometry as well as the metal layers used in the semiconductor process to realize the transformer baluns, as described in [35]. The performance metrics obtained in this work are listed in Table 3.5.

The performance metrics presented in Table 3.5 are therefore based on a SiGe BiCMOS semiconductor process as opposed to a standard CMOS implementation. In [35], comparisons between different up-conversion mixers in both BiCMOS and CMOS processes are presented. In most cases the gain (conversion gain) in the CMOS implementations are higher, whereas the BiCMOS (SiGe) process offers superior linearity (albeit at the cost of high power consumption).

Chen et al. [8] propose a 62–58 GHz linear up-conversion mixer adopting a "*transconductance stage consisting of a common-source path*" as well as a "*cross-coupled common-gate path*" for enhanced linearity. The transconductance stage in the work of [8] is also valuable to implement broad IF impedance matching. To minimize the physical size of the up-conversion mixer, "*a transformer is used as an inductive load stage, balun and output-matching network*". The performance metrics of the up-conversion mixer presented in [8] are summarized in Table 3.6.

Table 3.6 Performance metrics reviewed in this chapter and obtained by the up-conversion mixer presented in Chen et al. [8]

Performance metric	Specified unit	Value obtained (measured)
Operating frequency	GHz	62–85
Gain	dB	−4.3 @ 77 GHz
OIP3	dBm	–
NF	dB	–
1-dB compression	dBm	2.14
L-R isolation	dB	30
Power consumption	mW	10.8
LO power	dBm	–
Area	mm^2	0.3825

Finally, Mazor et al. [30] designed a linear 60 GHz SiGe up-conversion mixer in a 130 nm BiCMOS process. The mixer was implemented using a mixing core only strategy to improve linearity and without any LO buffers or transformer-matching network for IF. Separate bias control was used in the mixing process to enhance the LO suppression in-band. The performance metrics of the up-conversion mixer presented in [30] are summarized in Table 3.7.

A critical component in realizing an RF mixer to be used in applications such as mm-Wave 5G is the LO and integrating the LO on chip has several advantages compared to a discrete off-chip LO. At mm-Wave frequencies, discrete oscillators are difficult to realize because of their physical size in relation to the wavelength of the operating frequency. Furthermore, eliminating interconnects of several millimeters between on-chip and off-chip components reduces parasitic inductance from external wires. An mm-Wave oscillator also presents challenges and has limitations when implemented on chip; primarily the Q-factor of the inductive components have a proportional effect on the bandwidth of the oscillator. The following section highlights the design requirements of integrated mm-Wave oscillators and

Table 3.7 Performance metrics reviewed in this chapter and obtained by the up-conversion mixer presented in Mazor et al. [30]

Performance metric	Specified unit	Value obtained (measured)
Operating frequency	GHz	57–66
Gain	dB	−2.5
OIP3	dBm	7.5–10.5
NF	dB	–
1-dB compression	dBm	–
L-R isolation	dB	30
Power consumption	mW	27
LO power	dBm	0
Area	mm^2	0.22

the crucial considerations that must be highlighted to realize linear and low-power, high-frequency oscillators.

3.9 Integrated mm-Wave Oscillators

At this point it should be obvious that the local oscillator in any mm-Wave 5G-capable transceiver is another key component that needs to perform under stringent performance metrics. Integrating the LO on chip and in the same technology as the rest of the transceiver subsystems is becoming more important in realizing fully integrated 5G systems. Consequently, mm-Wave and THz oscillators have been realized in GaAs and InP [24], but with the rapid advances occurring in silicon-based technologies, practical implementations of mm-Wave oscillators in standard CMOS or BiCMOS with adequate output power to drive the frequency mixer(s) have become more commonplace. The inherent non-linearity of active components and the determinate *Q*-factor of passive devices both determine the overall performance of the LO and typically lead to output-power-limited implementations, if realized in CMOS or BiCMOS technology. The primary FOMs that describe oscillator performance are power consumption and phase noise. The following paragraph briefly describes phase noise in terms of oscillator performance.

3.9.1 Oscillator Phase Noise

From a practical perspective, phase noise is the ratio of the single-sideband noise power to the total signal power at a specified offset. The unit for phase noise is dBc/Hz, therefore the noise bandwidth is specified as 1 Hz [23]. The offset from the carrier is typically rated at 10 kHz, 100 kHz, 1 MHz or 10 MHz, depending on the frequency of the carrier signal. In mm-Wave 5G-capable transceivers, the offset frequency is normally specified at 10 MHz from the carrier frequency. Since phase noise is the proportion of power in the carrier frequency related to the power at an offset, it follows that the proportion will grow as the offset frequency reduces. Mathematically, phase noise is described using Leeson's model and adapted from the well-known feedback oscillator representation shown in Fig. 3.19.

The feedback oscillator presented in Fig. 3.19 comprises of "*an amplifier with transfer function A(jω) and a feedback network with transfer function H(jω)*" [26]. Sustained oscillations in the feedback oscillator will occur if the relationship $|A(j\omega)| \times |H(j\omega)| \geq 1$. A complete derivation of phase noise based on this model is not within the scope of this book and only the result is presented. The mathematical derivation of the Leeson formula is presented in [26] and it follows that phase noise $\mathscr{L}\{2\pi f_m\}$ is defined by

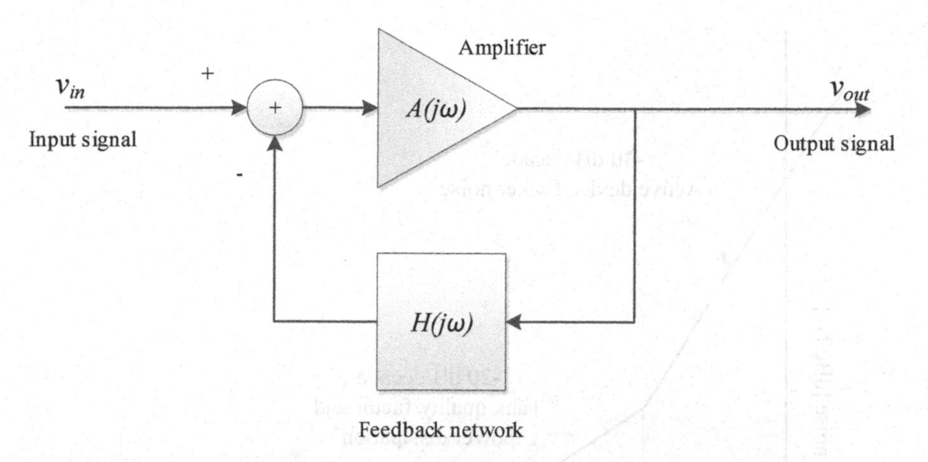

Fig. 3.19 Depiction of a feedback oscillator by means of Leeson's model to describe phase noise

$$\mathcal{L}\{2\pi f_m\} = 10\log\left[\left[\left(\frac{f_0}{2Q_L f_m}\right)^2 + 1\right] \times \frac{FkT}{P} \times \left(\frac{f_c}{f_m} + 1\right)\right] \tag{3.41}$$

where f_m is the offset frequency from the carrier (f_0), Q_L is the loaded Q-factor of the tank circuit in a LC oscillator, k is Boltzmann's constant (1.38×10^{-23} J/K), F is the noise factor, T is operating temperature, P is carrier power in mW and fc is the corner frequency for flicker noise [26]. This mathematical representation of phase noise can be represented in graphical form to indicate the influence of respective terms on the total phase noise. Such a graphical illustration is depicted in Fig. 3.20.

The overall FOM of an oscillator is derived from its phase noise and power consumption, such that

$$FOM = 10\log\left(\left(\frac{f_0}{f_m}\right)^2 \frac{1}{\mathcal{L}\{2\pi f_m\}P}\right) \tag{3.42}$$

where P is the dc power consumption of the oscillator in mW.

3.9.2 Oscillator Topologies

Various VCO topologies exist to realize low-noise and high-power oscillators for transceivers, but the challenges and limitations that mm-Wave operating frequencies incur on these circuits limit the number of topologies available to implement 5G-capable VCOs. Traditional crystal oscillators, which are essentially thin slices of "*quartz crystal with conducting electrodes on opposing sides* [20]", are typically used for applications in the mid-tier MHz range and rarely in the GHz domain. The

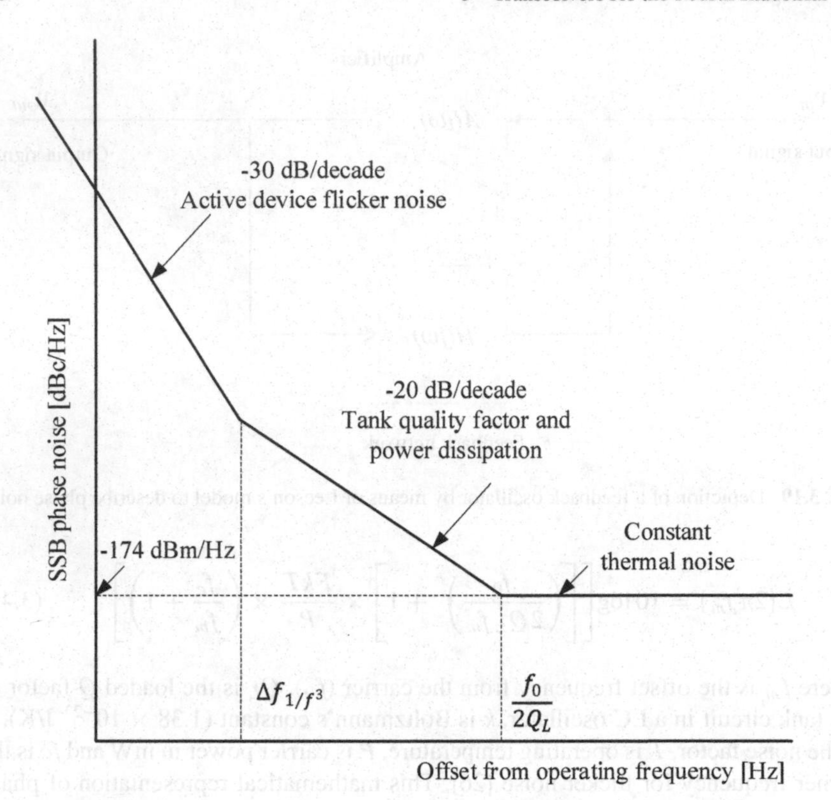

Fig. 3.20 Graphical representation of Leeson's equation to determine oscillator phase noise

Q-factor of crystal oscillators are superior to most discrete implementations; however, the frequency limitations in the GHz range eliminate crystal oscillators from mm-Wave transceivers. A discrete realization of VCOs is the ring oscillator, a ring of determinate equal stages, each stage consisting of an inverter and transmission line connecting succeeding stages. Furthermore, there is an active feedback link amid the first and final stages of the ring oscillator. An advantage for the ring oscillator is that it can simply be realized by means of standard cell modules and the transmission lines act as delay lines, eliminating the need for additional delay lines in the circuit. Ring oscillators also have a wide tuning range and are compatible with standard CMOS or BiCMOS technologies. However, the phase noise of ring oscillators increases dramatically at mm-Wave frequencies, since a specific quantity of energy is stored in the capacitive components for the period of each half-cycle, dissipating the same amount of energy during the subsequent half-cycle. Therefore, for a specified "*power dissipation in steady state, a ring oscillator suffers from a smaller maximum charge swing*" [17]. Since a ring oscillator is dependent on a string of standard cells, each additional component (albeit physically small in size), adds noise that decreases phase noise performance and reduces linearity. The most attractive alternative to mm-Wave oscillators (specifically VCOs) is the LC-tank VCO (or LC oscillator),

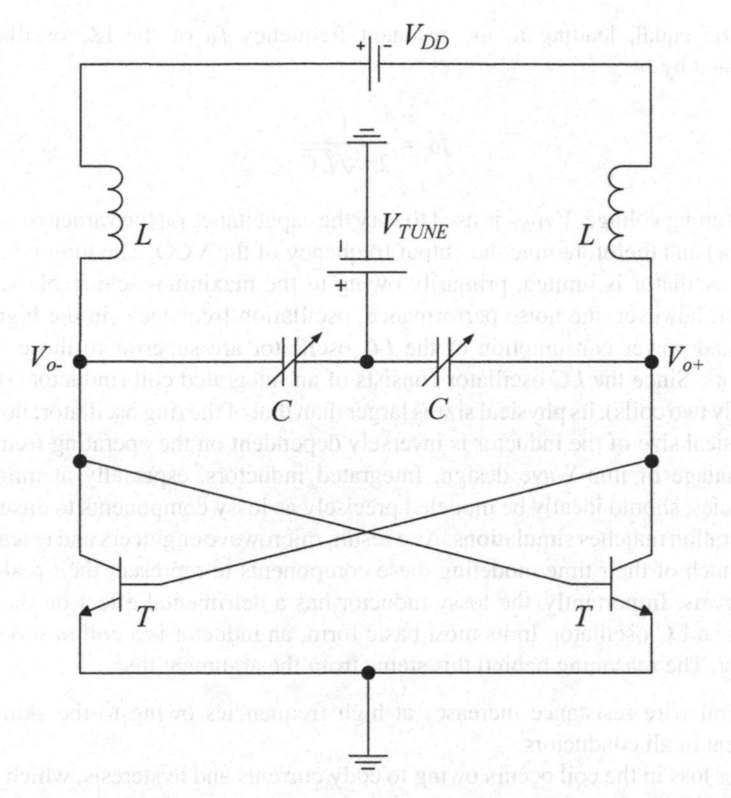

Fig. 3.21 Basic representation of an LC oscillator circuit, realizable in standard CMOS or BiCMOS and capable of generating mm-Wave 5G-capable oscillations

consisting of an inductor (L) and a capacitor (C) in parallel, which combine to realize a parallel resonance *tank* circuit. In its most basic form, the cross-coupled LC oscillator topology is represented in Fig. 3.21.

The LC oscillator shown in Fig. 3.21 represents two sides of the cross-coupled strategy, therefore a mirror image of a single-ended capacitive-inductive circuit. The topology has an active element, a transistor ($-R$) that compensates for the "*resistive losses of the inductor and capacitor in the tank circuit*", to maintain oscillation. To determine the condition at which oscillation occurs, the capacitive reactance

$$X_C = \frac{1}{2\pi f C} \tag{3.43}$$

and inductive reactance

$$X_L = 2\pi f L \tag{3.44}$$

should be equal, leading to the resonant frequency f_0 of the LC oscillator as determined by

$$f_0 = \frac{1}{2\pi\sqrt{LC}} \tag{3.45}$$

and the tuning voltage, V_{TUNE} is used to vary the capacitance on the varactor (variable capacitor) and therefore tune the output frequency of the VCO. The tuning range of the *LC* oscillator is limited, primarily owing to the maximum achievable varactor variation; however, the noise performance, oscillation frequency (in the high-GHz range) and power consumption of the *LC* oscillator are superior to those of ring oscillators. Since the *LC* oscillator consists of an integrated coil (inductor) on chip (typically two coils), its physical size is larger than that of the ring oscillator; however, the physical size of the inductor is inversely dependent on the operating frequency, an advantage of mm-Wave design. Integrated inductors, especially at mm-Wave frequencies, should ideally be modeled precisely as lossy components to ensure that real operation matches simulations. As a result, microwave engineers and researchers spend much of their time modeling these components to represent their real-world counterparts. Importantly, the lossy inductor has a detrimental effect on the phase noise of an LC oscillator. In its most basic form, an inductor is a coil in series with a resistor. The reasoning behind this stems from the argument that

- the coil wire resistance increases at high frequencies owing to the skin effect present in all conductors,
- power loss in the coil occurs owing to eddy currents and hysteresis, which is also dependent on the operating frequency,
- the coil has a finite resistance (assuming it is not a super-conductor), and
- radiation of power to surrounding components and structures incurs additional resistive losses.

To represent a practical inductor at high frequencies, an equivalent circuit that includes a finite series resistance as shown in Fig. 3.22 is typically used.

The effect of the finite series resistance R_s in Fig. 3.22 is noticeable when considering the *Q*-factor of the inductor, a characteristic that should ideally be maximized in the design of an oscillator. The *Q*-factor is determined by

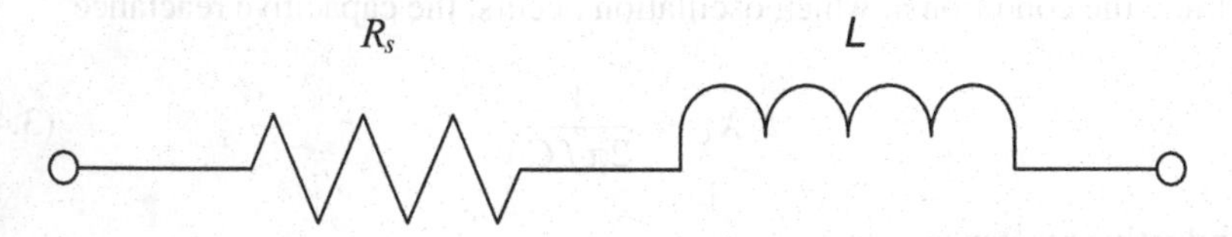

Fig. 3.22 The equivalent circuit model of an inductor (coil) at high frequencies, consisting of the coil and a finite series resistance

$$Q = \frac{\omega L}{R_s} \tag{3.46}$$

showing the inverse proportionality of the Q-factor on the series resistance and where f represents the operating frequency. As a result, the finite value of R_s in Ω should ideally be minimized, although typically a constant function of coil size (diameter and length) in the CMOS or BiCMOS technology. Furthermore, the finite resistance of the coil has an effect on the actual resonance frequency, derived in [5] such that

$$\omega_0 = \sqrt{\frac{1}{LC} - \frac{R_s^2}{L^2}} \tag{3.47}$$

which leads to a smaller resonant frequency compared to the lossless electronic circuit resonant frequency in (3.45). To obtain operating frequencies in the mm-Wave range as required by next-generation 5G-capable transceivers, CMOS oscillators with a high tuning range or low phase noise are required, in combination with low power consumption considering its IoT applications. At mm-Wave frequencies, however, the circuit modeling of inductors (and capacitors) becomes much more complex to account for parasitic effects present at these frequencies, considering integrated (on-chip) components. A complete equivalent circuit model of an on-chip inductor at mm-Wave frequencies is given in Fig. 3.23.

As shown in Fig. 3.23, the traditional model consisting of only the coil (inductor) L_S and finite series resistance R_S (due to eddy currents) is shown. However, since the physical component is realized by placing a metal line on a substrate, numerous additional parasitic effects need to be modeled to describe real-world performance. According to Fig. 3.23, C_P represents the parallel tank capacitor placed in parallel with the inductor. C_{OX1} and C_{OX2} are the parasitic oxide capacitance of the dielectric oxide layer between the substrate and the metal lines. The substrate material (typically silicon) incurs additional capacitance and resistance towards the actual metal lines, indicated by R_{S1}, R_{S2}, C_{S1} and C_{S2} in Fig. 3.23. The mathematical complexity of circuits operating at mm-Wave frequencies is therefore high and highlights the importance of component-level circuit modeling during the research and design phases of mm-Wave 5G-capable transceivers. Again, many of these parasitic effects can be minimized by component dimensions and geometry [25]. However, numerous parameters such as the resistivity of the conductive material, the thickness of the conductors, oxide thickness and substrate capacitance/conductance per unit area, are determined by the process technology.

3.9.3 CMOS mm-Wave Oscillators for 5G Systems

In terms of practical realizations of 5G-capable mm-Wave oscillators, specifically cross-coupled LC oscillators in CMOS or BiCMOS technology, numerous works have been presented that highlight the performance and practicality of using this

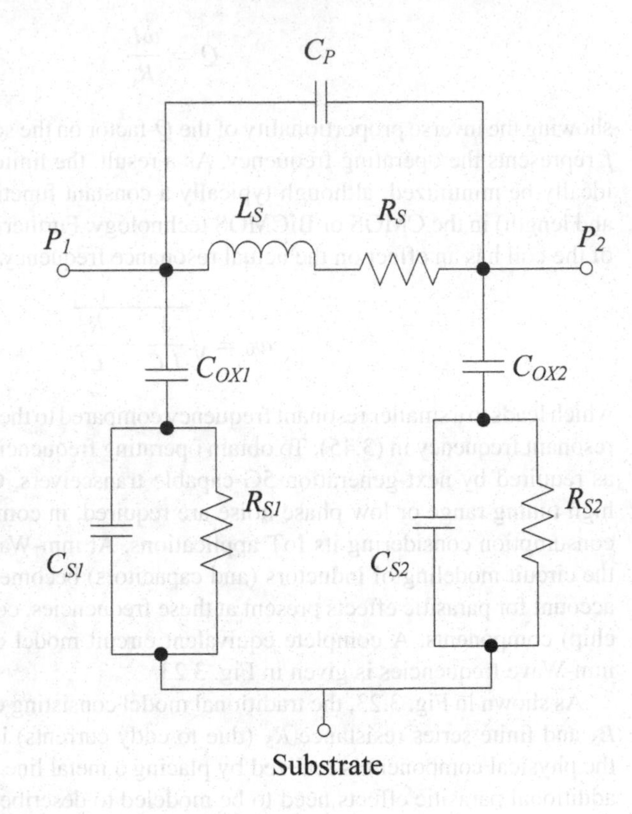

Fig. 3.23 The equivalent circuit model of an inductor (coil) at mm-Wave frequencies, consisting of the coil and a finite series resistance in combination with various other parasitic components

topology in 5G systems. In [10], a cross-coupled structure with a push-push circuit to enable frequency doubling is presented. The V-band VCO in [10] is tunable between 59 and 59 GHz with an output power of −10 dBm and phase noise of −103 dBc/Hz at a 10 MHz offset. Already in [36], Razavi proposed a mm-Wave (90 nm) CMOS oscillator using inductive feedback and realized a fundamental operating frequency of 128 GHz with 9 mW of power dissipation and a phase noise of −105 dBc/Hz at a 10 MHz offset. Motoyoshi et al. [31] also designed a 5G-capable mm-Wave cross-coupled LC oscillator with a fundamental operating frequency of 57.4 GHz and power consumption of only 130 μW in a 65 nm CMOS process, although phase noise performance was not given. Wang et al. [43, 44] presented a 213 GHz CMOS (65 nm) single-ended differential fundamental oscillator consuming 790 μW with phase noise of −93.4 dBc/Hz at a 1 MHz offset.

3.10 Conclusion

As the wireless communications milieu is changing in preparation for Industry 4.0, electronic systems also need to adapt to an increase in operating frequency. The unlicensed mm-Wave frequency band has been identified as the spectrum for future communications such as 5G; however, electronic circuits also need to adapt to facilitate this shift. These changes trickle down from the communication protocol standards, down to systems level and subsystem level all the way to component level. As a result, traditional architectures used in wireless transceivers need to be either adapted or re-imagined. The first step in achieving this is identifying the challenges, limitations and shortcomings of lower-frequency implementations used in mm-Wave or THz communications. Many of these shortcomings can be traced to component level, but others are also found higher up in the hierarchy.

In this chapter, the fundamentals of transceivers are critically reviewed in terms of their ability to operate at high frequencies. Understanding these fundamental principles aids researchers and designers of next-generation communication systems in identifying how these systems operate, as well as where limitations are most likely to be encountered. Identification of shortcomings is, however, only part of the journey in mitigating them and mm-Wave and THz operation at subsystem level should also be investigated. In this chapter, the first subsystem that is critically reviewed is the frequency mixer. The frequency mixer is responsible for translating between RF (mm-Wave) signals and lower frequency IF signals and is therefore a subsystem that must be capable of operating in the high-frequency band. Stringent specifications are placed on the frequency mixer and limitations on component level can influence the overall performance of an entire transmitter or receiver. This chapter aims to identify these at component level and also provides a review of implementations specifically for mm-Wave 5G-capable frequency mixers as proposed in literature, identifying the techniques used to mitigate some of the challenges of implementing this subsystem.

Using a similar strategy, the next chapter will research and investigate two subsystems that also need to be able to process high frequencies as part of a transceiver, the LNA and the PA. The LNA is the first point of contact in a receiver subsystem for the incoming RF (mm-Wave) signal and requirements of gain and low noise are typically stringent in these subsystems. The PA in the transmitter should be able to amplify (with high gain) high-frequency components to be transmitted across a channel, again with stringent requirements all the way to component level, where the choice of process technology plays a large role.

Appendix 1

The small-signal equivalent circuit of a MOSFET in Fig. 3.24 is used to derive the f_T of the transistor.

From Fig. 3.25, at the input node A, the current is given by

$$I_{in}(\omega) = j\omega\left(C_{gs} + C_{gd}\right)V_{gs}(\omega) \tag{3.48}$$

and at the output node B, the current is defined by

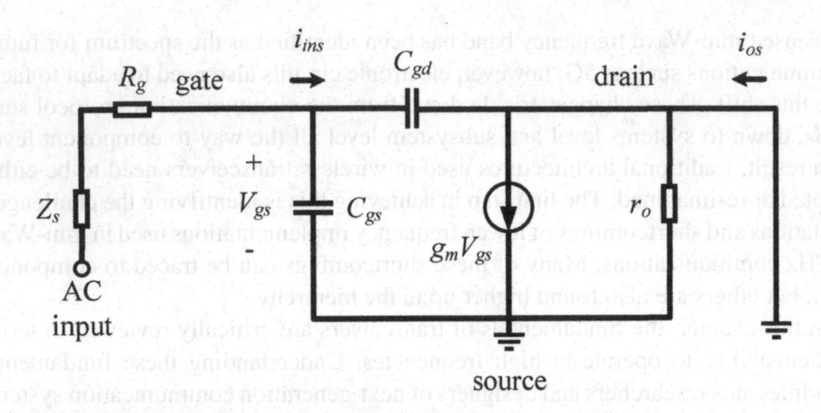

Fig. 3.24 The small-signal equivalent circuit of a MOSFET used to derive f_T

$$I_{out}(\omega) = \left(g_m - j\omega C_{gd}\right)V_{gs}(\omega). \tag{3.49}$$

The current gain can therefore be determined as

$$\left|\frac{I_{out}(\omega)}{I_{in}(\omega)}\right| = \left|\frac{\left(g_m - j\omega C_{gd}\right)V_{gs}(\omega)}{j\omega\left(C_{gs} + C_{gd}\right)V_{gs}(\omega)}\right| \tag{3.50}$$

which can be rewritten where the V_{gs} dependency falls away such that

$$\left|\frac{I_{out}(\omega)}{I_{in}(\omega)}\right| = \frac{\sqrt{g_m^2 - \omega^2 C_{gd}^2}}{\omega\left(C_{gs} + C_{gd}\right)} \tag{3.51}$$

and at low frequencies, therefore when $g_m \gg \omega C_{gs}$, the current gain becomes

$$\left|\frac{I_{out}(\omega)}{I_{in}(\omega)}\right| \approx \frac{g_m}{\omega\left(C_{gs} + C_{gd}\right)} \tag{3.52}$$

whereas at high frequencies, therefore when $g_m \ll \omega C_{gs}$, the current gain becomes

$$\left|\frac{I_{out}(\omega)}{I_{in}(\omega)}\right| \approx \frac{C_{gs}}{\left(C_{gs} + C_{gd}\right)}. \tag{3.53}$$

Theoretically, at f_T, the current gain should be unity, therefore

$$\omega_T = \frac{g_m}{C_{gs} + C_{gd}} \approx \frac{g_m}{C_{gs}} \tag{3.54}$$

or

$$f_T = \frac{g_m}{2\pi\left(C_{gs} + C_{gd}\right)}. \tag{3.55}$$

If the transistor gain g_m is substituted in terms of its gate bias voltage (V_T), where

$$g_m = \mu C_{ox}\frac{W}{L}\left(V_{gs} - V_T\right) \tag{3.56}$$

then ω_T becomes

$$\omega_T = \frac{3}{2}\frac{\mu\left(V_{gs} - V_T\right)}{L^2} \tag{3.57}$$

showing that decreasing the transistor gate length increases the cut-off frequency, or if substituting gm in terms of drain current and channel length

$$g_m = \sqrt{2I_D\mu C_{ox}\frac{W}{L}} \tag{3.58}$$

where I_D is the drain current, C_{ox} is the oxide capacitance and W and L are the width and length of the transistor, respectively. It follows that

$$\omega_T = \frac{\sqrt{2I_D\mu C_{ox}\frac{W}{L}}}{C_{gs}} \tag{3.59}$$

showing that an increase in bias current also increases the cut-off frequency. To derive $f_{\max}$, the maximum frequency where the power gain is unity, consider (3.55) and the input and output impedance of the MOSFET, where

$$R_{in} = R_g + \frac{1}{j\omega C_{gs}} \approx R_g \tag{3.60}$$

since at high frequencies

$$\frac{1}{j\omega C_{gs}} \approx 0. \tag{3.61}$$

Furthermore, the output impedance is simplified as

$$R_{out} = r_o//\frac{C_{gs} + C_{gd}}{g_m C_{gd}}. \tag{3.62}$$

The power gain must be calculated under conjugate match at the input and the output, with reference to Fig. 3.25.

From Fig. 3.25a, a conjugate match, therefore

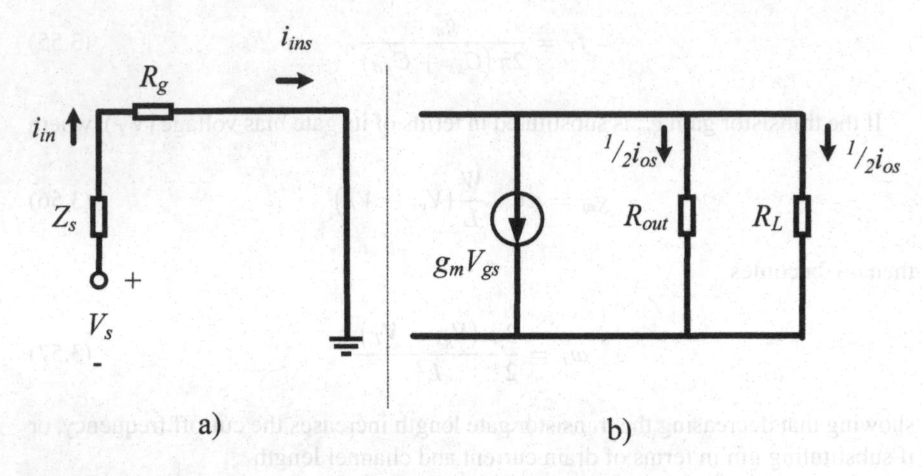

Fig. 3.25 Conditions for a conjugate match at the **a** input and the **b** output

$$Z_s = R_g \tag{3.63}$$

is achieved if

$$i_{in} = i_{ins} = \frac{V_s}{2R_g} \tag{3.64}$$

and from Fig. 3.25b, for

$$R_L = R_{out} \tag{3.65}$$

the equation

$$i_o = \frac{i_{os}}{2} \tag{3.66}$$

must be satisfied. The power gain can be calculated by

$$G_P = \frac{\frac{1}{2} i_o^2 R_{out}}{\frac{1}{2} i_{in}^2 R_{in}} \tag{3.67}$$

which can be rewritten in terms of the cut-off frequency such that

$$G_P = \frac{1}{4}\left(\frac{f_T}{f}\right)^2 \frac{R_L}{R_g} \tag{3.68}$$

and when the power gain is unity, therefore $|G_P| = 1$, and rewriting to make the operating frequency f the subject, it is found that

$$f = \frac{1}{2} f_T \sqrt{\frac{R_L}{R_g}} = f_{\max}. \quad (3.69)$$

By substituting $R_L = R_{out}$ (conjugate match), which is equal to (3.62) calculated, such that

$$R_L = \frac{1}{\frac{1}{r_o} + (2\pi f_T C_{gd})} \quad (3.70)$$

and $f_{\max}$ becomes

$$f_{\max} = \frac{1}{2} \frac{f_T}{\sqrt{2\pi f_T R_g C_{gd} + \frac{R_g}{r_o}}} \quad (3.71)$$

and with $R_g \ll r_o$, $f_{\max}$ is simplified to

$$f_{\max} \approx \sqrt{\frac{f_T}{8\pi R_g C_{gd}}}. \quad (3.72)$$

For bipolar transistors, the following small-signal model is used to derive f_T (Fig. 3.26).

The following variables for a bipolar transistor should be noted:

$$C_{gs} \to C_\pi \quad (3.73)$$

$$C_{gd} \to C_\mu \quad (3.74)$$

$$V_{gs} \to V_{be} \quad (3.75)$$

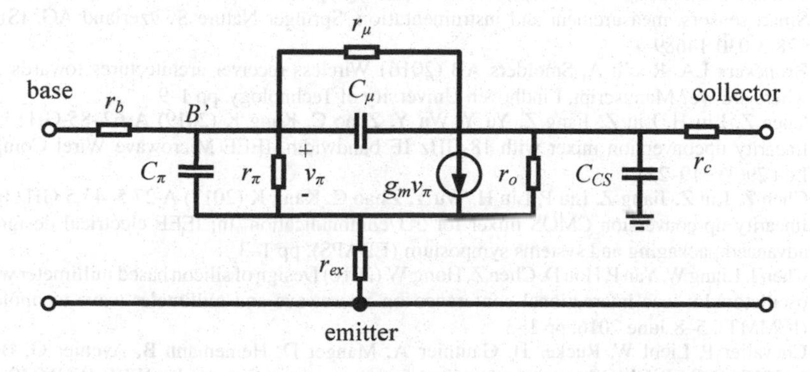

Fig. 3.26 The small-signal equivalent circuit of a bipolar transistor used to derive f_T

$$r_o \to \infty \tag{3.76}$$

therefore, the cut-off frequency for a bipolar transistor becomes

$$f_T = \frac{g_m}{2\pi(C_\pi + C_\mu)} \tag{3.77}$$

and (3.71) shows in a similar derivation that

$$f_{\max} = \frac{1}{2}\frac{f_T}{\sqrt{2\pi f_T R_b C_{bc} + \frac{R_g}{r_o}}} \tag{3.78}$$

and taking into account (3.76), $f_{\max}$ simplifies to

$$f_{\max} \approx \sqrt{\frac{f_T}{8\pi R_b C_{bc}}}. \tag{3.79}$$

References

1. Abidi AA (1995) Direct-conversion radio transceivers for digital communications. IEEE J Solid-State Circuits 30(12):1399–1410
2. Al-Araji SR, Gosling W (1973) A direct conversion v.h.f. receiver. Radio Electron Eng 43(7):442–446
3. Armstrong EH (1924) The super-heterodyne—its origin, development, and some recent improvements. Proc Inst Radio Eng 12(5):539–552
4. Bhana VB (2017) A slow-wave CMOS delay line filter for mm-Wave applications. Master's dissertation, University of Pretoria, November 2017
5. Boon CC, Do MA, Yeo KS, Ma JG, Zhao RY (2004) Parasitic-compensated quadrature LC oscillator. IEE Proc Circuits, Dev Syst 151(1):45–48
6. Božanić M, Sinha S (2019) Systems-level packaging for millimeter-wave transceivers. In: Smart sensors, measurement and instrumentation. Springer Nature Switzerland AG. ISBN: 978-3-030-14689-4
7. Bronckers LA, Roc'h A, Smolders AB (2016) Wireless receiver architectures towards 5G: where are we? Manuscript, Eindhoven University of Technology, pp 1–9
8. Chen Z, Liu H, Liu Z, Jiang Z, Yu Y, Wu Y, Zhao C, Kang K (2019) A 62–85-GHz high linearity upconversion mixer with 18-GHz IF bandwidth. IEEE Microwave Wirel Compon Lett 29(3):219–221
9. Chen Z, Liu Z, Jiang Z, Liu P, Liu H, Wu Y, Zhao C, Kang K (2017) A 27.5–43.5 GHz high linearity up-conversion CMOS mixer for 5G communication. In: IEEE electrical design of advanced packaging and systems symposium (EDAPS), pp 1–3
10. Chen J, Liang W, Yan P, Hou D, Chen Z, Hong W (2016) Design of silicon based millimeter wave oscillators.In: ieee international conference on microwave and millimeter wave technology (ICMMT), 5–8 June 2016, pp 1–3
11. Chevalier P, Liebl W, Rucker H, Gauthier A, Manger D, Heinemann B, Avenier G, Bock J (2018) SiGe BiCMOS current status and future trends in Europe. In: IEEE BiCMOS and compound semiconductor integrated circuits and technology symposium (BCICTS), pp 64–71

12. Chi B, Song Z, Jia H, Kuang L, Lin J, Wang Z (2018) CMOS circuit techniques for mm-Wave communications. In: IEEE MTT-S international wireless symposium (IWS), pp 1–3
13. Choi C, Son JH, Lee O, Nam I (2017) A +12 dBm OIP3 60-GHz RF downconversion mixer with an output-matching, noise- and distortion-canceling active balun for 5G applications. IEEE Microwave Wirel Compon Lett 27(3):284–286
14. Dinc T, Chakrabarti A, Hrishnawamy H (2016) A 60 GHz CMOS full-duplex transceiver and link with polarization-based antenna and RF cancellation. IEEE J Solid-State Circuits 51(5):1125–1140
15. Frenzel L (2013) What's the difference between the third-order intercept and the 1-dB compression points? Retrieved 13 June 2019 from http://www.electronicdesign.com
16. Friis HT (1944) Noise figure of radio receivers. IEEE 32(7):419–422
17. Hajimiri A, Limotyrakis S, Lee TH (1999) Jitter and phase noise in ring oscillators. IEEE J Solid-State Circuits 34(6):790–804
18. Hajimiri A (2007) mm-Wave silicon ICs: challenges and opportunities. In: Invited paper: IEEE 2007 custom integrated circuits conference (CICC), 22-1-1, pp 741–748
19. Hedayati MK, Abdipour A, Shirazi RS, Ammann MJ, John M, Cetintepe C, Staszewski RB (2019) Challenges in on-chip antenna design and integration with RF receiver front-end circuitry in nanoscale CMOS for 5G communications systems. IEEE Access 7:43190–43204
20. Inzelt G (2008) Conducting polymers: a new era in electochemisty. Springer Science & Business Media. ISBN 3540759301
21. Jia H, Chi B, Kuang L, Yu X, Chen L, Zhu W, Wei M, Song Z, Wang Z (2015) Research on CMOS mm-Wave circuits and systems for wireless communications. China Commun 12(5):1–13
22. Kim Y, Tam S, Itoh T, Chang MF (2017) A 60-GHz CMOS transceiver with on-chip antenna and periodic near field directors for multi-Gb/s contactless connector. IEEE Microwave Wirel Compon Lett 27(4):404–406
23. Lambrechts JW, Sinha S (2016) Microsensing networks for sustainable cities. Springer International Publishing, Switzerland. ISBN 978-3-319-28358-6
24. Lambrechts JW, Sinha S (2017) SiGe-based re-engineering of electronic warfare subsystems. Springer International Publishing, Switzerland. ISBN 978-3-319-47402-1
25. Lambrechts JW, Sinha S (2019) Last mile internet access for emerging economies. Springer International Publishing, Switzerland. ISBN 978-3-030-20956-8
26. Leeson DB (1966) A simple model of feedback oscillator noise spectrum. Proc IEEE 54(2):329–330
27. Li Y (editor), Lie DYC, Li C, Zhao D, Fager C (guest editors) (2018) RF front-end circuits and architectures for IoT/LTE-A/5G connectivity. Wireless communications and mobile computing open access article
28. Marcu C, Chowdhury D, Thakkar C, Park J, Kong L, Tabesh M, Wang Y, Afshar B, Gupta A, Arbabian A, Gambina S, Zamani R, Alon E, Niknejad AM (2009) A 90 nm CMOS low-power 60 GHz transceiver with integrated baseband circuitry. IEEE J Solid-State Circuits 44(12):3434–3447
29. Marki F, Marki C (2010) Mixer basics primer. A tutorial for RF and microwave mixers. Marki Microwave Inc
30. Mazor N, Sheinman B, Katz O, Levinger R, Bloch E, Carmon R, Ben-Yishay R, Elad D (2017) Highly linear 60-GHz SiGe downconversion/upconversion mixers. IEEE Microwave Wirel Compon Lett 27(4):401–403
31. Motoyoshi M, Kameda S, Suematsu N (2018) 57 GHz 130 μW CMOS millimeter-wave oscillator for ultra low power sensor node. In: 11th global symposium on millimeter waves (GSMM), pp 1–3
32. Moulthrop AA, Muha MS, Silva CP (2005) On the frequency response symmetry of diode mixers. IEEE Trans Instrum Meas 4(6):2458–2461
33. Niknejad AM (2005) Introduction to mixers. University of California, Berkeley. Retrieved 5 June 2019 from http://rfic.eecs.berkeley.edu

34. Pang J, Wu R, Wang Y, Dome M, Kato H, Huang H, Narayanan AT, Liu H, Liu B, Nakamura T, Fujimura T, Kawabuchi M, Kubozoe R, Miura T, Matsumoto D, Li Z, Oshima N, Motoi K, Hori S, Kunihiro K, Kaneko T, Shirane A, Okada K (2019) A 28-GHz CMOS phased-array transceiver based on LO phase-shifting architecture with gain invariant phase tuning for 5G new radio. IEEE J Solid-State Circuits 54(5):1228–1242
35. Qayyum JA, Albrecht JD, Ulusoy AC (2019) A compact V-band upconversion mixer with –1.4-dBm OP1dB in SiGe HBT technology. IEEE Microwave Wirel Compon Lett 29(4):276–278
36. Razavi B (2008) A millimeter-wave circuit technique. IEEE J Solid-State Circuits 43(9):2090–2098
37. Razzel C (2013) System noise-figure analysis for modern radio receivers: Part 1, calculations for a cascaded receiver. Application Note: Maxim Integrated—A1131
38. Reynaert P, Steyaert W, Standaert A, Simic D, Kaizhe G (2017) mm-Wave and THz circuit design in standard CMOS technologies: Challenges and opportunities. In: IEEE Asia Pacific microwave conference (APMC), pp 85–88
39. Singh S, Kumar N (2018) Design of wideband millimeter wave mixer in CMOS 65 nm for 5G wireless. In: International conference on advances in computing, communications and informatics (ICACCI), pp 717–721
40. Swift G, Molinski TS, Lehn W (2001) A fundamental approach to transformer thermal modeling. I. Theory and equivalent circuit. IEEE Trans Power Deliv 16(2):171–175
41. Teppati V, Tirelli S, Lovblom R, Fluckiger R, Alexandrova M, Bolognesi CR (2014) Accuracy of microwave transistor f_T and f_{MAX} extractions. Electron Dev 61:984–990
42. Terrovitis MT, Meyer RG (1999) Noise in current-commutating CMOS mixers. IEEE J Solid-State Circuits 34(6):772–783
43. Wang H, Chen J, Do JTS, Rashtian H, Liu Z (2018) High-efficiency millimeter-wave single-ended and differential fundamental oscillators in CMOS. IEEE J Solid-State Circuits 53(8):2151–2163
44. Wang X, Zhu X, Yu C, Li C, Liu P, Shi X (2018). Wideband transceiver front-end integrated with Vivaldi array antenna for 5G millimeter-wave communication systems. In: IEEE international symposium on antennas and propagation & USNC/URSI national radio science meeting, pp 405–406
45. Yang X, Matthaiou M, Yang J, Wen CK, Gao F, Jin S (2019) Hardware-constrained millimeter-wave systems for 5G: challenges, opportunities, and Solutions. IEEE Commun Mag 57(1):44–50

Chapter 4
Transceivers for the Fourth Industrial Revolution. Millimeter-Wave Low-Noise Amplifiers and Power Amplifiers

Abstract In this chapter, a review is presented on receiver subsystem-level of the low noise amplifier (LNA) in a millimeter-wave (mm-Wave)-compatible fifth-generation (5G) transceiver to identify the challenges and limitations of this subsystem at microwave operation (referring specifically in this context to centimeter-wave (3–30 GHz) and mm-Wave (30–300 GHz) operation. In this chapter, the expressions mm-Wave and microwave operation are used interchangeably and this includes mm-Wave frequencies. An overview of the considerations of the power amplifier (PA) in the transmitting front-end is correspondingly presented in this chapter, with a similar analysis of its microwave operation for 5G communications. Although the 5G specification allows for various options for carrier modulation, power levels, data rates and other capabilities as reviewed in Chap. 1 of this book, many of the performance characteristics are derived from the transmitter and receiver front-ends, more specifically the quality of the operation of the LNA and the PA. The LNA is responsible for receiving a weak, noisy signal and (ideally linearly and efficiently) amplifying this signal to a usable level without adding noise. Numerous LNAs have been used that are specifically designed for lower-GHz operation (such as for the 2.4 GHz and 5 GHz bands), but difficult operation in the mm-Wave 5G domain increases the complexity of these circuits significantly. Not only is it necessary for the architecture and topology of the LNA to be optimized for mm-Wave operation; the process technology and transistor type should also be considered based on their merits that are applicable and required for a specific application. Improvements on technologies such as "*complementary metal-oxide semiconductor (CMOS), bipolar CMOS silicon germanium (SiGe), silicon-on-insulator and gallium arsenide field-effect transistors*" are being researched to improve LNA performance from process level through inherent characteristics of the materials, such as leakage currents and electron mobility. References in this chapter to the process technologies are made for the subsystems being reviewed, and a detailed summary of the benefits and drawbacks of the various processes is presented in Lambrechts and Sinha [16]. In this chapter, a review of the fundamentals of LNAs and PAs is presented, along with an analysis on circuit level of the architectures that are typically used for high-frequency operation.

W. Lambrechts and S. Sinha, *Millimeter-wave Integrated Technologies in the Era of the Fourth Industrial Revolution*, Lecture Notes in Electrical Engineering 679,
https://doi.org/10.1007/978-3-030-50472-4_4

The scope of this chapter is limited to describing the performance aspects of the LNA and the PA in terms of their architecture. This technical overview allows the reader to understand the performance limitations when designing transceiver subsystems for 5G communications; the overview does not aim to derive all performance metrics, as this has been reported in various works, referred to throughout this chapter. The subsequent chapters of this book focus on a techno-economic perspective of 5G the fourth industrial revolution, concentrating on emerging markets. Thorough understanding of the limitations and complexities of microwave circuit design is encouraged to avoid underestimating the skills required from researchers and engineers to implement and sustain the technology in these markets. Together with Chap. 2 of this book, the subsystems that are required to process high-frequency signals within a mm-Wave 5G communications system are therefore identified and reviewed. This provides the necessary background to implement these types of systems in preparation for big data communications.

4.1 Introduction

The performance of a millimeter-wave (mm-Wave) fifth-generation (5G) transmitter or receiver front-end has similar performance metrics and design considerations to those of lower frequency communication systems. However, at mm-Wave frequencies many of the challenges and limitations are *amplified* in terms of complexity. Communications at mm-Wave frequencies are at the mercy of the technology—the limitations of the active and passive components, of parasitic effects that are often considered negligible in lower frequency systems, as well as propagation limitations due to the inverse relationship between frequency and distance.

On the receiver front-end, the low noise amplifier (LNA) is responsible for amplifying the weak incoming signal short of additional noise that is generated and summed to the signal. On the transmitter front-end, the power amplifier (PA) is responsible for amplifying the signal to be transmitted through the channel, and the quality of this signal effectively determines the distance that it can travel and still be detected by a receiver. This has cost implications for the system if *too many* base stations are needed along a route. Both these subsystems are therefore crucial in a mm-Wave 5G communication system and require special attention during the design and implementation phases. In this chapter, these considerations are reviewed and presented to identify the challenges and limitations of mm-Wave communications. The LNA is reviewed with respect to the topologies and their alterations to offer mm-Wave capability, and refers directly to the input and output matching, noise and gain characteristics as performance metrics of these subsystems. The PA is reviewed likewise and focus is placed on its primary performance metrics such as linearity, gain and efficiency. These analyses and reviews do not aim to offer a step-by-step guide to design and implement these subsystems, as this is outside the presented scope of this chapter. Numerous references are supplied during the reviews that provide additional design-specific discussions and understanding on the microelectronic design of these circuits.

For both subsystems, the fundamentals are briefly reviewed to identify the importance of each subsystem in a mm-Wave 5G communications system. The following section is devoted to a review of the fundamentals of the LNA followed by its performance metrics specifically within the microwave domain.

4.2 The Fundamentals of LNAs

The LNA is arguably the most critical subsystem of the receiver front-end and defines many facets of the receiver performance. There are various LNA topologies and architectures, each with its own merit, but the chosen topology should be optimized for the application, therefore the environment in which the system will operate and the radio frequency (RF) spectrum. This is especially true of mm-Wave 5G-capable receivers and requires additional consideration in terms of gain and noise performance, as well as linearity, since the active and passive components in the chosen technology might incur inherent limitations on certain topologies. To determine the limitations and performance characteristics that increase in complexity as operating frequency increases, the fundamental operation of the LNA is briefly reviewed. The simplified representation of a direct-conversion receiver architecture as presented in Chap. 2 of this book is referenced and presented in Fig. 4.1, with the placement of the LNA highlighted.

The direct-conversion receiver architecture (specifically indicating microwave operation through the input signal f_{RF}) in Fig. 4.1 receives an incoming wireless RF signal at the receiving antenna. This (weak and noisy) signal is initially amplified by an LNA, therefore the (input-matched to the antenna) LNA is fundamentally a vital interface between the information input signal and the rest of the receiving subsystem. For this reason, the LNA is deemed arguably one of the most crucial

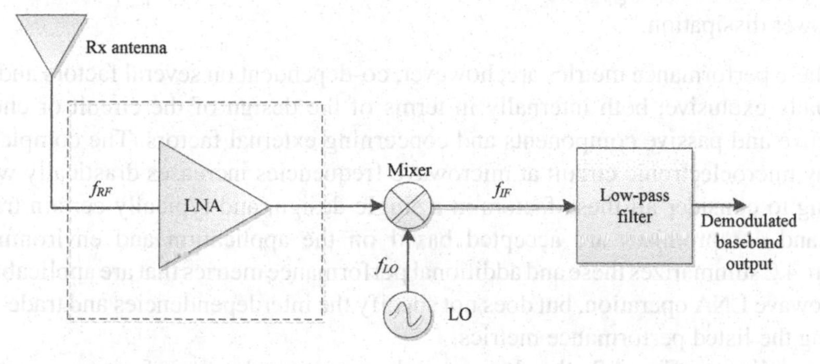

Fig. 4.1 A simplified representation of a direct-conversion receiver architecture showing the placement of the LNA as the first point of contact with the received RF signal

performance-critical components in the receiver, with stringent requirements. The signal that enters the LNA from the antenna already has a fair amount of noise, picked up during transmission through the channel, and the LNA should ideally not enlarge the noise of the signal. The performance metrics of the LNA define the quality and ability of the LNA to perform the required tasks.

The performance metrics of the LNA are therefore crucial to categorize and grasp the interdependence of several factors in the internal and external environment of the LNA. At microwave operation the LNA cannot be treated as an independent subsystem and numerous internal and external factors affect its performance. The following section reviews the internal, external and systems-level performance metrics of LNAs in terms of its operation at microwave frequencies for 5G communication.

4.3 Microwave LNA Performance Metrics

The LNA, which like frequency mixers was reviewed in Chap. 2 of this book, ideally should have superior performance in terms of low noise and low interference of the microwave signal transmitted through the medium, which is especially crucial in low-power high-frequency uses like the internet of things (IoT) within the 5G infrastructure. In addition, several device- and component-level as well as board-level and external factors influence the accuracy and efficiency of LNA performance. Typical performance metrics of the LNA that are usually specified either in research reports published in the body of knowledge, or on datasheets for off-the-shelf components, broadly include (not listed in order of importance)

- power gain (uniformity and constancy in a temperature band),
- noise figure (NF)/noise factor,
- linearity,
- stability,
- impedance matching, and
- power dissipation.

These performance metrics are, however, co-dependent on several factors and not mutually exclusive, both internally in terms of the design of the circuit or choice of active and passive components and concerning external factors. The complexity of any microelectronic circuit at microwave frequencies increases drastically when aiming to consider all these factors in a single design, and typically certain trade-offs and compromises are accepted based on the application and environment. Figure 4.2 summarizes these and additional performance metrics that are applicable to microwave LNA operation, but does not specify the interdependencies and trade-offs among the listed performance metrics.

According to Fig. 4.2, the device- and component-level performance metrics include the strategy and architecture of the feedback and stability networks in the design of the LNA. Performance metrics then become more complex to optimize

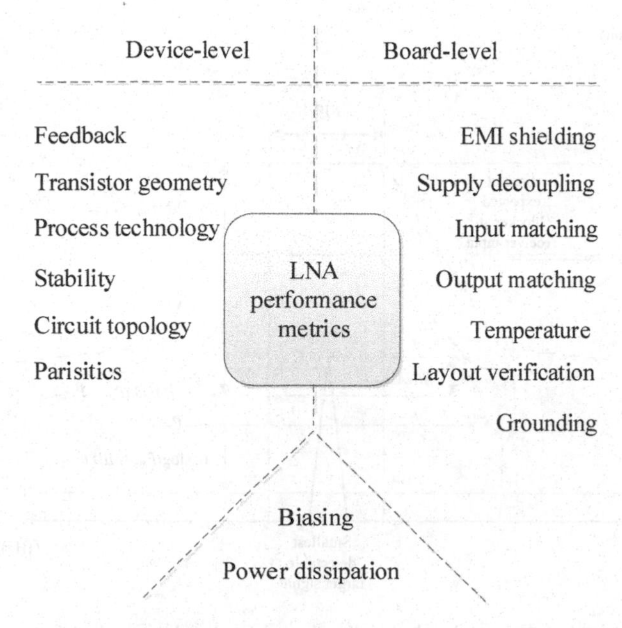

Fig. 4.2 Device- and component-level as well as board-level and external performance metrics of LNAs

at frequencies approaching mm-Wave and recognizing that these circuits are typically designed on chip, as opposed to discrete components, factors such as transistor geometry, process technology and package parasitics must be considered. Circuit topology can be altered and modified based on the limitations of the process technology and requires thorough circuit simulations and post-layout verification. On board level, therefore the interface between the LNA and its external connections, numerous factors influence the performance of the LNA, including temperature of operation and variations, input and output matching layout and grounding, electromagnetic shielding and power supply decoupling. A performance variable that depends on both the internal and external factors of the LNA and is typically defined at systems level is the biasing and power dissipation, which is also dependent on the process technology used to integrate the LNA on chip. The receiver sensitivity, essentially the product of all internal and external factors and a fair indication of the total link budget in a communications structure, is detailed at systems level in Das [5] and adapted in Fig. 4.3. This adaptation serves as a good indication of how certain performance metrics of the LNA affect the overall systems-level implementation.

At systems level, the receiver sensitivity as summarized in Fig. 4.3 is used to identify LNA performance optimization in a particular environment. Performance requirements become especially complex and stringent in microwave operation. Based on Fig. 4.3, the following analysis describes the overall link budget of the receiving subsystem (essentially an analysis of the spurious-free dynamic range (SFDR)): From the input power perspective, primarily, the thermal noise density

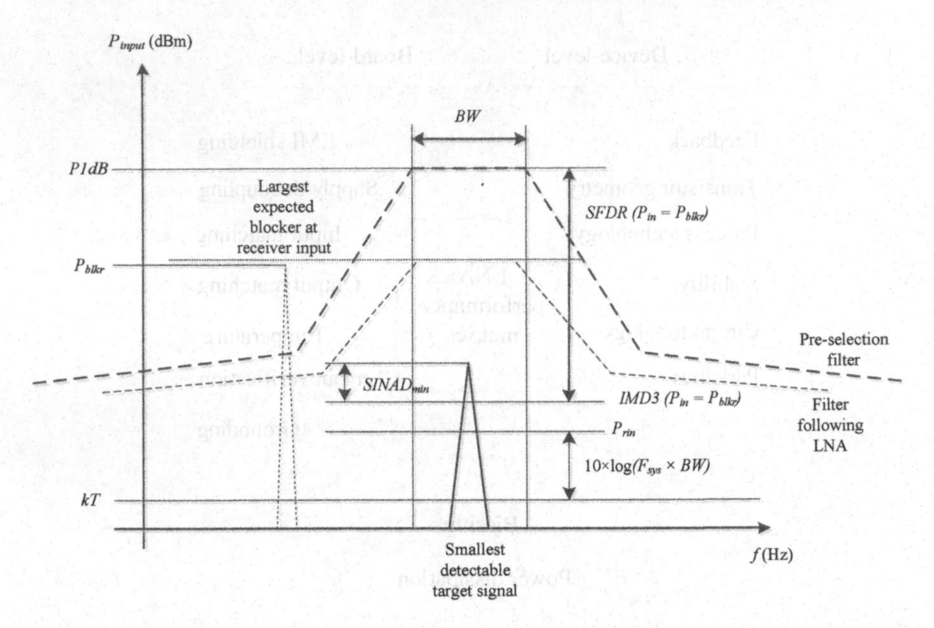

Fig. 4.3 Device- and component-level as well as board-level and external performance metrics of LNAs. Adapted from Das [5]

(kT) defines the smallest detectable signal of any receiver. Specific to the receiver, P_{rin} defines the noise floor, a function of the bandwidth (BW) requirement of the circuit and the noise in the system. P_{1dB} defines the signal power and the input and output that correspond to a 1 dB gain compression (see Chap. 2 of this book). Figure 4.3 serves as a good reference for LNA design and defines the areas where optimization can be done when considering the link budget of the signal. Interdependencies are identified and used in research to focus on specific areas of the LNA and the way in which overall improvements will affect the subsystem operation. The parameters identified in Fig. 4.3 are also referred to throughout this chapter when reviewing the circuit topologies and component choice of LNAs. This chapter also aims to identify circuit-level limitations and proposed enhancements of the performance metrics of microwave LNAs, particularly ones realized in complementary metal-oxide semiconductor (CMOS) and bipolar CMOS (BiCMOS) processes.

Before analyzing the typical LNA topologies and architectures that are capable of microwave operation, the following section expands on the performance considerations by reviewing design considerations of the LNA. These design considerations are specifically tailored to microwave operation but have also been considered for lower frequency systems, although microwave systems apply similar principles but with additional complexity.

4.4 LNA Design Considerations

The LNA is not only a function of its circuit topology (as reviewed in the following section) but also depends on various external factors to operate at its required performance, as stated in the preceding section. As pointed out in the abstract of this chapter, the process technology ("*silicon (Si), silicon germanium (SiGe), silicon-on-insulator (SOI) or gallium arsenide (GaAs)*" and therefore *not* power transistor technologies "*such as gallium nitride (GaN) or indium phosphide (InP)*") is typically the first decision when choosing to implement an LNA on chip, and many design considerations are based on the specifications of the chosen process. Both active and passive components in a technology process have limitations, advantages and disadvantages and should be considered before starting the design phase. In Lambrechts and Sinha [16], numerous process technologies are reviewed and compared with respect to their material performance. The technologies reviewed in Lambrechts and Sinha [16] include

- "*Si,*
- *SiGe,*
- *GaN,*
- *silicon carbide,*
- *InP, and*
- *GaAs*".

The performance parameters of active components that are reviewed in Lambrechts and Sinha [16] include

- "*electron bandgap,*
- *electron mobility,*
- *power density,*
- *breakdown voltage,*
- *thermal conductivity, and*
- *cut-off frequency*".

Using a performance comparison such as the one presented in Lambrechts and Sinha [16] serves as a good starting point in choosing the specific technology when microwave amplification is required, as in the case of 5G LNAs. Once this step is complete, there are further considerations that influence the performance of the LNA, which are especially important to consider at microwave frequencies. Among these are:

- External connections (tracks) from chip level to an external printed circuit board require careful consideration of the printed circuit board (PCB) material and their geometry (especially length). In the GHz range of frequencies, track losses (transmission line losses, skin effect) at the input and the output of the LNA are major factors and require additional modeling and post-layout simulations to mitigate. The parasitic capacitance or inductance generated as a result is typically regarded as an additional passive component that needs to be designed for in the final layout.

- At GHz frequencies, capacitors and inductors with high quality factors should ideally be used for matching at the input and the output of the LNA. High-quality factors essentially lead to better noise performance and maintain a low NF into and out of the LNA. A performance versus cost analysis is required, since high-quality factor passive components are typically expensive to manufacture.
- For stable and consistent microwave operation, power supply bypassing is crucial and bypass capacitors should be chosen such that their impedance is at a minimum at the operating frequency to ensure maximum decoupling performance. As frequencies rise, the required capacitance decreases substantially and on-chip decoupling is often necessary. Again, parasitic components should be considered and PCB layout becomes crucial to avoid additional parasitic passive components on tracks and other external connections.
- If physical cables are required to interconnect subsystems, circuit balancing should be maintained to decrease susceptibility to noise. Cable selection and calibration are required in addition to account for any amplitude or phase shifts inherent to mm-Wave propagation as a function of insertion loss and cable length. High-performance RF cables also drive the cost of mm-Wave and 5G systems and should be considered in advance and during the transceiver design phases.

Internally, and referring to CMOS, BiCMOS or high electron mobility transistors (HEMT) on-chip LNAs, there are various characteristics that require additional attention, understanding and simulations if operated in the microwave spectrum. Essentially, noise performance and power gain are two of the most important characteristics of the LNA, as these determine its capacity to intensify the weak signal without accumulation extra noise. Additional performance characteristics that affect the LNA performance are outlined in the LNA performance metrics section of this chapter. Noise and power gain matching, and techniques to decrease noise and improve linearity, are discussed in further detail below.

In terms of noise and power gain matching, the key design considerations are optimum noise matching to obtain the minimum achievable NF (NF_{min}). Furthermore, conjugate impedance matching for maximum available power gain becomes a trade-off for noise matching. In CMOS, it is possible to achieve a state where both matching circumstances are adjacent and the key aim in CMOS LNAs is to achieve simultaneous noise and power matching as closely as possible.

To achieve peak power gain matching in an LNA, the input impedance of the LNA need a resistive part and the matching network changes, from the perspective of the source impedance, this resistive component to a real term, typically 50 Ω or 75 Ω. The chosen topology, for example the common-gate or common-source topology, then has different techniques to achieve this matching. The common-gate-based LNA has the advantage that the resistive term is already a portion of the input impedance to the source of the input transistor, as discussed in the following section. In a common source topology (or if a cascode configuration is used) the input impedance becomes purely capacitive at low frequencies and therefore a resistive part should be introduced to the input terminal. This can be achieved either through resistive feedback (which will add additional thermal noise and increase the NF) or through a parallel resistance in the gate of the input transistor. Another method, one that is often used, is adding a degenerating inductance in the source of the common-source transistor, which does not add any additional noise to the system. A degenerating inductance in the source of the common-source input transistor yields a resistive term in the input impedance of the transistor. Furthermore, this inductive component increases the possibility of achieving simultaneous power and noise matching. Because of its advantages in terms of power gain and noise matching, this method will be discussed in the following section.

The second additional characteristic of LNA design and performance is improving the noise performance as well as linearity of the receiver. Noise canceling techniques for CMOS-based LNAs, especially for ultra-wideband applications, have received much attention in literature and in practical implementations. It is advantageous to incorporate such techniques in both narrowband and wideband application. In narrowband LNAs, the $NF_{\min}$ is already achievable with proper matching; however, for wideband applications, the frequency response in the operating band places limitations on obtaining $NF_{\min}$. Essentially, noise-canceling techniques take advantage of signals propagating in the circuit and arriving at specific points from different paths. Noise arriving at these specific points is 180° out of phase and consequently attenuated at the output, while the signal is being amplified. These schemes are typically feed-forward-based schemes but become less effective at high frequencies (above approximately 10 GHz) owing to the stringent phase accuracy requirements [25]. Externally, noise that originates from instruments and even inside the chip that couples through the substrate affects the noise performance of the LNA. These noise sources require additional decoupling and on-chip layout techniques to minimize the effects on the receiver performance. Finally, nonlinearities of the CMOS-based LNA are inherently an effect of the transistor biasing affecting its conductance and transconductance characteristics.

There are various other design considerations when choosing or designing an LNA capable of mm-Wave 5G operation. Among these considerations are:

- If the active devices (transistor) have poles in frequencies well above feedback loop bandwidth, it is often decided to use feedback techniques to match input terminals of the chosen topology and transfer the optimum noise impedance to a anticipated point, especially in common-source configurations. Advances in transistor cut-off frequency (f_T) have allowed feedback techniques to be used, even in mm-Wave LNAs. Feedback additionally reduces the nonlinearity of the circuit and improves the IIP3 point (discussed in Chap. 2 of this book).
- It is important to consider adequate electrostatic discharge (ESD) protection in CMOS circuits because of its inherent high input impedance and low gate breakdown voltages. These ESD circuits, when implemented in high-frequency mm-Wave (and lower RF) circuits, should ideally be simplistic and not lead to performance degradation of the LNA. Various techniques of ESD protection are available and depend on the operating frequency, with dual-diode implementations

or silicon-controlled rectifier-based protection circuits often implemented for RF applications.

- Another important characteristic, especially with renewable energy sources becoming more prevalent in electronic applications, is power dissipation associated with the physical area used on chip when designing an LNA. Low-power RF operation is a requirement for most receiver front-ends and the popularity of mobile 5G transceivers has recently driven these requirements. Chip area has always been an important consideration for two reasons, the first being the cost of chip real estate and the second being driven by miniaturization at subsystem level. Passive components such as inductors and capacitors are known to take up large physical area on chip; however, these components do offer higher quality factors when compared to discrete off-chip solutions, leading to an inherent tradeoff when designing circuits that implement passive components (such as the common-source with inductive degeneration). It has been shown that power consumption [14] is primarily driven by the gain requirements of the active components. This requires careful consideration to limit overall dissipation. 5G architecture has been criticized for excessive power dissipation [14] as a function of its modulation techniques. The need for additional base stations due to the inherent free-space loss of mm-Wave frequencies adds complexity, power requirements and additional constraints to these deployments.

An important element to take away from this section is that the design considerations, both internally and externally, of the LNA for an mm-Wave front-end become extremely complex and a strong function of the environment in which they will be used. It becomes difficult to choose a topology, process technology and optimization technique and to perform system-level tasks efficiently in any environment and under most circumstances. For this reason, understanding the challenges and limitations of the LNA becomes crucial in determining the viability and sustainability of a chosen front-end. The following section reviews the LNA topologies that were highlighted in this section and specifically underlines the design considerations for optimal noise and power gain performance. The section does not review input and output matching techniques, but emphasizes the component-level characteristics that influence the matching networks.

In Reddy and Choi (undated), a practical summary of design considerations of the LNA circuit is presented. The key points are listed below:

- Possibly the most crucial consideration in LNA design, before topology and matching are considered, is the transistor(s) used. The geometry (aspect ratio) and process-specific parameters should be carefully considered before attempting to implement an LNA, which might possibly be inadequate for mm-Wave operation owing to limitations of the active devices.
- Oscillation has to be avoided. In-depth stability analysis to determine whether the amplifier is inherently stable is crucial. To verify the stability of a transistor, the Rollet's stability factor analysis [27] can be used, where (using *S*-parameter analysis)

$$K = \frac{1 - |S_{11}|^2 - |S_{22}|^2 + |\Delta|^2}{2|S_{21}S_{12}|} \quad (4.1)$$

where

$$\Delta = |S_{11}||S_{22}| - |S_{12}||S_{21}| \quad (4.2)$$

and stability is achieved when $K < 1$ and $|\Delta| < 1$.

- Transistor biasing and hence the steady-state operating point of the transistor need to be chosen with respect to the entire operating bandwidth of the circuit.
- Load line analysis to set up the bias point for the transistor to operate at the chosen operating point is required.
- An acceptable NF and adequate power gain should be achieved through input and output matching of the LNA.

From the numerous considerations that are applicable to mm-Wave LNAs and have to be accounted for in 5G front-ends, the chosen circuit topology and its implementation determine the maximum performance in both power gain and noise. Typical LNA topologies have been identified in this chapter (common-gate, common-source, cascode as well as variants on each) and these topologies are used in mm-Wave 5G applications, as shown in Appendix 1 of this chapter.

Considering LNAs implemented for lower frequencies as well as more modern microwave LNAs, certain topologies are used, depending on the application and the requirement of the system. The various topologies used to implement LNAs should be considered and analyzed on component level to determine the performance with respect to the parameters outlined in Fig. 4.3.

4.5 LNA Topologies Applicable to 5G mm-Wave Communication

The LNA topologies that are typically used in HEMT, CMOS and/or BiCMOS processes are described and analyzed on component level in this section. The topologies that are discussed include

- common-gate,
- common-source,
- cascode, and
- differential topologies.

The performance characteristics of each of these topologies vary in terms of its gain and noise performance, typically a trade-off, and depend on the application of the LNA. In mm-Wave 5G receivers, transistor performance is crucial and due to the inherent free-space losses of mm-Wave signals leading to incoming RF signals that are both weak and noisy. As a result, matched LNA design at these frequencies are complex and typically require more than one stage to satisfy both gain and noise requirements. The first topology to be discussed is the common-gate, an architecture that allows for effective input impedance matching due to a resistive input impedance when observing into the source node of the MOSFET. In an IC implementation, the LNA is often connected off-chip and typically matched to a 50 Ω or 75 Ω impedance (from the antenna), and the output is not automatically matched if driving the frequency mixer input. For the topology analysis, only CMOS MOSFETs are considered and heterojunction bipolar transistors (HBTs) in technologies such as SiGe are not reviewed. Component-level analysis on SiGe HBTs are presented in depth in Lambrechts and Sinha [16].

4.5.1 LNA Topologies: The Common-Gate

The common-gate LNA has been widely used in wireless communications; however, in view of recent operating frequency requirements in the mm-Wave spectrum, as required for 5G systems, there are drawbacks to this configuration in terms of *NF*. One of the primary advantages of the common-gate topology, which still drives arguments for its use in wideband LNAs, is its relatively simple input matching realization. Since the input resistance of this configuration is at the source terminal of the input transistor, it is inversely proportional to the internal current gain (g_m) of the transistor, thus having wideband capabilities. The wideband noise performance of the common-gate topology therefore becomes more comparable to other topologies (discussed in this section) that are typically operated in narrowband applications. In this chapter analysis of the common-gate LNA, identification of the limitations of mm-Wave operation and recognition of the difficulties of implementing this LNA configuration in 5G systems are also considered. The simplified common-gate LNA configuration (essentially a current buffer with no current gain and only voltage gain) is given in Fig. 4.4 and the equivalent small-signal model of this configuration is presented in Fig. 4.5.

In Figs. 4.4 and 4.5, the input signal v_S (typically generated by the receiver antenna) is coupled to the source of the input transistor and the gate is grounded. The output is obtained at the drain of the transistor as the voltage generated by the drain resistor R_D, as shown in Figs. 4.4 and 4.5. For an in-depth component-level analysis on small signal models of both MOSFETs and bipolar transistors, refer to Gray et al. [9]. To achieve the desired input impedance of a common-gate LNA input stage, the bias current, transistor aspect ratio ($^W/_L$) and the overdrive voltage V_{ov} are adjusted. These parameters should be adjusted such that the resistance looking into the transistor, thus $1/g_m$, is close to the termination impedance, which is typically 50 Ω or 75 Ω. As a result, $1/g_m$ should ideally be adjusted to approximately 20 mS or 13 mS, depending on the configuration. At high frequencies, the common-gate LNA does not undergo the effects of the Miller capacitance, since the gate of the input transistor

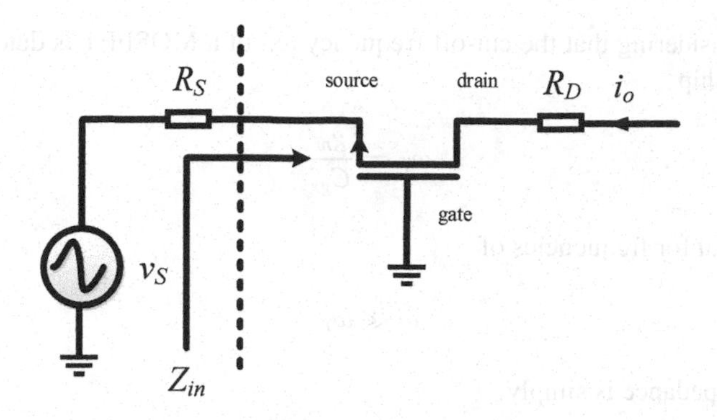

Fig. 4.4 Simplified representation of the common-gate LNA configuration

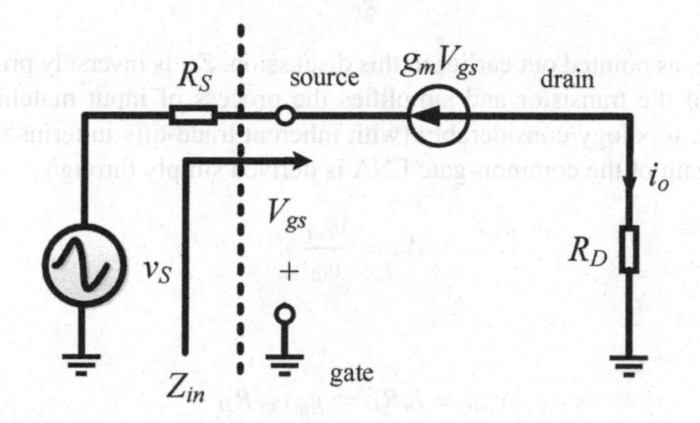

Fig. 4.5 Equivalent small-signal model of the common-gate LNA configuration

is grounded and this is an advantage of the common-gate configuration. Reasonable reverse isolation can therefore be achieved by a single transistor stage. The input impedance Z_{in} considering observing into the source of the transistor is derived simply through its gain and the parasitic gate-source capacitance, resulting in

$$Z_{in} = \frac{1}{j\omega C_{gs} + g_m} \tag{4.3}$$

where $\omega = 2\pi f$ is the operating frequency and C_{gs} is the gate-source capacitance of the transistor. If it is considered that typically

$$\omega C_{gs} \ll g_m \tag{4.4}$$

and therefore

$$\omega \ll \frac{g_m}{C_{gs}} \tag{4.5}$$

and also considering that the cut-off frequency ω_T of a MOSFET is determined by the relationship

$$\omega_T = \frac{g_m}{C_{gs}} \tag{4.6}$$

it follows that for frequencies of

$$\omega \ll \omega_T \tag{4.7}$$

the input impedance is simply

$$Z_{in} \approx \frac{1}{g_m} = R_S \tag{4.8}$$

and therefore, as pointed out earlier in this discussion, Z_{in} is inversely proportionate to the gain of the transistor and simplifies the process of input matching for the common-gate topology considerably (with inherent trade-offs in terms of gain and noise). The gain of the common-gate LNA is derived simply through

$$A_v = \frac{v_{out}}{v_{in}} \tag{4.9}$$

where

$$v_{out} = i_o R_D = g_m v_{gs} R_D \tag{4.10}$$

and

$$v_{in} = v_{gs} \tag{4.11}$$

thus

$$A_v = \frac{g_m v_{gs} R_D}{v_{gs}} \tag{4.12}$$

and is therefore simply equal to

$$A_v = g_m R_D \tag{4.13}$$

which shows the proportionality to the drain resistance R_D. To determine the noise factor F (recall that the noise figure is denoted as NF) of the LNA (first at low frequencies), the total output noise in relation to the noise due to the source resistor is analyzed and is derived in Gray et al. [9]. In Gray et al. [9] it follows that the noise factor is equal to

$$F = 1 + \frac{\overline{i_{nd}^2}\left(\frac{1}{1+g_m R_S}\right)^2}{\overline{e_{ns}^2}\left(\frac{g_m}{1+g_m R_S}\right)^2} \quad (4.14)$$

where i_{nd} is the drain current noise and e_{ns} is the output noise current from the source resistance. This representation of the noise factor can be simplified to

$$F = 1 + \frac{\gamma}{\alpha} \frac{1}{g_m R_S} \quad (4.15)$$

and if a conjugate power match is achieved, therefore if $1/g_m = R_S$, the noise factor F is simplified to

$$F = 1 + \frac{\gamma}{\alpha} \quad (4.16)$$

where γ and α are described in Chap. 2 of this book. Although this performance is adequate for broadband operation at lower frequencies, the 5G milieu requires much better noise performance from the LNA, and modifications to the common-gate LNA exist. However, rather than modifying a topology with inherent limitations, an alternative to the common-gate LNA, the common-source (more ideally the common-source with inductive source degeneration) can be used and is reviewed in the following section.

4.5.2 LNA Topologies: The Common-Source

To improve on *NF* degradation experienced in the common-gate LNA topology owing to the noise resistance in the signal path, the common-source (and the common-source with inductive source degeneration) LNA topologies are often used. These circuits are especially popular in narrowband high-frequency applications. The typical common-source configuration is presented in Fig. 4.6.

The common-source topology in Fig. 4.6 is resistively loaded at the input terminal, therefore at the gate of the input transistor. The equivalent small-signal model of the common-source LNA is presented in Fig. 4.7 and used to analyze the alternating current (AC) characteristics of the circuit. It is at this point assumed that the direct current (DC) bias characteristics through load line analysis to bias the transistor in the saturation region has already been achieved.

From the equivalent small-signal model of the common-source LNA presented in Fig. 4.7, the output voltage can be derived as

$$v_{out} = -g_m v_{gs}(r_o \| R_D) \quad (4.17)$$

and therefore the (unloaded) voltage gain can be represented as

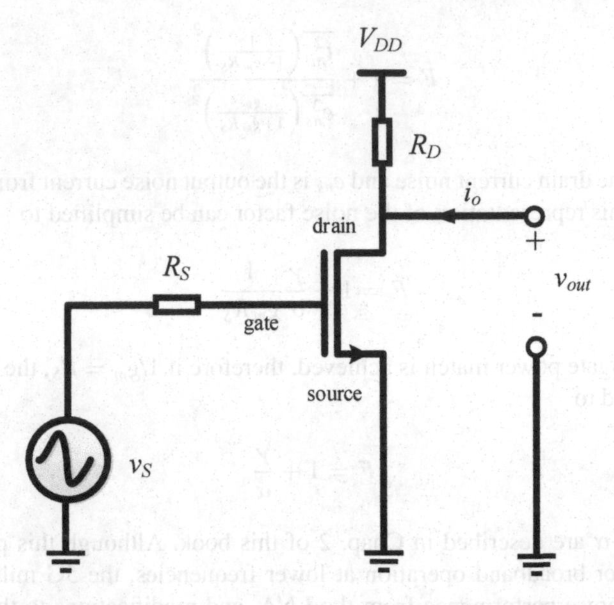

Fig. 4.6 The common-source LNA topology (not showing biasing network on transistor gate)

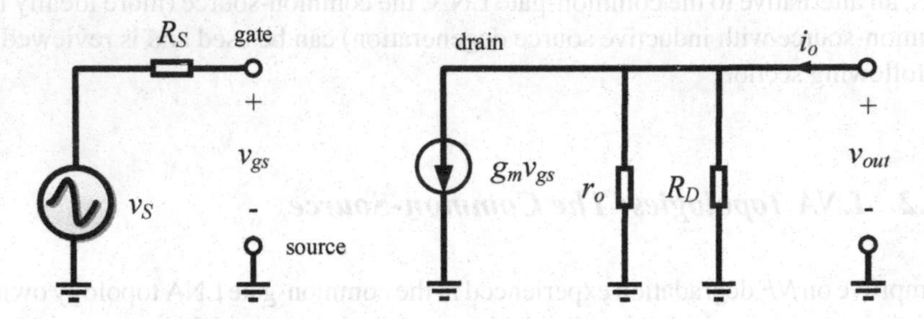

Fig. 4.7 The equivalent small-signal model of the common-source LNA topology

$$A_{vo} = \frac{v_{out}}{v_{gs}} = -g_m(r_o \| R_D) \tag{4.18}$$

and can be amplified by improving the transistor gain (a higher current) proportionally. The drain resistor R_D can also be enlarged to increase the gain of this configuration; however, the limitation is to keep the transistor in the saturation region, such that

$$V_{DS} = V_{DD} - I_D R_D > V_{DS,sat} \tag{4.19}$$

so for a fixed current, an inherent limitation is set by R_D. The input impedance of the common-source LNA, from the equivalent small-signal model in Fig. 4.7, is derived from

$$R_{in} = \frac{v_s}{i_s}. \tag{4.20}$$

However, $i_s = 0$ on the gate of a MOSFET, therefore the input impedance is ∞. The output impedance is simply a parallel combination between R_D and r_o of the transistor, such that

$$R_{out} = r_o \| R_D \tag{4.21}$$

and can be adjusted through R_D, again considering the limits placed on ensuring operation of the transistor in the saturation region. From a transistor-design perspective, the gain, input impedance and output impedance can be adjusted by considering that

$$g_m = \sqrt{2 I_D \frac{W}{L} \mu_n C_{ox}} \tag{4.22}$$

where I_D is the drain current of the transistor, W and L are the width and length of the transistor, respectively (adjustable during the design phase), μ_n is the electron mobility (process-dependent), and C_{ox} is the oxide capacitance (also process-dependent). Furthermore, the output resistance of the transistor is a proportional relationship that can also be described by

$$r_o \propto \frac{L}{I_D} \tag{4.23}$$

and therefore can also be controlled in some part by the geometry of the active device and varying the current in the drain. The transconductance gain of the common-source LNA at medium frequencies is derived in Gray et al. [9] and is given as

$$G_m = \frac{1}{1 + j\omega C_{gs}(R_S + R_g)} \tag{4.24}$$

and the derivation of the high-frequency noise factor [8] yields

$$F = 1 + \frac{R_g}{R_S} + \frac{\gamma}{\alpha}\left(\frac{\omega}{\omega_T}\right)^2 g_m R_S \tag{4.25}$$

where the low frequency noise factor, as derived in Lee and Perrot [18], is given as

$$F = 2 + 4\frac{\gamma}{\alpha}\frac{2}{g_m R_S}. \tag{4.26}$$

This derivation outlines the considerations that can be varied to attain the required performance, a necessary step in any LNA design and especially true in microwave operation. Table 4.1 lists the potential variations in gain, input impedance and output

Table 4.1 Variations in gain, input impedance and output impedance of the common-source LNA configuration that can be achieved through transistor-level and supply-current variations

Device parameter	Circuit parameter		
	Gain A_{vo}	Input resistance R_{in}	Output resistance R_{out}
Supply-current increase	Decrease	∞	Decrease
Transistor width (W) increase	Increase	∞	n/a
Transistor length (L) increase	Increase	∞	Increase

impedance of the common-source LNA configuration that can be achieved through transistor-level and supply-current variations.

A relatively minor adjustment of the common-source topology yields superior performance; however, as in most electronic circuits, it also presents drawbacks. Essentially, a source-terminal inductor can be added to the common-source topology, leading to several enhancements in performance, but at the expense of, for example, additional size on chip and therefore cost of manufacturing and miniaturization restrictions. The common-source topology with inductive source degeneration is reviewed in the following section.

4.5.3 LNA Topologies: The Common-Source with Inductive Source Degeneration

"*Source degeneration introduces negative series feedback and reduces the gain of the LNA*" (Daniel and Terrovitis [3]). As an advantage, it offers one additional "*degree of freedom*" to optimize the input impedance of the circuit and allows adjustment of the ideal source impedance with the consequence of the lowest NF (essentially through lower gain). Inductive source degeneration does not add additional (thermal/resistive) noise to the common-source LNA as a resistive degeneration alternative would. This inductive degeneration technique is implemented to control the real part of the input impedance by varying the reactance on the gate of the input transistor. The "*common-source with inductive source degeneration topology*" (Daniel and Terrovitis [3]) (also showing an inductive component on the input terminal to compensate for the input capacitance) is presented in Fig. 4.8.

From the common-source with inductive source degeneration topology as presented in Fig. 4.8 the 50 Ω (or 75 Ω) source resistance is connected across the input terminal (gate) of the LNA and therefore provides input matching. The bandwidth of this configuration is also affected by the gate-source capacitance C_{gs} of the transistor and the capacitance can be large, leading to bandwidth degradation. The inductance L_g placed in series with the input terminal effectively removes the gate-source capacitance of the transistor at the resonant frequency, thus making the input impedance at the input terminal real, denoted as R_{in} (as opposed to the typical

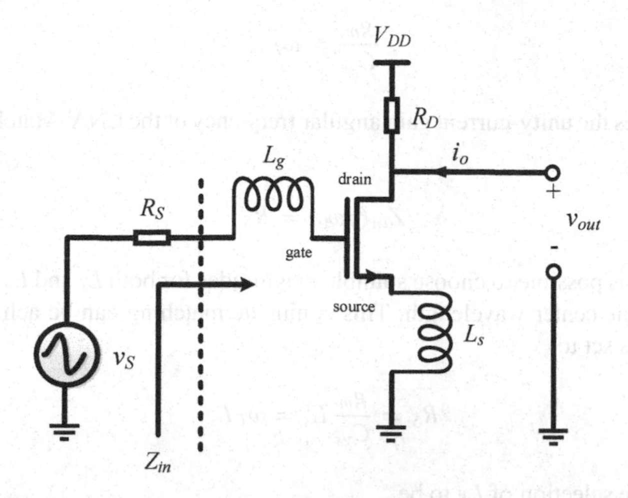

Fig. 4.8 Common-source with inductive degeneration LNA topology, Daniel and Terrovitis [3], including an inductive component on the input (gate) terminal

complex term of Z_{in}) [17]. To analyze the performance of the said topology, the equivalent small signal model is presented in Fig. 4.9.

The input impedance of the equivalent small-signal model in Fig. 4.9 is derived in Das [5] as

$$Z_{in} = j\omega(L_g + L_s) + \frac{1}{j\omega C_{gs}} + \frac{g_m}{C_{gs}} L_s \tag{4.27}$$

where

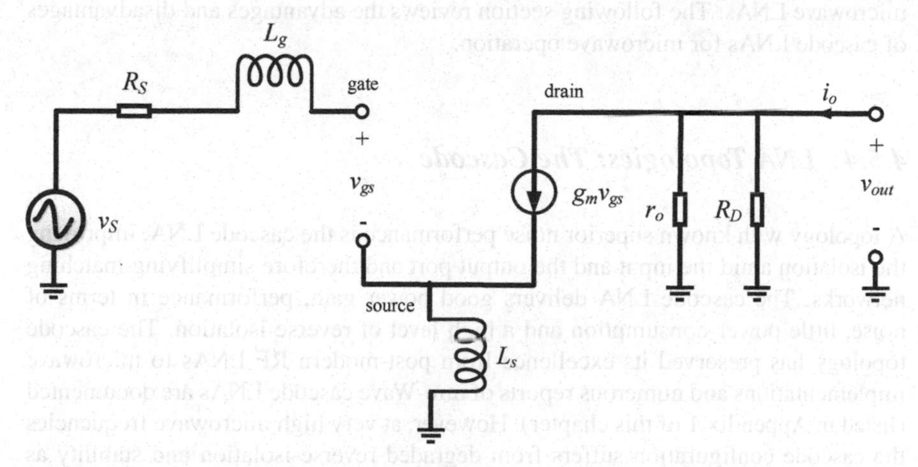

Fig. 4.9 The equivalent small-signal model of the common-source LNA configuration with inductive source degeneration

$$\frac{g_m}{C_{gs}} = \omega_T \tag{4.28}$$

which denotes the unity-current gain angular frequency of the LNA. Matching occurs when

$$Z_{in}(j\omega_0) = R_S \tag{4.29}$$

and hence, it is possible to choose suitable magnitudes for both L_s and L_g to resonate with C_{gs} at the center wavelength. This conjugate matching can be achieved if the value of R_S is set to

$$R_S = \frac{g_m}{C_{gs}} L_s = \omega_T L_s \tag{4.30}$$

leading to the selection of L_S to be

$$L_S = \frac{R_S}{\omega_T} \tag{4.31}$$

and if this value becomes practically too small, a capacitor can be inserted in parallel with L_S to reduce ω_T artificially. Therefore, the common-source topology with inductive source degeneration takes advantage of the gain characteristics of the common-source topology and adds additional degrees of freedom to adjust the input matching and therefore improve on noise performance. However, the addition of (at least) one inductive component is a disadvantage in terms of space on chip (leading to higher cost of integration), but this topology is typically still preferred above the common-gate and traditional common-source topologies. To increase reverse isolation as well as improve noise performance and power gain, cascode topologies are used in microwave LNAs. The following section reviews the advantages and disadvantages of cascode LNAs for microwave operation.

4.5.4 LNA Topologies: The Cascode

A topology with known superior noise performance is the cascode LNA, improving the isolation amid the input and the output port and therefore simplifying matching networks. The cascode LNA delivers good power gain, performance in terms of noise, little power consumption and a high level of reverse isolation. The cascode topology has preserved its excellence from post-modern RF LNAs to microwave implementations and numerous reports of mm-Wave cascode LNAs are documented (listed in Appendix 1 of this chapter). However, at very high microwave frequencies the cascode configuration suffers from degraded reverse-isolation and stability as a result of a feed-forward path created amid the input and output port owing to

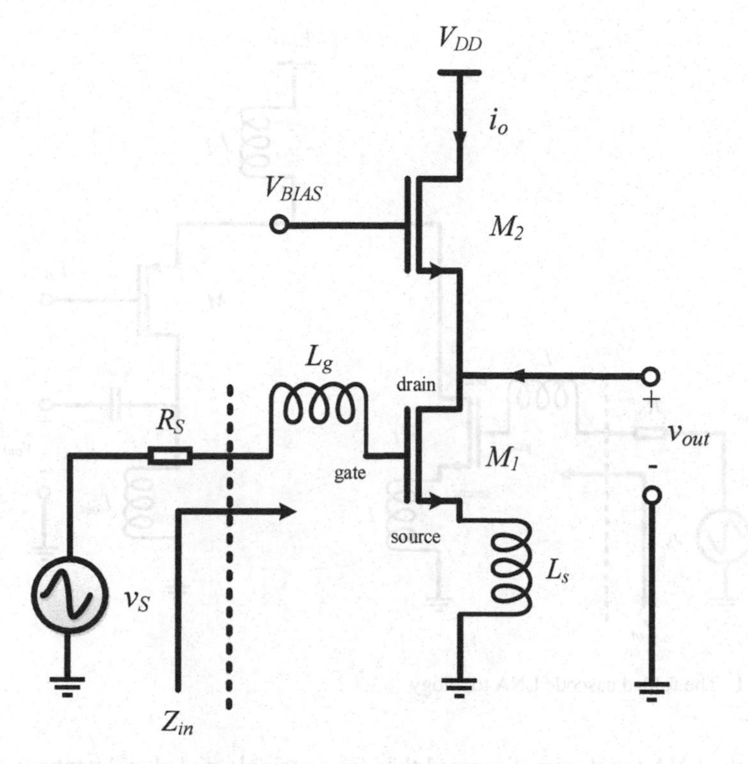

Fig. 4.10 The cascode LNA topology

an intrinsic gate-drain capacitance. A typical representation of the cascode LNA is shown in Fig. 4.10.

As presented in Fig. 4.10, the cascode LNA comprises a "*common-source topology with inductive source degeneration*" (Daniel and Terrovitis [3]) with a current-biasing common-gate network through M_2. This configuration therefore adopts a current-sharing technique between the transistors to intensify the gain but not at the overhead of power dissipation and therefore noise performance. The cascode topology also has higher output impedance and the voltage gain of the cascode arrangement is similar to that of the common-source LNA. The performance of the noise and gain of the cascode LNA, however, also reduces at extreme microwave wavelengths owing to the "*parasitic admittance at the drain-source common node that increases as the operating frequency increases*" (Daniel and Terrovitis [3]). The resulting lower impedance seen at the source of M_2 leads to higher noise at its train terminal. The folded cascode LNA topology extends the cut-off wavelength of M_1 through folding M_2. The typical circuit representation of the folded cascode LNA topology is presented in Fig. 4.11.

The folded cascode topology shown in Fig. 4.11 eliminates the parasitic capacitances of M_1 through resonance with the inductive component L_d at the supply. This technique suppresses the noise contribution of M_2 at the output and mitigates substrate losses at this node.

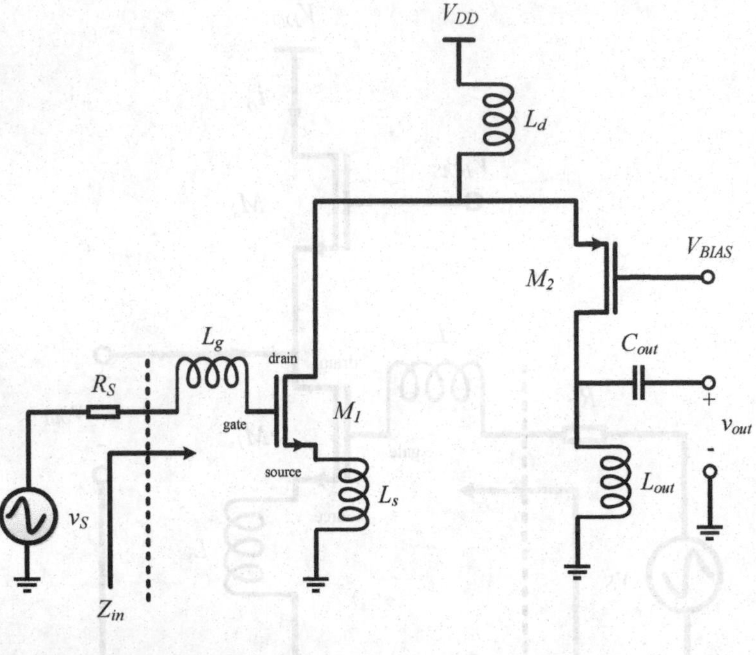

Fig. 4.11 The folded cascode LNA topology

All the LNA topologies discussed thus far are single-ended architectures, therefore only accepting a single input and providing one output. The performance of a single-ended architecture still has inherent limitations that cannot be overcome through component placement and cascode topologies. For further improvement on performance, a differential topology is required to *cancel* out noise that is inherent to the system. The benefits and weaknesses of the differential LNA are reviewed in the following section.

4.5.5 LNA Topologies: Differential

The single-ended LNA topologies reviewed thus far all require additional balun circuitry to translate the single-ended output to a differential output for the mixer (see Chap. 3 of this book). To realize the output of the LNA in a differential form, differential topologies are often preferred [2]. The differential LNA is constructed by two single-ended cascode LNAs, for example a common-gate and common-source topology in cascode configuration [2]. To optimize the gain, noise and reverse isolation performance at the output of the LNA, an output buffer is added. Figure 4.12 represents a typical differential LNA configuration.

A differential LNA as shown in Fig. 4.12 implements the premise of applying gain to the difference amid the two input signals, as opposed to applying it to one signal

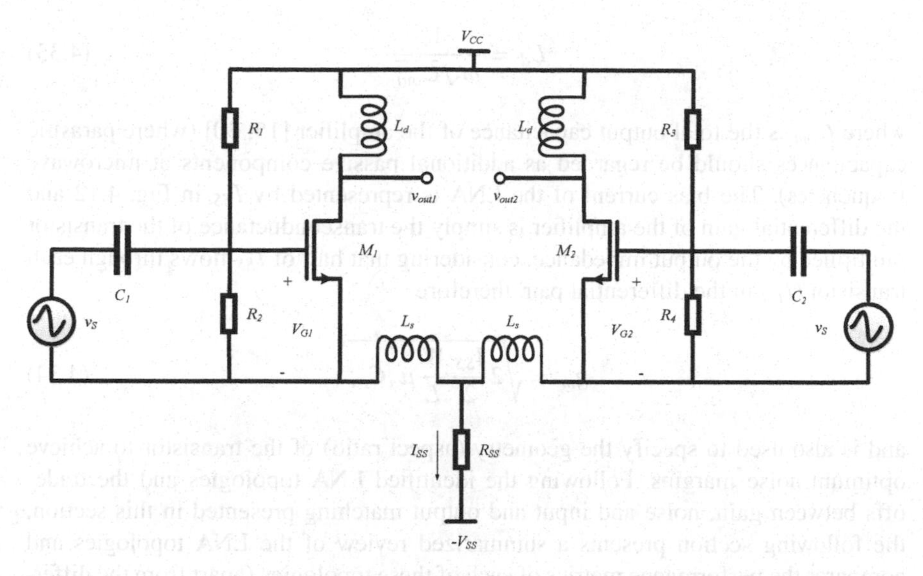

Fig. 4.12 Differential LNA topology showing the inductive component on the transistor drain and source and biasing networks on the transistor gates

at a time. This essentially translates that the differential configuration eliminates noise or interference that is existent (and similar) in the two input signals, naturally. Furthermore, the differential increase overpowers common-mode voltages/currents, for example a DC component that is existent in the two input signals. The gain is therefore only applied to the signal of interest and not to the noise or DC components. As a result, which is especially advantageous, the mm-Wave on-chip LNA design obviates the requirement for large DC-blocking capacitors and the output of the LNA. The differential LNA, however, does require a higher component count, including the larger passive (inductive) components, and these components (as well as transistors and resistors) should ideally be symmetrical and matched, increasing the cost and complexity of these circuits. These drawbacks are, however, often not considered serious, especially taking into account the superior performance of a differential LNA compared to the single-ended configurations.

The differential LNA in Fig. 4.12 already represents an amplifier where the drain, gate and source inductors are included, as reviewed in the single-ended representations of the LNA (essentially to match the input and output without adding resistive noise). The magnitudes of L_g and L_d are calculated by

$$L_g = \frac{QR_S}{\omega} \tag{4.32}$$

where Q is the quality factor of the inductor and $\omega = 2\pi f$ is the operating frequency in rad/s. L_d is determined through

$$L_d = \frac{1}{\omega\sqrt{C_{out}}} \tag{4.33}$$

where C_{out} is the total output capacitance of the amplifier [15, 30] (where parasitic capacitances should be regarded as additional passive components at microwave frequencies). The bias current of the LNA is represented by I_{SS} in Fig. 4.12 and the differential gain of the amplifier is simply the transconductance of the transistor multiplied by the output impedance, considering that half of I_{SS} flows through each transistor (I_D) in the differential pair, therefore

$$g_m = \sqrt{2\frac{I_{SS}}{2}\frac{W}{L}\mu_n C_{ox}} \tag{4.34}$$

and is also used to specify the geometry (aspect ratio) of the transistor to achieve optimum noise margins. Following the identified LNA topologies and the trade-offs between gain, noise and input and output matching presented in this section, the following section presents a summarized review of the LNA topologies and compares the performance metrics of each of these topologies (apart from the differential topology, thus only the single-ended implementations). This comparison can be used to identify the preferred architecture if considering the requirements that the subsystem should fulfill for a specific application. The following section additionally lists some comparative parameters of the technologies available to implement LNAs on-chip.

4.6 LNA Topologies: Comparison and Tradeoffs

4.6.1 Comparison of LNA Topologies

Adapted from Das [5], Table 4.2 provides a summarized list of the performance comparison of the common LNA topologies. The list provides device-level tradeoffs among the single-ended common-gate, common-source and cascode topologies; the differential circuit is not added to this list. The trade-offs and characteristics considered include the NF and the power gain, considered as the two parameters that directly affect performance variables such as linearity, bandwidth, stability, reverse isolation and sensitivity to tolerances (in the environment and in the manufacturing process).

As shown in Table 4.2, adapted from Das [5], it can be argued that the cascode amplifier is the most versatile of the three (single-ended) topologies. In terms of gain and bandwidth, it performs significantly better than the common-gate and common-source amplifiers, with little degradation in terms of its NF (which can be analyzed from the examples in Appendix 1 of this chapter). If superior NF is crucial and gain requirements can be relaxed, the common-source configuration (in conjunction with the inductive source degeneration) performs well. It is important to note that

Table 4.2 Comparison of the common-gate, common-source and cascode LNA topologies in terms of various performance metrics

Performance metric	Common-gate	Common-source	Cascode
Noise figure	Increases rapidly with frequency	Low	Medium
Power gain	Low	Moderate	High
Linearity	High	Moderate	High
BW	Medium/wide	Narrow	Wide
Stability	High	Compensation needed	High
Reverse isolation	High	Low	High
Overall sensitivity to tolerances	Small	Large	Small

Adapted from Das [5]

the superior NF is a result of the permissible transistor sizing and results in larger sensitivity to bias conditions, temperature and component-level tolerances. All these factors can be managed externally, through feedback or during the design and layout phases of the LNA, but could add to the cost and complexity of the system.

4.6.2 LNA Transistors: FET Versus BJT

The choice of transistor type for LNA design affects the performance and is typically application-specific, considering the trade-offs and environment in which the LNA will operate. There are two choices of transistor when designing an LNA, *either a field-effect transistor (FET) or a bipolar junction transistor (BJT)*, each with its own advantages and disadvantages for specific applications. Some trade-off considerations with respect to FET and BJT transistors if used in LNA design are given in the summary in Table 4.3.

As summarized in Table 4.3, there are distinct differences in terms of noise performance, gain and biasing requirements between FET and HBT implementations when used in LNAs. In general, the pHEMT FET exhibits higher power consumption but with a higher OIP3, whereas BiCMOS transistors will have a lower power consumption but owing to the shot noise component, exhibit lower OIP3. In terms of performance characteristics, Das [5] presents a summarized breakdown of technology-based performance of LNAs, adapted here and listed in Table 4.4.

As shown in Table 4.4, using GaAs (pHEMT) transistors in LNA design is preferred when a low NF is crucial, since this process involves a thinner heterojunction amid the doped AlGaAs and the undoped GaAs layer. This process also has higher linearity when compared to SiGe BiCMOS implementation. In terms of usability and frequency capability, GaAs and SiGe are comparable; however, the dynamic range of SiGe is limited because of its low breakdown voltage. SiGe does

Table 4.3 Comparison of LNA typical performance parameters if realized in GaAs (pHEMT) and SiGe heterojunction bipolar transistor (HBT) (BiCMOS)

FET (pHEMT)	BJT (HBT)
Gate resistance and thermal noise higher	Higher shot noise levels
Gate geometry determines noise performance	Optimal NF bias considerably lower than maximum power gain bias
Optimal NF bias and maximum power gain bias very close	Higher real input admittance for minimum NF [Y_{opt}]
Lower real input admittance for minimum NF [Y_{opt}]	Thermal runaway to be controlled through additional ballast resistor
ESD protection required to avoid gate damage	Higher transconductance
High low-frequency gain (prone to oscillation)	

Adapted from Das [5]

Table 4.4 Comparison of LNA typical performance parameters, if realized in GaAs (pHEMT), GaAs (HBT) and SiGe (BiCMOS)

Performance parameter	Unit	GaAs (pHEMT)	GaAs (HBT)	SiGe BiCMOS
Noise figure	dB	≥0.4/0.5	≥1.4	≥0.9
Power gain	dB	12–21	8–13	10–17
OIP3	dBm	≤41	≤30	≤31
Breakdown voltage	V	≈15	≈16.5	≈4
Typical inductor quality factor	–	≈15	≈15	5–10
f_T/f_{max}	–	High	High	High
Advantages	–	Good linearity Low NF	Reasonable performance at low cost	High integration potential Low-cost Inherent ESD immunity

Adapted from Das [5]

have advantages in terms of integration and cost and can be combined with standard CMOS designs on chip without requiring any additional interconnect circuitry. The advantages and limitations of each technology should ideally be analyzed based on the specific application. With respect to 5G and mm-Wave operation, both high frequency and low noise performance are crucial, but because of high inherent transmission losses in free space at mm-Wave, power gain and cost (since numerous base stations are needed to amplify signals along its path continuously) are equally important, and the tradeoffs need to be considered in every respect.

The following sections follow a similar analysis of the transmitter front-end subsystem, the PA, required to amplify the signal and achieve high power gain while remaining as linear as possible—another inherent tradeoff that can be *managed* through topology and component-level analysis. The review on the microwave PA is again not an exact guideline on designing and implementing the subsystem, but rather aims to identify the complexities of microwave operation.

4.7 The Fundamentals of PAs

In its most simplistic form, the PA (like most amplifiers) converts a DC power supply to AC power and generates an enlarged form of the signal at its input terminal. There is therefore a difference between the AC input power and the AC output power and the power gain is added by the PA. In terms of a communications transmitter front-end, the PA is the final stage "*before the signal is transmitted by the antenna*" [22], as indicated in Fig. 4.13.

The performance of an RF PA "*can often dominate the overall transmitter performance as its power-added efficiency (PAE) dictates the power and heat dissipation for the entire transmitter*" [22]. For massive multiple-input multiple-output (MIMO) antennas at microwave frequencies, 5G transceivers will need additional PAs to be combined in the RF front-end components and therefore the topology/architecture and efficiency of the 5G RF PA are crucial. According to Lie et al. [22] and Shakib et al. [28], successful 5G PA implementation requires critical analysis and design in terms of fundamental performance metrics, which include

- "*output power P_{out}*
- *power gain G_p*
- *PAE, and*
- *linearity*".

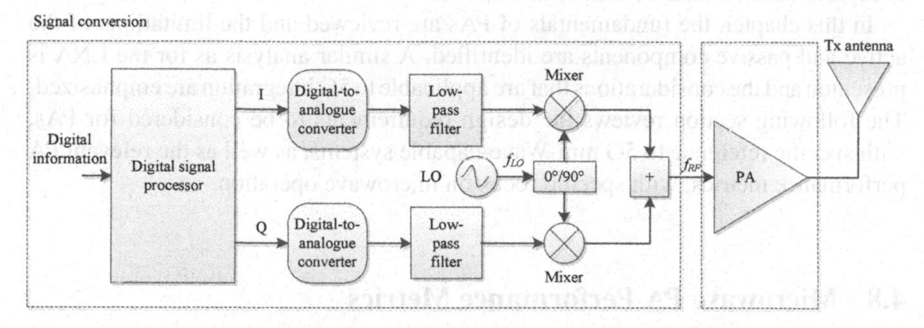

Fig. 4.13 A simplified representation of a direct-conversion transmitter design showing the placement of the PA as the last point of contact with the outgoing RF signal, before it reaches the antenna

These four metrics are specifically performance-based and form part of the design phase of the PA, typically with trade-offs between parameters that should be prioritized in terms of the application and environment where the PA is to be used. Other metrics that result from the design of the PA should also be considered, including

- reliability,
- cost, and
- form factor.

These listed metrics are becoming increasingly important, especially since 5G mm-Wave communications are specifically targeted at the mobile world, therefore smartphones and IoT devices. The reliability, cost and form factor are therefore crucial not only to allow the technology to be used in power-limited mobile devices, but also to increase the profitability of manufacturers. With regard to the *technical* specifications, the broadband modulation of 5G transmissions requires high power efficiency and linearity from the integrated PAs. Furthermore, since 5G networks will employ the phased array antenna strategy to steer multiple beams, transmission tasks will have to be divided among multiple beams and the PA must be capable of adapting its power requirements to the variations. The PA must be capable of adjusting its power consumption dynamically based on the number of beams (with an inverse proportionality) to which it is transmitting. 5G operates in more bands (and wider apart) compared to 4G and 3G and therefore requires more RF switching, filtering and power amplification elements within the PA, further increasing its complexity and dependence on the performance requirements. 4G PAs have been relying (successfully) on GaAs as the underlying technology, since this technology supports the high voltages (power transistors) required for power amplifiers. Above the sub-6 GHz band and especially within the mm-Wave spectrum, however, arrangements of high-frequency technologies such as SOI, GaN and SiGe are researched and considered to replace traditional GaAs implementations.

In this chapter, the fundamentals of PAs are reviewed and the limitations of the active and passive components are identified. A similar analysis as for the LNA is presented and the considerations that are applicable to 5G integration are emphasized. The following section reviews the design requirements to be considered for PAs, with specific reference to 5G mm-Wave-capable systems, as well as the relevant PA performance metrics, with specific focus on microwave operation.

4.8 Microwave PA Performance Metrics

In this segment, the performance metrics of the PA that delivers a relatively worthy outline of the performance of the PA is presented. Again, there are numerous internal and external effects that have an effect on the overall PA performance and considering all these (mostly parasitic) effects increases the complexity of the design significantly. The performance metrics considered in this section include the output power, power gain, PAE and linearity—all parameters that should be designed for based on the application and environment where the microwave PA is to be used.

4.8.1 Output Power

The output power of the PA (P_{out}) is the power that is delivered to the load, which in most cases for wireless communications systems is the antenna that will transmit the signal through the air (the channel). The saturated continuous wave output power (denoted as P_{sat}) is a function of the maximum supported voltage of the transistor (process-specific and directly relating to the applied DC power supply voltage) and the maximum current that can flow through the transistor, $I_{\max}$, which can be adjusted by adjusting the geometry of the transistor during the design phase (assuming high-frequency implementation that is on chip and not with discrete components). Importantly, the output power of the PA should be kept lower than P_{sat} to avoid clipping of the output signal, which will result in degraded performance and essentially distorted output signals. The output power distributed to the load can be conveyed mathematically by considering the signals presented in Fig. 4.14.

If it is assumed that in Fig. 4.14 a typical matched output load with real resistance R is used; the expression of the output power is given as

$$P_{out} = \frac{\left(\frac{V_{p\text{-}p}}{2}\right)^2}{2R} \tag{4.35}$$

where $V_{p\text{-}p}$ is the voltage from the trough of the output waveform to the crest and the output power is typically specified in dBm. The output power rating of PAs also differs between applications. If gain and linearity were not crucial (for example in frequency modulation transmissions) then the rated output power would refer to the

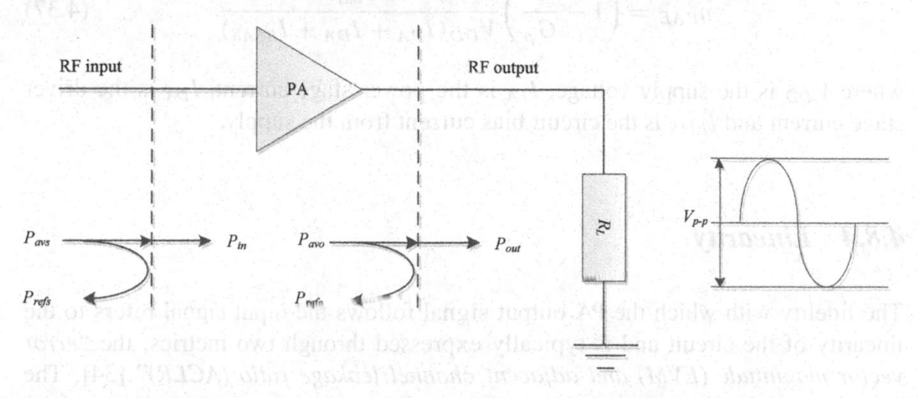

Fig. 4.14 Simplified representation the input and output signals of the PA used to define the output power as well as the power gain

saturated output power. For an ideally linear amplifier, the output power is rated along the linear portion of the input-output curve.

4.8.2 Power Gain

The power gain of an amplifier is, simply put, the proportion of the output- to input-power and is typically expressed in dB, or dBm is more common in RF and microwave amplifiers. The power gain considers the input and output matching conditions and can be determined practically by an RF power meter.

4.8.3 Power-Added Efficiency

The PAE of a PA relates the amount of power that is supplied to the PA through its supply voltage and current to the RF output power. The PAE is an important metric, since the PA typically draws high currents from its supply, and at low efficiency, this current will not generate RF power but heat instead. The PAE is therefore also crucial in low-power wireless mobile devices that have limited power supply. The PAE is essentially the input power subtracted from the output power, divided by the DC supply power, and it is proportionate to the power dissipated in the power transistors. Mathematically, PAE is expressed as

$$\eta_{PAE} = \frac{P_{out} - P_{in}}{P_{DC}} \tag{4.36}$$

and can be expanded to

$$\eta_{PAE} = \left(1 - \frac{1}{G_p}\right) \frac{P_{out}}{V_{DD}(I_{PA} + I_{DA} + I_{BIAS})} \tag{4.37}$$

where V_{DD} is the supply voltage, I_{PA} is the power stage current, I_{DA} is the driver stage current and I_{BIAS} is the circuit bias current from the supply.

4.8.4 Linearity

The fidelity with which the PA output signal follows the input signal refers to the linearity of the circuit and is typically expressed through two metrics; the "*error vector magnitude (EVM) and adjacent channel leakage ratio (ACLR)*" [34]. The

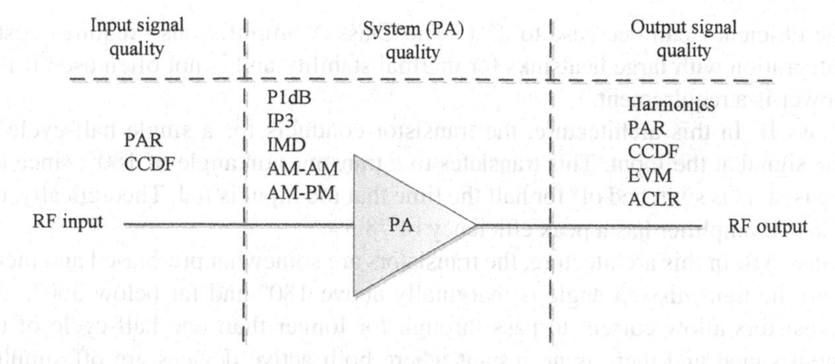

Fig. 4.15 Simplified representation of the input and output of the PA to define degradation of the input signal due to non-linear effects

linearity therefore essentially defines how the quality of the output signal is maintained through the PA and Fig. 4.15 represents the effects on the input signal throughout the system.

As presented in Fig. 4.15, the quality of the input is defined by its peak-to-average ratio (PAR) and its complementary cumulative distribution function (CCDF). In the system (PA), non-linear effect metrics that change the quality of the input signal refers to its P1dB, IP3, intermodulation distortion, and its narrowband amplitude-amplitude and amplitude-phase components. The output signal are to be degraded and analysis is done on its harmonics, PAR, CCDF, EVM and ACLR, to compare the final quality to that of the input signal.

The following section compares the typical PA topologies as well as advanced topologies used specifically for RF (microwave) communications and specifically for 5G low-power devices that require linear and efficient transmissions from limited power supplies.

4.9 PA Topologies

Various PA topologies are available, each with its own merit, drawbacks and complexity. The typical PA topologies used (for both low- and high-frequency operation) are not reviewed in detail in this book, since each topology is well known and numerous references exist that outline each topology. The performance metrics of the PA in general is, however, discussed in this section with reference to different available topologies. For convenience, a list of the typical PA topologies is provided:

- Class A. In this architecture, the active device is always switched on and the current towards the output drifts for the full phase of the signal on the input. This results in total immunity to crossover distortion and high fidelity but at the cost of increased power consumption and poor efficiency. Theoretically, 50% is the maximum power efficiency of the Class A amplifier and in practical scenarios the efficiency can decrease to 25%. The Class A amplifier also requires costly integration with large heatsinks for thermal stability and is not often used if low power is a requirement.

- Class B. In this architecture, the transistor conducts for a single half-cycle of the signal at the input. This translates to a transmission angle of 180°; since the transistors is switched off for half the time that the input is fed. Theoretically, the Class B amplifier has a peak efficiency of 78.5%.
- Class AB. In this architecture, the transistors are somewhat pre-biased and therefore the transmission angle is marginally above 180° and far below 360°. The transistors allow current to pass through for longer than one half-cycle of the input signal and there is no instant where both active devices are off simultaneously. As a result, crossover distortion is mitigated. In terms of distortion, the Class AB is superior when related to the Class B, but its effectiveness is somewhat lower, at a theoretical maximum of 70%.
- Class C. In this architecture, comparable to the Class B, the transistor allows current to pass through for a smaller amount of one half-cycle of the input, therefore smaller than 180° and typically between 80° and 120°. Consequently, the efficiency is increased at the expense of increased distortion. The theoretic peak efficiency of the Class C is 90%. The Class C PA is not regularly implemented in audio amplification due to the high levels of distortion, but it has many RF applications in oscillators and amplifiers where the distortion has a more limited effect on the output (due to additional tuned circuits).
- Class D. In this architecture, the active power-handling transistors operate as binary switches (as opposed to linear gain devices) and little to no power is consumed in the zero-input condition. As a result, the Class D amplifier is among the most efficient configurations in terms of power efficiency, with a theoretical efficiency of 100% and around 90% in practice. Higher efficiency also leads to lower thermal power dissipation and therefore this architecture consumes less power than the previously mentioned architectures, an advantage in power-limited mobile devices.
- Class E. In this architecture a single-pole transferring component as well as a specifically tuned reactive system amid a switching circuit and the load is used for operation that is efficient. To improve the RF efficiency and linearity of the traditional classes of amplifiers discussed above (A, B, C and D), harmonic tuning (active matching) of the output networks has become more commonplace, especially at microwave frequencies. The switching elements in the amplifier configuration are only operated at the points of zero current or zero voltage, when switching occurs. This technique minimizes the amount of power that is lost within the switch itself and mitigates the effect of the switching time of the device in relation to the operating frequency.
- Class F. In this architecture, further improvements to RF amplifier efficiency are implemented through an extra resonant network located in the bias subsystem. This network is resonant at the third harmonic component of the drain voltage waveform. Output filter design is therefore crucial in the Class F amplifier to prevent peak harmonics from reaching the load.

The typical PA architectures listed above are well-known theoretical implementations that have been used in numerous PA output stages. In terms of mm-Wave 5G communication systems, variations of these amplifier topologies are used to satisfy the stringent criteria of low-power operation at microwave frequencies, which is a misnomer, as it is known that RF PAs are inevitably power-hungry [31]. As a result, an analysis of each of these configurations is not presented in this book, and the next section evaluates the advanced adaptations and trends of PA implementations for 5G communication. In Vasjanov and Barzdenas [31], a detailed review of each of these PA architectures is presented, and these reviews are adapted and summarized below. The architectures reviewed in Vasjanov and Barzdenas [31] include

- "*envelope elimination and restoration*
- *envelope tracking*
- *linear amplification using non-linear components*", as well as the
- Doherty topology, and
- the distributed amplifier.

The first advanced PA architecture used in 5G communication that is adapted from Vasjanov and Barzdenas [31] and reviewed is the envelope elimination and restoration (EER) amplifier.

4.9.1 Envelope Elimination and Restoration

Generally, *envelope tracking* enhances the efficiency of traditional EER amplifiers and incorporates "*a modulator for shaping the PA power supply according to the baseband (low-frequency) signal*" [31]. Figure 4.16 indicates this process, adapted from Vasjanov and Barzdenas [31].

The technique presented in Fig. 4.16 offers a dynamically fluctuating source to the RF PA to improve its overall effectiveness. Essentially, the overall efficiency $\eta_{overall}$ is described as

$$\eta_{overall} = \eta_{envelope\_amp} \times \eta_{RFPA} \tag{4.38}$$

indicating that the overall efficiency is a function of the efficiency of the envelope tracking amplifier as well as the efficiency of the RF PA. The modulation on the supply is possible through a linear regulator or a DC-DC switching amplifier. In Vasjanov and Barzdenas [31] several architectures of the envelope tracking amplifier are presented, along with their benefits and weaknesses related to frequency operation, power consumption and efficiency. Another advanced architecture for 5G communications is the "*linear amplification using a non-linear components (LINC)*" approach, as discussed in the following section.

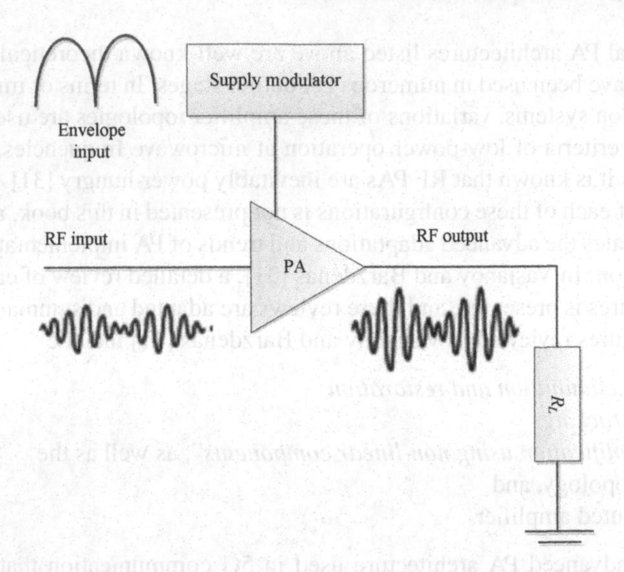

Fig. 4.16 Simplified representation of envelope tracking implemented in an EER PA

4.9.2 Linear Amplification Using Non-linear Components

The linear amplification using the LINC approach is an extension into the microwave domain of the traditional out-phasing approach to improve efficiency and linearity of amplitude-modulated transmitters. The out-phasing approach is depicted in Fig. 4.17.

The out-phasing approach as shown in Fig. 4.17, driven by two signals, combines the outputs of the non-linear PAs with constant amplitudes but with altered phases consistent with the envelope of the RF input. Even if each amplifier operates effectively, the total efficiency is still a function of the accessible combiner output-power. In theory, the LINC architecture can achieve 100% efficiency, although in practice

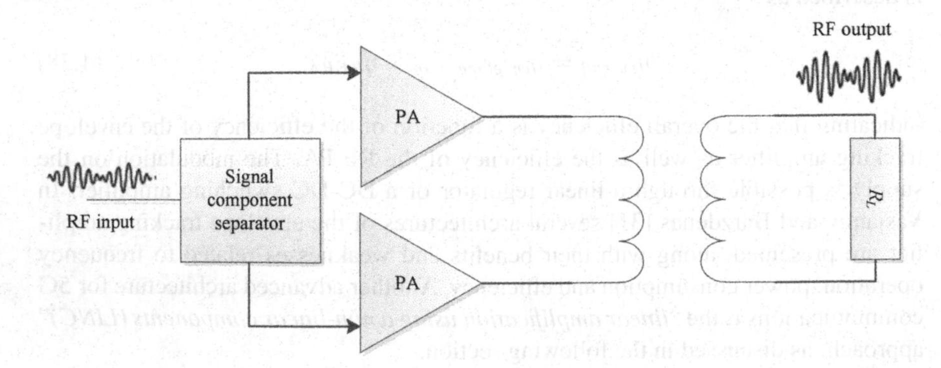

Fig. 4.17 Simplified representation of the out-phasing approach as used in traditional amplitude modulated transmitters. Adapted from Vasjanov and Barzdenas [31]

the PAE is determined by

$$\eta_{PAE} = \frac{2(\cos\varphi)^2}{\sqrt{\left(2(\cos\varphi)^2\right)^2 + \left(\sin 2\varphi - \sin 2\varphi_{comp}\right)^2}} \quad (4.39)$$

where φ is the out-phasing angle and φ_{comp} is the compensation angle. In Vasjanov and Barzdenas [31], a summary of reports that implement the LINC approach is presented. The PAE of these amplifiers ranges between 16 and 62%, where it is noted that the efficiency is strongly dependent on the digital algorithms implemented to achieve efficient LINC, limiting the ability to interpret the effect of the approach on the PA. Finally, and arguably the most popular implementation of microwave PAs, is the Doherty architecture, as reviewed in the following section.

4.9.3 Doherty RF PA

The Doherty RF PA is grounded on the active load notion [31] capable of dynamically modulating the load impedance for maximum efficiency for various input signals [36]. The simplified architecture of the Doherty RF PA is presented in Fig. 4.18.

The typical Doherty RF PA shown in Fig. 4.18 comprises one main carrier amplifier, as indicated in Fig. 4.18, with an output load that is modulated through a

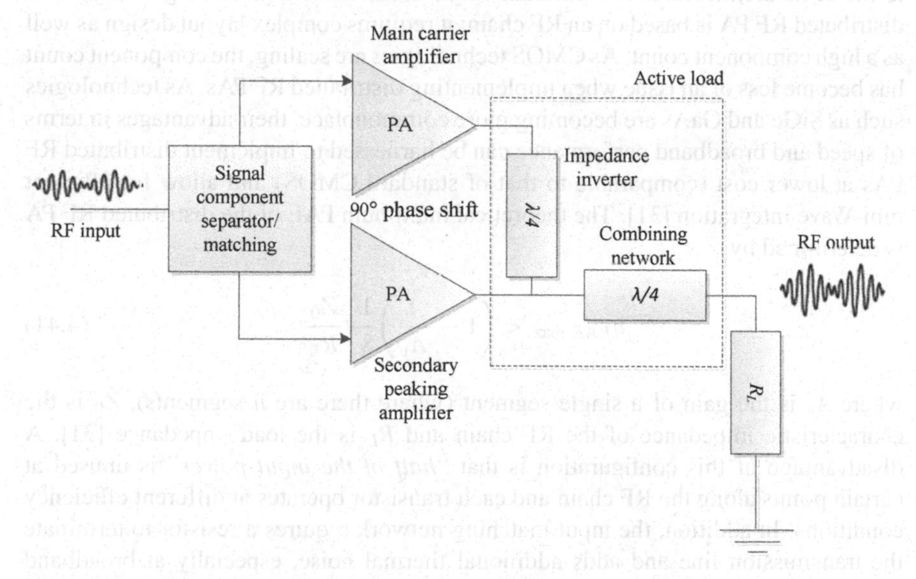

Fig. 4.18 Simplified representation of the Doherty RF PA approach. Adapted from Vasjanov and Barzdenas [31]

secondary peaking amplifier, also indicated in Fig. 4.18. The output impedance determines the active load concept and therefore receives a lot of research attention to improve its performance. According to Vasjanov and Barzdenas [31], the PAE of the Doherty RF PA is determined by

$$\eta_{PAE} = \frac{P_{out} - P_{in}}{\sum_{n=1}^{m} V_{DDn} I_{DQn}} \tag{4.40}$$

where P_{out} is the output power of the PA and P_{in} the input power, V_{DDn} is the power supply voltage of the nth PA in the Doherty arrangement and I_{DQn} is the quiescent current that is expended by the secondary PA, where m is the total number of corresponding branches within the configuration [31]. Vasjanov and Barzdenas [31] list the papers that have reported the Doherty RF PA for 5G communications and the performance attained in these papers. The overall PAE obtained in these papers ranges between 21 and 51% and all reports are related to portable devices, therefore with limited input power available.

4.9.4 *Distributed RF PA*

The distributed (traveling wave) RF PA is often used to enhance the speed and bandwidth of communication front-ends. It can be separated into two key classes in terms of its architecture: conventional single-stage and cascaded single stage. The distributed RF PA is based on an RF chain; it requires complex layout design as well as a high component count. As CMOS technologies are scaling, the component count has become less of an issue when implementing distributed RF PAs. As technologies such as SiGe and GaAs are becoming more commonplace, their advantages in terms of speed and broadband performance can be harnessed to implement distributed RF PAs at lower cost (comparable to that of standard CMOS) and allow for efficient mm-Wave integration [31]. The theoretical maximum PAE of the distributed RF PA is determined by

$$\eta_{PAE,\max} < \left(1 - \frac{1}{A_v}\right)\frac{1}{8} n \frac{Z_0}{R_L} \tag{4.41}$$

where A_v is the gain of a single segment (where there are n segments), Z_0 is the characteristic impedance of the RF chain and R_L is the load impedance [31]. A disadvantage of this configuration is that "*half of the input power*" is unused at certain points along the RF chain and each transistor operates at different efficiency conditions. In addition, the input matching network requires a resistor to terminate the transmission line and adds additional thermal noise, especially at broadband microwave frequency operation.

According to Vasjanov and Barzdenas [31], mm-Wave PAs are typically implemented in the listed advanced RF topologies at frequencies below 25 GHz. Single- or two-stage stacked methodologies in either single-ended or differential systems that typically function in the non-linear areas (using for example Class E/F PAs) are used for operation above 25 GHz. However, this statement is only based on three reported PAs and further investigation and research into mm-Wave PAs continue.

The following section lists and describes the fundamental PA design considerations for microwave operation with specific reference to 5G applications.

4.10 Microwave PA Design Considerations

A common theme in wireless communications is the requirement for RF power transistors and amplifiers to achieve high efficiency, since the transmitting subsystem in any transceiver relates directly to the overall operating cost. Incremental improvements in efficiency have numerous advantages, such as reduced power consumption and potentially fewer transmitting devices to deliver a specific RF signal to receivers and for technologies within the fourth industrial revolution (Industry 4.0), such as the IoT, these advantages add up. Power efficiency is a function of component-level advances, for example RF power transistors, on integrated level, within the RF integrated circuit as well as on monolithic microwave integrated circuits, and academic institutions and researchers are also studying techniques to improve power efficiency in transmitters. The inception of 5G as a commercially available technology integrated into modern mobile phones is predicted to grow the market in research and applications of PAs. Low-power microwave PAs are cornerstones in 5G front-ends and their high complexity is directly proportional to increasing operating frequency.

The high integration requirements of microwave PAs typically "*favor silicon-based technologies for 5G mobile products, even though technologies such as GaAs and GaN offer superior performance*" [22]. Furthermore, successful MIMO realizations are complex to achieve, as linearity is compromised at microwave frequencies with additional heat issues degrading performance. Most modern PAs are implemented and manufactured in III-V semiconductor technologies that offer greater performance in terms of its highest frequency, breakdown sturdiness and quicker to reach a point where it can be sold. In Lie et al. [22] numerous PA technologies for optimal performance in 5G technologies are researched and compared and this comparison shows that most silicon-based CMOS PAs are popular, with SOI and SiGe stacked PAs demonstrating the highest output power capabilities. These output power capabilities are, however, still not able to equal those of III-V semiconductors GaAs and GaN, typically used for high-power military or aerospace applications, whereas commercial 5G can benefit from the better frequency performance of silicon-based technologies.

4.11 Conclusion

This chapter researches the performance challenges and restrictions of the LNA and PA in a mm-Wave-capable communications system, specifically with reference to 5G communications that are likely to turn into the de facto protocol normal for Industry 4.0. The chapter firstly discusses the LNA, including its fundamental principles, microwave performance metrics, design considerations and the topologies that are capable of mm-Wave operation. The chapter also provides a list of 5G LNA implementations (see Appendix 1 of this chapter) that have been reported in the body of knowledge and compares the performance of these works in terms of the performance metrics presented in this chapter. A similar analysis is done of the mm-Wave 5G PA, where the fundamental operation is reviewed and followed by its microwave performance metrics, topologies and design considerations. A summary of reported mm-Wave 5G PAs in the body of knowledge is also presented in Appendix 2 of this chapter and relates the performance of these works in terms of the reviewed performance metrics presented in the chapter.

This chapter concludes the technical analysis and review of the mm-Wave subsystems that are required for a 5G communications system. These reviews are specifically focused on identifying challenges and limitations of mm-Wave subsystems, since the subsequent chapters of this book are focused on the techno-economic perspective of mm-Wave 5G communications in emerging markets. It is, however, also necessary to grasp the technical aspects of high-frequency communications in order to make informed decisions on implementing the technology holistically. If the challenges and limitations are underestimated, the economic consequences could have a negative effect and investments and capital could be lost owing to uninformed decisions.

Appendix 1

See Tables 4.5 and 4.6.

Appendix 2

See Tables 4.7 and 4.8.

Table 4.5 A summary of published mm-Wave LNAs in CMOS technology

Reference	Topology	Frequency	Power	IP1dB	Gain	NF	Area	Node
Lo and Kiang [24]	CG-2CS	14.3–29.3	13.9	–	9.9	4.3–5.8	0.54	180
Qin and Xue [25]	Cascaded CG/CS	7.6–29.0	12.1	–	10.7	4.5–5.6	0.23	65
Qin and Xue [26]	1CC	15.8–30.3	12.4	–	10.2	3.3–5.7	0.18	65
Huang et al. [13]	3CC	35–43	28.8	–	14.3	3.8	0.252	65
Yu et al. [35]	4CS	54.4–90.0	19.0	–	17.7	5.4–7.4	0.28	65
Pepe and Zito [28]	6CC	91.0	36.0	–	32.0	5.3	0.21	28
Vigilante and Reynaert [32]	4CS	68.1–96.4	31.3	–	29.6	6.4–8.2	0.25	28
Doan et al. [6]	3CC	51–65	54	−18	14.6	8.8	1.3	130
Lo et al. [23]	3CC	50–58	72	−23	24.7	7.1	0.46	130
[34]	2CC	62	10.5	–	12.2	6	0.53	90
Wu and Chen [33]	3CS	65–72	5.4	−17	10.9	5.1	0.38	130

Adapted from Hasani [10] and publications listed from Hedayati et al. [11]

Table 4.6 Parameter units and/or descriptions for Table 4.5

Parameter	Unit/description
Topology	nCC/nCS/nCG (n-stage cascode/common-source/common-gate)
Frequency	GHz
Power	mW
IP1dB	dBm
IIP3	dBm
Gain	dB
NF	–
Area	mm^2
Year	–
Node	nm CMOS, unless specified otherwise

Table 4.7 A summary of published mm-Wave PAs in CMOS technology

Ref.	Topology	Frequency	Supply	P_{sat}	Peak power gain	PAE, max	Area	Node
Hu et al. [12]	Doherty	28	1.5	+16.8	18.2	20.3	1.76	130 SiGe
Lee et al. [19]	Split-parallel cascode	28	1.1	+12.2	14.0	17.8	0.14	65
Lv et al. [26]	Doherty/Class AB	3.3–3.8/5.8	−2.4/−4.6	+42.6/+41	12/10.5	–	–	250 GaN
Ali et al. [1]	Class F	29	1.1	+14.75	10	46.4	0.12	65
Li et al. [21]	Doherty	5.1–5.9	–	–	14.4–17.3	43.2	3.88	250 GaN
Shakib et al. [29]	Two-stage transformer coupled	30	1.0	+14	15.7	35.5	0.16	28

Table 4.8 Parameter units and/or descriptions for Table 4.7

Parameter	Unit/description
Frequency	GHz
Supply	V
P_{sat}	dBm
Peak power gain	dB
PAE, max	%
Area	mm^2
Node	nm CMOS, unless specified otherwise

References

1. Ali SN, Agarwal P, Gopal S, Mirabbasi S, Heo D (2019) A 25–35 GHz neutralized continuous Class-F CMOS power amplifier for 5G mobile communications achieving 26% modulation PAE at 1.5 Gb/s and 46.4% peak PAE. IEEE Trans Circuits Syst I Regul Pap 66(2):834–847
2. Almusallam S, Ashkanani A (2019) Differential amplifier using CMOS technology. Int J Eng Res Appl 9(2) (Series-1):31–37
3. Daniel L, Terrovitis M (1999) A broadband low-noise-amplifier. EECS217-Microwave circuit design projects, University of California at Berkeley
4. Das T (2013) Practical considerations for low noise amplifier design. White Paper: Freescale Semiconductor. Rev. 0, 5/2013

5. Doan CH, Emami S, Niknejad AM, Broderson RW (2005) Millimeter-wave CMOS design. IEEE J Solid-State Circuits 40(1):144–155
6. Fan X (2007). High performance building blocks for wireless receiver. Multi-stage amplifiers and low noise amplifiers. Submitted to the Office of Graduate Studies of Texas A&M University in partial fulfillment of the requirements for the degree of Doctor of Philosophy, December 2007
7. Gray PR, Hurst PJ, Lewis SH, Meyer RG (2001) Analysis and design of analog integrated circuits, 4th edn. Wiley
8. Hasani JY (2008) Design of radiofrequency front-end module for "Smart Dust" sensor network. Doctoral Dissertation, Université Joseph-Fourier, Grenoble
9. Hedayati MK, Abdipour A, Shirazi RS, Centipede C, Staszewski RB (2018) A 33-GHz LNA for 5G wireless systems in 28-nm bulk CMOS. IEEE Trans Circuits Syst II Express Briefs 65(10):1460–1464
10. Hu S, Wang F, Wang H (2019) A 28-/37-/39-GHz linear Doherty power amplifier in silicon for 5G applications. IEEE J Solid-State Circuits 54(6):1586–1599
11. Huang BJ, Lin KY, Wang H (2009) Millimeter-wave low power and miniature CMOS multicascode low-noise amplifiers with noise reduction topology. IEEE Trans Microwave Theory Techn 57(12):3049–3059
12. Johnson D (2018) The 5G dilemma: more base stations, more antennas—less energy? Retrieved 5 Aug 2019 from http://spectrum.ieee.org
13. Keim R (2016) The basic MOSFET differential pair. Retrieved 9 Aug 2019 from http://allaboutcircuits.com
14. Kusama MS, Shanthala S, Raj CP (2018) Design of common source low noise amplifier with inductive source degeneration in deep submicron CMOS processes. Int J Appl Eng Res 13(6):4118–4123
15. Lambrechts JW, Sinha S (2017) SiGe-based re-engineering of electronic warfare subsystems. Springer International Publishing. ISBN 978-3-319-47402-1
16. Lee S, Kang S, Hong S (2019) A 28-GHz CMOS linear power amplifier with low output phase variation over dual power modes. IEEE Microwave Wirel Compon Lett 29(8):551–553
17. Lee HS, Perrot MH (2005) High speed communication circuits and systems. Low noise amplifiers. Massachusetts Institute of Technology
18. Li S, Hsu SSH, Zhang J, Huang K (2018) Design of a compact GaN MMIC Doherty power amplifier and system level analysis with X-parameters for 5G communications. IEEE Trans Microw Theory Tech 66(12):5676–5684
19. Lie DYC, Mayeda JC, Li Y, Lopez J (2018). A review of 5G power amplifier design at cm-Wave and mm-Wave frequencies. Wirel Commun Mobile Comput, Article ID 6793814, 1–16
20. Lo YT, Kiang JF (2011) Design of wideband LNA using parallel-to-series resonant matching network between common-gate and common-source stages. IEEE Trans Microwave Theory Technol 59(9):2285–2294
21. Lo C, Lin C, Wang H (2006). A miniature V-band 3-stage cascode LNA in 0.13 μm CMOS. In: IEEE solid-state circuits conference, pp 1254–1263, 6–9 Feb 2006
22. Lu F, Xia L (2008) A CMOS LNA with noise cancellation for 3.1–10.6 GHz UWB receivers using current-reuse configuration. In: Proceedings of the 4th IEEE international conference on circuits and systems for communications (ICCSC 2008), pp 824–827
23. Lv G, Chen W, Liu X, Feng Z (2019) A dual-band GaN MMIC power amplifier with hybrid operating modes for 5G application. IEEE Microwave Wirel Compon Lett 29(3):228–230
24. Pepe D, Zito D (2015) 32 dB gain 28 nm bulk CMOS W-band LNA. IEEE Microwave Compon Lett 25(1):55–57
25. Qin P, Xue Q (2017) Compact wideband LNA with gain and input matching bandwidth extensions by transformer. IEEE Microwave Compon Lett 27(7):657–659
26. Qin P, Xue Q (2017) Design of wideband LNA employing cascaded complimentary common gate and common source stages. IEEE Microwave Compon Lett 27(6):587–589
27. Rollet J (1962) Stability and power-gain invariants of linear twoports. IRE Trans Circuit Theory. 9(1):29–32

28. Shakib S, Dunworth J, Aparin V, Entesari K (2019) mmWave CMOS power amplifiers for 5G cellular communication. IEEE Commun Mag 57(1):98–105
29. Shakib S, Park H, Dunworth J, Aparin V, Entesari K (2016) A highly efficient and linear power amplifier for 28-GHz 5G phased array radios in 28-nm CMOS. IEEE J Solid-State Circuits 51(12):3020–3036
30. Sumathi M, Malarviszhi S (2011) Performance comparison of RF CMOS low noise amplifier in 0.18-μm technology scale. Int J VLSI Des Commun Syst 2(2):45–54
31. Vasjanov A, Barzdenas V (2018) A review of advanced CMOS RF power amplifier architecture trends for low power 5G wireless networks. Electronics 7(11):1–17
32. Vigilante M, Reynaert P (2016) A 68.1-to-96.4 GHz variable-gain low-noise amplifier in 28 nm CMOS. In: Proceedings of the IEEE international solid-state circuits conference (ISSCC), San Francisco, CA, USA, pp 360–362
33. Wu C, Chen P (2007) A low power V-band low noise amplifier using 0.13 μm CMOS technology. In: Proceedings of the IEEE international conference on electronics, circuits and systems (ICECS 2007), pp 1328–1331, Dec 2007
34. Yao T, Gordon M, Yau K, Yang KT, Voinigescu SP (2006) 60-GHz PA and LNA in 90-nm RF-CMOS. In: Proceedings of the IEEE radio frequency integrated circuits symposium (RFIC2006), p 4
35. Yu Y, Liu H, Wu Y, Kang K (2017) A 54.4–90 GHz low-noise amplifier in 65-nm CMOS. IEEE J Solid-State Circuits 52(11):2892–2904
36. Zhao C, Park B, Cho Y, Kim B (2017) Analysis and design of CMOS Doherty power amplifier using voltage combining method. IEEE Access 5:5001–5012

Chapter 5
Preparing Emerging Markets to Participate in a New Era of Communication. A Technical and Economic Perspective

Abstract In this digital era and with the fourth industrial revolution (Industry 4.0) imminent and earmarked to change many facets of the digital economy, fast and reliable internet and connectivity in general are possibly some of the most important *products* required. As the digital service sector evolves, certain key characteristics remain a commonality, such as adequate digital infrastructure, technology-literate end users, innovative skilled entrepreneurs, and business environments that encourage creative thinking. There are numerous economic advantages to upskilling the technical (digital) ecosystem of a country or region and benefits will increase over the long term. However, some emerging markets are struggling to develop adequate infrastructure to participate in a digital economy and have pressing issues that further stifle techno-economic growth. As fifth-generation (5G) and millimeter-Wave communications are opening up new ways of connectivity, from high bandwidth, low latency connectivity to structured and tailored service offerings with dynamic access (therefore potentially dynamic pricing), this is probably the best time for emerging markets to invest in their digital infrastructure (and even to consider abandoning old infrastructure). This chapter investigates the techno-economic perspectives of unequal markets and reviews causes, effects, and potential solutions to modernizing policies and regulations to encourage and incentivize participation in the digital economy.

5.1 Introduction

The technical and economic potential of fifth-generation (5G) and millimeter-Wave (mm-Wave) communications in emerging markets is largely untapped, but there is still time to enter this market, participate in Industry 4.0 and share the economic benefits that it will present. Mobile broadband is the key to getting people in emerging markets online, according to the World Economic Forum [31]. The potential for high penetration could have a fundamental impact on fields such as education and

W. Lambrechts and S. Sinha, *Millimeter-wave Integrated Technologies in the Era of the Fourth Industrial Revolution*, Lecture Notes in Electrical Engineering 679,
https://doi.org/10.1007/978-3-030-50472-4_5

healthcare, but populations in these regions must be skilled and trained to meet the demands. There are also challenges in bringing people online in emerging markets that are less relevant in developed countries, and the World Economic Forum [31] lists some of these, with a detailed discussion also available in Lambrechts and Sinha [16]. These challenges are among others:

- There is a major digital divide in emerging economies between the wealthy urban population and rural districts.
- Fixed-line infrastructure in these regions is outdated, is not maintained, or does not exist and this limits the potential solutions to cost-effectively bringing people online to mobile broadband (where current infrastructure typically only supports 2G and 3G at best, according to the World Economic Forum [31]).
- Typically, there is low average revenue per user and this discourages investments by mobile operators to capitalize in these areas, especially rural areas.
- Mobile devices are not manufactured in these areas, government tax and import duties are high and the price of these devices for the users is thus unreasonably high and unaffordable in already impoverished areas.
- As also reviewed in Lambrechts and Sinha [16], the internet predominantly uses English, Chinese, Spanish and other languages spoken by large population groups. In emerging markets, people do not necessarily speak any of these languages and the internet is inaccessible to them, even if they are brought online.

Although these are big challenges that have many facets with which emerging markets are struggling, for example corruption and limited financial resources, modern versatile technology such as 5G and mm-Wave networking brings about opportunities to develop new infrastructure without having to support legacy systems. Mobile operators have the opportunity to implement and adapt technologies that suit their needs, not the other way round. In the World Economic Forum [31] discussion, a summary of the challenges versus potential opportunities is outlined. The analysis is adapted and presented in Fig. 5.1.

The World Economic Forum [31] also gives several examples of where these opportunities have been capitalized on, which include solutions in countries such as Brazil and India. In certain ways, emerging markets can therefore accelerate the development of their digital service market and focus on new technologies with the emphasis on technologies that allow such opportunities and are sustainable for future expansion. The digital services sector in these countries should therefore invest in technologies that address infrastructure needs and the countries should also (according to the World Economic Forum [31])

- pursue public-private investment partnerships,
- encourage the development of local services using digital technology,
- address local challenges and issues through these digital services,
- develop innovative funding mechanisms, and
- develop innovative market access mechanisms.

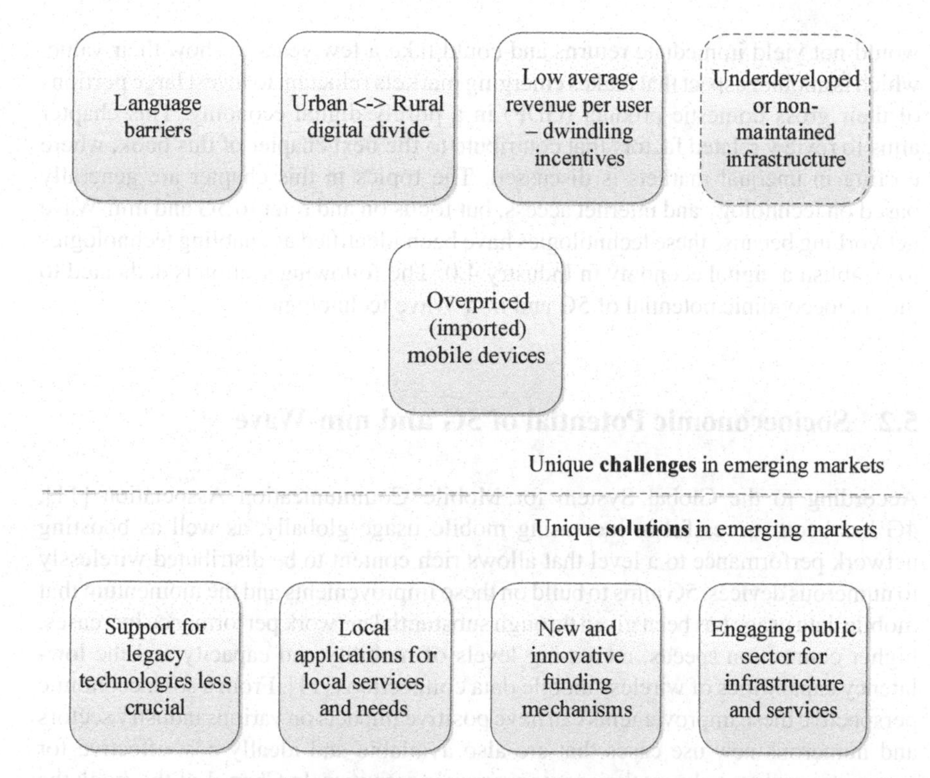

Fig. 5.1 Unique challenges and opportunities of internet connectivity in emerging markets, as outlined by the World Economic Forum [31]

To develop such goals, a clear understanding and direction for the future must be in place and aim to overcome current uncertainty as well as any disagreements nationally and internationally. Despite the numerous goals and directions globally and the differences between these for emerging markets and developed countries, the World Economic Forum [31] identifies three pillars required to achieve these goals, regardless of the means of implementation, namely

- commitment to actions that endorse long-term growth of the digital economy,
- removing any impediments to the expansion of digital infrastructure, and
- modernizing policies and regulations to encourage investment and innovation through the internet ecosystem.

For these pillars to be addressed, unequivocal devotion from government and public-private enterprises is required and financial and intellectual resources must be allocated to such a vision. Again, in many emerging markets, this is not particularly practical, as resources are limited and there are more pressing issues that must be addressed. The socioeconomic benefits of such a decision and investment might not be clear from the beginning and stakeholders should be informed and should understand what the long-term effects would be. Furthermore, such investments would not yield immediate returns and could take a few years to show their value, which is another aspect that makes emerging markets reluctant to invest large portions of their gross domestic product (GDP) in a purely digital economy. This chapter aims to review related factors that contribute to the next chapter of this book, where costing in unequal markets is discussed. The topics in this chapter are generally based on technology and internet access, but focus on and refer to 5G and mm-Wave networking because these technologies have been identified as enabling technologies to establish a digital economy in Industry 4.0. The following section is dedicated to the socioeconomic potential of 5G and mm-Wave technologies.

5.2 Socioeconomic Potential of 5G and mm-Wave

According to the Global System for Mobile Communication Association [11], 4G has been successful in increasing mobile usage globally, as well as boosting network performance to a level that allows rich content to be distributed wirelessly to numerous devices. 5G aims to build on these improvements and the momentum that mobile data usage has been given through substantial network performance increases, higher connection speeds, increasing levels of mobility and capacity and the low-latency capabilities of wireless mobile data connections [11]. From a socioeconomic perspective these improvements can have positive impacts on various industry sectors and numerous new use cases that are also available and ideally cost-effective for emerging markets to boost their socioeconomic activities. In Chap. 1 of this book the importance of efficient and fair spectrum allocation for 5G services has already been discussed and this chapter focuses on the benefits that these processes could have if prioritized and implemented effectively. Furthermore, using mm-Wave as the backbone of 5G communications has several advantages, not only in terms of network performance, but also in creating applications and use cases that make efficient use of the spectrum and bandwidth advantages within mm-Wave. Direct comparisons between mm-Wave 5G communications and its contribution to GDP and tax revenue are reviewed in this chapter, since these factors have a direct impact on socioeconomic potential.

In the opening sections of the report published by GSMA [11] an estimation that is directly attributed to the global impact that the mm-Wave spectrum could have on GDP and tax revenue (assuming its timely distribution in 2019–2020) is presented. The estimation is done for a 15-year period between 2020 and 2034 and in summary identifies the following potential global impacts:

- At the end of this period, the mm-Wave spectrum will have increased global GDP (with a conservative assumption) by US $565 billion.
- During the same time global tax revenue due to the mm-Wave spectrum could be US $152 billion or much higher.

- Early adopters of this technology will show larger economic benefits at the end of this period; however, the rate of contribution in markets that lag in adoption will surpass that of the early adopters.

The third finding listed above has an interesting prospective for emerging markets, since it can almost already be assumed that numerous emerging markets will be slower to adopt the technology compared to developed countries that have already started rolling out 5G. A topic that will be discussed separately in this chapter, but briefly touched on here, is the adoption rate of 5G in emerging economies. Recent developments in emerging economies such as South Africa already show that 5G and its potential are being taken seriously primarily because of its socioeconomic benefits in the long run. Local (to South Africa) mobile providers such as Rain realize the potential of 5G and non-equivocally state that 5G will comprise a larger portion of their overall revenue in the future compared to fiber and 4G offerings [22]. Rain already, as of October 2019, has over 250 sites that offer 5G connectivity throughout South Africa and aims to have over 700 sites by the end of 2020. This example shows that when the global potential and impact of 5G are referred to throughout this chapter, the assumption is not that most of the (positive) statistics refer to developed countries with state-of-the-art infrastructure; emerging markets are aware of the socioeconomic benefits of 5G too.

In referring to the GSMA [11] report on the socioeconomic benefits of mm-Wave and 5G, a convenient and concise regional breakdown is presented, referring specifically to Asia-Pacific and the Americas, Europe, and Sub-Saharan Africa. The importance of this breakdown is again seen when focusing on emerging markets and the estimation that the late adopters will also benefit greatly from 5G and mm-Wave, both from separate technologies and from their integral relationship. Based on the GSMA [11] report, the following results with respect to the estimated regional breakdown of socioeconomic benefits of 5G and mm-Wave are highlighted in relation to the overall theme of this book:

- After removing from the list the early adopters in the Asia-Pacific region, therefore China, Japan, New Zealand, Australia, and the Republic of Korea, a fifth of the contribution to GDP and tax revenue is achieved by the rest of the markets.
- One-tenth of the GDP and tax revenue contribution in the Americas region is made up by Latin-America and Caribbean countries.
- The gap between the early adopters and Sub-Saharan Africa will start closing up significantly from 2026 onwards.

From these findings it is evident that the GDP and tax revenue contributions of emerging markets are not negligible and although late adopters such as Sub Saharan Africa might not reap the benefits at first, the estimated time for these markets to start seeing socioeconomic benefits is not far away. Furthermore, as mm-Wave and 5G are considered relatively new and immature technology, these markets will have the benefit of adopting strategies, ecosystems and use cases that would essentially have matured by 2026 onwards, making adoption a non-linear process. Additional to the socioeconomic benefits that mm-Wave and 5G could potentially present to emerging markets in terms of pure GDP growth and tax revenue, the added benefits that will inevitably follow are equally important in their impact on the wellbeing and prospects of emerging markets. Added benefits include, as outlined in the GSMA [11] report and listed here,

- better health for the local population (essentially through improved access to healthcare) leading to a longer lifespan and higher productivity during the active (working) years of the population,
- a positive shift in independence and autonomy in emerging markets that typically rely on outsourcing and foreign skills,
- a significant reduction in pollution as awareness through technology increases,
- similarly, an increase in the accessibility of education through technology that will have numerous ripple effects contributing to the socioeconomic status of any country,
- improved public safety and emergency response through monitoring and reacting to incidents with the aid of technology (especially relevant to communications technology under the 5G, mm-Wave, or internet of things (IoT) umbrella), and
- technology-assisted public transport such as scheduling and communicating downtime to local residents, which will have the positive effect of decreasing commute times to and from the workplace.

These positive effects of improved communications technology, and more importantly, cost-effective and accessible communications technology, are dependent on applied use cases, increased access and infrastructure and skills development. GSMA [11] foresees the role of mm-Wave and 5G to be especially prevalent in industrial automation and remote object manipulation, which fits in with the argument that connectivity is at the core of Industry 4.0—the fundamental and underlying shift in global operations that allows emerging markets to participate in GDP growth and tax revenue. Finally, again referring to the GSMA [11], several interesting use cases that the GSMA deems enabling use cases for mm-Wave and 5G technology are presented and analyzed. The following list only summarizes these concepts and detailed descriptions can be found in the GSMA report [11]. The identified use cases that highlight the socioeconomic potential of mm-Wave and 5G technology from a global perspective (therefore not just for developed or emerging markets but as an enabler for both) include:

- High-speed broadband access through a wireless mobile backhaul network dependent on mm-Wave will expand remote access, especially in rural and underdeveloped areas.
- Fast deployment and temporary connectivity to supply interim solutions as infrastructure is developed will give network and internet access to critical services in healthcare, public safety and disaster relief.
- High-speed and ubiquitous connectivity between humans and machines/devices will play its part in industrial automation and sharing of resources during outages or local unrest.

- Remote object manipulation is essentially the broad term for working remotely on numerous types of applications and use cases and allowing socioeconomic stimulation to a population that is not able to commute freely, where working conditions are dangerous, or specialists are situated far away, allowing them to offer their skills remotely.
- Virtual meetings and virtual/augmented reality all require high-speed low-latency connectivity to be effective and the applications and use cases for these technologies are explored more deeply as issues of connectivity are resolved.
- Transport technology is being refreshed through broadband access on transport (such as trains and airplanes) as well as the use of connected/autonomous vehicles. From the perspective of a developed market, transport technology has a different list of benefits when compared to that of emerging markets. However, from the perspective of safety and accessibility, many of the advantages will be transferred to emerging markets as matured technologies are exported to these nations. For example: autonomous vehicles will provide luxurious self-driving vehicles and the same technology can be used for emergency response in hazardous areas.

In conclusion to this section that highlights the (estimated) socioeconomic potential of mm-Wave and 5G, primarily referring to the GSMA [11] for content, the following remarks are made:

- Late adopters of mm-Wave and 5G are not doomed never to realize the full potential of high-speed, low-latency, ubiquitous, and cost-effective communications; in fact, adopting the technology closer to maturity seems to be beneficial.
- GDP growth and increased tax revenue are the two primary measures of the success and adoption rate of next-generation communications technology and will affect the socioeconomic status of the said markets.
- Emerging markets and late adopters will still make up a fair ratio of global GDP and tax revenue and the market will not be completely dominated by a select few early adopters,
- Numerous ripple effects in various sectors (healthcare, safety etc.) will be noticeable as mm-Wave and 5G technology matures.

It is, however, also important to recognize that although the general trend for late adopters of mm-Wave and 5G technology could have a smaller effect on exclusion over the long term, this does not mean that these markets should not prioritize efforts to grow local accessibility to the technologies. In fact, as the example shows, emerging markets such as South Africa have already started rolling out 5G services domestically, albeit on a smaller scale and with limited resources; the importance of the technologies have been realized and positive private/public backing endeavors show this. Socioeconomic growth also goes hand in hand with productivity and therefore overall economic growth, and in a Bureau of Communications and Arts Research [1] report some key findings show important considerations for emerging markets entering Industry 4.0 built with mm-Wave and 5G. The following section briefly summarizes some of the relevant findings in BCAR [1] in association with the topics discussed in this chapter.

5.3 The Role of 5G on Productivity and Economic Growth

The report generated by BCAR [1] focuses on productivity as a function of economic growth, specifically since growth in productivity is defined in BCAR [1] as the increase in output achieved per unit of input and this is a key driver of income growth over the long term. External factors such as an ageing population have an effect on productivity growth, especially in the labor sector. Previous industrial revolutions provided high levels of productivity growth and as Industry 4.0 is approaching, these traditional models of productivity growth (leading to income growth) are slowing down. New models are needed to sustain and grow income levels in future and BCAR [1] identifies 5G as a key driver through digital transformation.

5G offers the prospect for innovative and enhanced products and services through its fundamental principles of high efficiency. BCAR [1] therefore estimates that 5G is expected to have an economic impact past short-term profitability, proposing that 5G is an investment in economic growth. The scale, however, of the economic impact of 5G depends on several factors identified in BCAR [1] and although specific to the economy of Australia, these factors can be generalized for other markets as well. The factors listed in BCAR [1] that affect the scale of the economic impact of 5G include:

- Is 5G a paradigm shift in communications technology or merely an improvement on earlier standards such as 4G? Much of this will also depend on the use cases and applications implemented with 5G as the enabling technology. This chapter will also focus on the accessibility and cost of 5G, especially since 5G offers network slicing, which could be a key factor in separating 5G to be a paradigm shift.
- Regulatory factors play a large role in the speed and efficiency of rolling out 5G in domestic markets. In general, communications authorities take a long time to allocate and approve spectrum to service providers and if these regulators drag their feet, the economic benefit of having 5G in the early stages could deteriorate, although it should pick up again over the long term, according to the GSMA [11].
- The adoption rate of 5G in households and businesses will also to some extent determine its impact on economic growth; however, adoption rates are typically low when initial investment is high, and there are numerous players that should work together to make the technology cost-effective and accessible (communications regulators, service providers, and device manufactures).

The safety, quality and reliability of 5G-enabled applications and devices are also crucial factors that must be standardized as quickly as possible for 5G uptake to grow exponentially, especially in the first phases of adoption. If consumers do not trust the technology based on these factors, resistance to uptake will inevitably follow. Assuming the path of least resistance to standardizing, deploying, and using mm-Wave and 5G technology and services, productivity and economic growth seem to be expected. BCAR [1] refers to IHS [14] and provides a concise list of the global impact of 5G on various industries, specifically with respect to economic growth (estimated activity by 2035). The global economic impact is divided into categories of industry sectors, ranging from arts and entertainment to manufacturing. This impact is estimated as low to high based on three technological focus areas of 5G, namely

- enhanced mobile broadband, essentially referring to stable and expandable 5G infrastructure that is capable of servicing devices with high-speed and low-latency internet connectivity,
- massive IoT and the effect that high-speed low-latency (and low power) ubiquitous communications could have on the sector, and
- the impact that 5G will have on mission-critical services, therefore any failure or disruption in a network having direct effects on business operation.

Concerning the theme of this book and the potential of 5G to give emerging markets the opportunity to take part in Industry 4.0 and grow economies and the socioeconomic status of these countries, some of the results in IHS [14] that fit in with this theme are evaluated below:

- In agriculture, forestry and fishing, it is likely that 5G will have a large impact on all three levels, improved mobile broadband, massive IoT and mission-critical services. Agriculture is still one of the largest economy drivers in many emerging markets and therefore, if the impact of 5G on these sectors is estimated to be large, it should affect the economic wellbeing of many emerging markets.
- In education, massive IoT and mission critical-services are estimated to have a low impact on economic and productivity growth. However, enhanced mobile broadband is identified as a very high impact factor. Especially in emerging markets where skills development for Industry 4.0 is crucial, the high impact of enhanced mobile broadband cannot be overlooked and should ideally be prioritized as a key long-term sustainability factor.
- Professional services, similar to education, are unlikely to be largely influenced in terms of massive IoT and mission-critical services enhancements due to 5G, but will also experience high impact regarding the enhanced mobile broadband capabilities of 5G. This is another aspect that concerns emerging markets, since during skills development remote access to professional services is required.
- Finally, manufacturing will experience a high impact in all three categories, improved mobile broadband, massive IoT and mission-critical services. Industry 4.0 will revolutionize manufacturing in general and for emerging markets to participate, communications infrastructure must be established.

The BCAR [1] analyzes a number of additional industry sectors and the economic effects of 5G, which pertain more specifically to developed countries (at first), such as real estate and hospitality. It will take some time for these industries to contribute in real terms to productivity and economic growth in emerging markets where agriculture, education, professional services and manufacturing are typically the foundations that contribute to long-term sustainability. Of equal importance is the tests to study the impact of technology on productivity, since the approach, strategy, and accuracy of such tests will give a better indication of the advantages that 5G could provide to productivity and the economy.

Similar to many technologies in their early stages, there is a degree of uncertainty about the extent and scope of the impact of the technology. BCAR [1] provides a modelling approach to analyze the potential impact of 5G on productivity and economic benefits based on a number of fair assumptions due to uncertainty. For example, if productivity increases in response to successful rollout and uptake of 5G, it is assumed that resources within the economy will be reallocated and have a further impact on the economy, therefore leading to a *snowball* effect. The measured impact is still unclear and the level of uncertainty about the long-term impact is high. BCAR [1] therefore accepts an order of magnitude impact on both productivity and tax income, identifying the key areas of uncertainty as follows:

- Will 5G entail incremental improvements on earlier technologies or be considered a general-purpose technology (GPT), a similar observation as in the GSMA [11] report.
- The required inputs to realize a 5G infrastructure are different for all regions and especially dissimilar when considered in developed countries and emerging markets.

To quantify these uncertainties effectively, BCAR [1] endeavors to model the impact of 5G on productivity and the economy through a range of scenarios where each scenario uses a varied set of assumptions, all of which are essentially modelled around three primary factors, namely

- cost,
- outputs, and
- timing.

The cost factor, as described in BCAR [1], estimated only the cost of building 5G networks and purchasing the required spectrum and indicated that the cost varies significantly depending on whether it is calculated on a per capita or per area basis. This discrepancy appears if a region (or a country) is either densely populated or comprises vast rural areas where the population density is low. BCAR [1] subdivides the cost factor into low, medium and high cost primarily based on geographically linked per capita quantities and uptake (therefore investment from consumers in the technology once rolled out). It is clear that although population density and therefore per capita cost assumptions can be made fairly accurately, investment from consumers remain a large uncertainty. The second figure of merit is outputs; a comparison based on historical data of how incremental technology updates and GPT have affected productivity and economies. The focus here is on the rate of growth versus the sustainability of growth (therefore the period in which the technology enables positive growth). Typically, incremental technology updates provide a higher output rate and GPT provides longer periods of progressive growth, and these data are incorporated in the modeling approach proposed by BCAR [1]. Finally, productivity is also affected by the timing when a technology is released. There is a tradeoff and typically impact studies are crucial to determine how long it will take to develop a sustainable infrastructure, either building on previous infrastructures or having to develop a new one, and the time it will take for technology uptake and the realization of a profit from the said technology. BCAR [1] used two different cases in terms of timing, the first being a fast rollout assuming that benefits will follow shortly, and the second a staggered approach where both investments and outputs increase slowly during the beginning phases of the rollout, with an upsurge during the middle phases and again slower during the later and final stages of the rollout.

The four broad scenarios in BCAR [1] considering the cost, output, and timing factors above are therefore specifically defined as

- shorter output effect with a quick rollout,
- shorter output effect with a lagged rollout,
- longer output effect with a quick rollout, and
- longer output effect with a lagged rollout.

The results of these scenarios are based on the effect on the communications sector multifactor productivity per annum by 2030 and 2050, as well as the effect on the total economy multifactor productivity and additional per capita GDP relative to a baseline for the same periods. BCAR [1] analyzes these results in detail, but a brief overview of the results is adapted and presented in Fig. 5.2.

From Fig. 5.2 it is noticeable that the largest productivity gains can be experienced from a quick rollout and longer GPT benefits, whereas the smallest productivity gains are estimated if rollout is delayed and a shorter (incremental) benefit is experienced from 5G, thus making it only an optimization of earlier generations (4G) in terms of its speed and latency. It is therefore still possible for internet service providers (ISPs) to construct use cases that take advantage of 5G in ways that 4G is not capable of doing

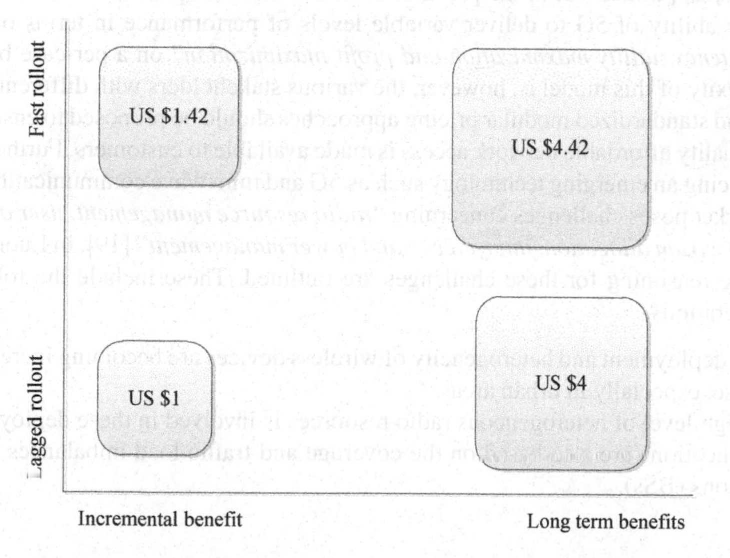

Fig. 5.2 An overview of the modeled results on productivity gains on real GDP per capita in 2050 presented in BCAR [1], based on the cost, outputs and timing of rolling out 5G and mm-Wave. The US dollar values are normalized in this adaptation and accurate estimates are available in BCAR [1]

(network slicing for example), but it remains early to predict the productivity gains of 5G accurately in the long to medium term. Another comment made in BCAR [1] is that looking at previous generations of mobile technology, these networks and infrastructures are typically not built quickly and are especially dependent on geographical area and current infrastructure. The conservative view in terms of rollout time is therefore to use at least a medium to long rollout period and have conservative estimates on productivity over a longer period, which could improve.

The speed and latency benefits of 5G and considering that 5G can be considered a GPT make it a strong contender in future network integrations and do not place it merely as an incremental improvement on 4G. The uses of "*economic and pricing policies for resource management*" [19] present an opportunity for both emerging markets and developed nations to place 5G wireless and mm-Wave technologies strategically as a long-term productivity benefit technology. The following section reviews a survey presented by Luong et al. [19] and builds on the observations made in this work.

5.4 Resource Management and Pricing Models of 5G and mm-Wave Networks

In Luong et al. [20] a review is presented on uses of "*economic and pricing theory for resource management in the evolving 5G*" and mm-Wave milieu. At the core of this analysis, as pointed out in the previous section of this chapter, are network slicing and the ability of 5G to deliver variable levels of performance in terms of "*data rate, latency, utility maximization and profit maximization*" on a per-case basis. A complexity of this model is, however, the various stakeholders with different objectives and standardized modular pricing approaches should be proposed to ensure that high-quality affordable network access is made available to customers. Furthermore, introducing an emerging technology such as 5G and mm-Wave communications into the market poses challenges concerning "*radio resource management, user association, spectrum allocation, interference and power management*" [19]. In Luong et al. [19] the reasoning for these challenges are outlined. These include the following considerations:

- The deployment and heterogeneity of wireless devices are becoming increasingly dense, especially in urban areas.
- A high level of heterogeneous radio resources is involved in these deployments.
- Implications are also based on the coverage and traffic load imbalances of base stations (BSs).

- Handover between transceivers occurs at very high frequencies.
- Stringent constraints are placed on both fronthaul and backhaul networks.
- The objectives of the numerous stakeholders vary vastly.

Optimal resource allocation in previous generation mobile networks is typically done by a centralized entity, which performs these tasks for the entire network. These strategies are typically difficult to implement in a way that optimizes economic applications and from a technical perspective are complex and efficiency is limited in medium-to-large scale networks. 5G does offer efficient strategies of network optimization on a use-case basis, but few of the traditional methods can be reused and the risks of non-standardization are also identified in Luong et al. [19]. The involvement of numerous stakeholders and entities, each with different objectives, further complicates the scenario and economic and pricing approaches are being identified and proposed to mitigate this. These models analyze complex interactions among both the stakeholders and entities and observe, learn and predict the current status or planned actions to propose decisions that satisfy use-cases—reminiscent of machine-learning. Luong et al. [19] report that through the investigation of economic and pricing models, several advantages over previous generation mobile networks can be identified. These advantages of economic and pricing models are outlined in detail in Luong et al. [19] and are briefly summarized as:

- dynamic user association schemes,
- resource utilization and dynamic demands,
- share spectrum once profitability is achieved,
- lower levels of computational complexity to distributed dynamic solutions to mitigate signal interference,
- tiered pricing based on user requirements, and
- lowering the burden of front- and backhaul links and contributing to spectrum resources through small cells.

These advantages of economic and pricing models listed in Luong et al. [19] typically refer to specific models and the advantages that each model presents to the profitability of the mobile operator, resource management in a crowded spectrum, and affordability to the customer. Although these factors do work in unison, in emerging markets, a large degree of the implementation and sustainability of 5G and mm-Wave technology is dependent on pricing, as opposed to spectrum interference (at least during the initial phases of deployment). Luong et al. [19] also present a list of economic and pricing models and analyses each in terms of their advantages and disadvantages. These models are not specific to emerging markets or developed countries and offer a general approach. In this section, these models are identified and the merits and caveats of these models in terms of their validity in emerging markets are briefly proposed.

5.5 Economic and Pricing Models of 5G

According to Luong et al. [19], and expanded on in this section, the identified economic and pricing models of 5G that will be analyzed in terms of their viability in emerging markets are

- game theory (non-cooperative),
- the Stackelberg game,
- auction,
- network utility maximization (NUM)-based pricing, and
- the Paris metro pricing (PMP) model.

Each of these pricing models is based on a specific subset of goals, for example profit maximization or maximization of network resources, and has a degree of computational complexity. It is also important to investigate these models from the perspective of the potential clients and the typical requirements in a network connection that differ vastly among rural, suburban and urban areas and between emerging markets and developing markets.

5.5.1 Non-cooperative Game Theory

Game theory, an applied subdivision of mathematics, refers to strategic interaction among multiple decision-makers (players), where each player has an inclination organizing between numerous options captured in an independent purpose for that specific actor (*player*). The player attempts to either make the most of or minimalize towards the goal. According to Basar [2], if the player aims to maximize, the goal is a utility (benefit) function. Conversely, if the player aims to minimize, the objective function is a cost (loss) function [2]. The game can refer to any set of functions aimed at achieving a common goal. The terms can be applied to define the scenario of competing for telecommunication networks, for example.

If the game is non-trivial, then the goal of a player is subject to on the selection factors of no less than one player, usually of all the players involved and the choices of all players have an effect on the actions of each player. Therefore, even in a non-cooperative environment, player decisions are reciprocal. *Cooperative game theory* can be achieved in such a scenario if all players collectively agree on a set of actions and decisions that benefit all players equally. This is, however, typically not the case, since the primary objective of most games is to win. If little to no cooperation is allowed among the players, the scenario is defined as a non-cooperative game theory. In Basar [2] in-depth analysis of non-cooperative game theory is presented in terms of its main elements and equilibrium solution concepts. The detail of defining the concept of non-cooperative game theory mathematically is not reiterated in this section; an adaptation for 5G wireless communication as presented in Luong et al. [19] is rather highlighted. Referring to non-cooperative game theory, Luong et al. [19]

highlight that typically, mobile network operators aim to maximize their own profits without forming coalitions with one another. This is a fair observation; although coalitions in emerging markets could have widespread benefits for consumers and shared infrastructure, most bottom-line decisions are made to take profits into account.

The non-cooperative game theory model can be applied if, for example, in a market of *N* competing mobile network operators (MNOs), each MNO competes for buying spectrum at a specific price and sells this spectrum to its users. The strategy of the MNOs is to select a spectrum price that maximizes their utility—the utility in this scenario would refer to the profits of buying and selling spectrum. Each MNO will identify its optimum approach, which take full advantage of profits within a set of identified approaches. The set of best approaches is the *Nash equilibrium* if the MNO can increase utility by varying its specific approach, assuming the other MNO approaches remain the same. However, the presence and distinctiveness of the Nash equilibrium must first be tested and verified when deciding on prices founded on the non-cooperative theory. Basar [2] defines several strategies to ascertain the presence and distinctiveness of the Nash equilibrium.

According to Luong et al. [19], the non-cooperative game theory model is commonly used in environments where there is considerable competition (developed countries) or if there are limited resources (emerging markets). In 5G specifically, the model can be applied for spectrum trading with competitive MNOs or for data rate allocation typically for multimedia applications. Importantly, the Nash equilibrium can only be obtained if all players make their decision at the same time and if this is not the case, therefore a single player makes a strategy decision before any of the other players, the Stackelberg game theory can be applied, as reviewed in the next section.

5.5.2 Stackelberg Game

In the Stackelberg game, the currently leading participant *moves* (strategizes) first and subsequent players move (choose their optimal strategy) sequentially after this. The objective of this *game* is to find the Stackelberg equilibrium once the leading player, for example an MNO, has decided on its strategy to maximize utility through buying and selling spectrum and the second MNO follows after this.

There are certain constraints and assumptions when following the Stackelberg game model. Among these are accepting that the leading player knows beforehand that the followers are observing its actions. The followers must also not be able to commit to a future non-Stackelberg leader's actions and the leading player must know this. As described in Luong et al. [19], backward induction is typically used to compute the Stackelberg equilibrium. Essentially, how this works is that owing to the

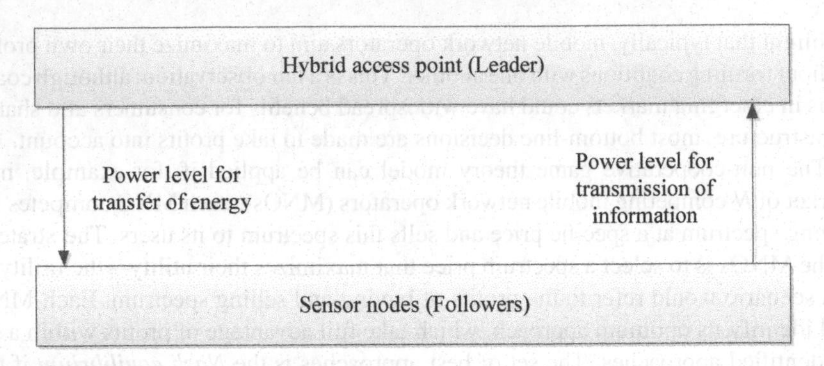

Fig. 5.3 In Su et al. [26], the Stackelberg game is used to allocate resources within a wireless IoT system where the hybrid access points are defined as leaders and sensor nodes are defined as followers

first-move advantage of the leading player, it has placed itself in a favorable position. Therefore, the utility of the forerunner at the Stackelberg equilibrium is assured to be a lesser amount than the Nash solution. According to Luong et al. [20], this characteristic of the Stackelberg game makes it appropriate for supply administration in 5G. Su et al. [26], for example, implement the Stackelberg game for resource distribution in wireless powered IoT structures to formulate the relationship between hybrid access points and wireless sensor nodes. It follows in Su et al. [26] that the power regulation in the IoT structure is then conveyed as an averaged Stackelberg game model with the goal to minimalize communication cost for respective sensors based on optimal power distribution. Su et al. [26] define the hybrid access points as the leaders and the sensor nodes as the followers and build the model based on the representation in Fig. 5.3.

According to Fig. 5.3, adapted from Su et al. [26], the leader (hybrid access point) can considerably influence the performance of the sensor nodes, the primary decision for defining the hybrid access point as the forerunner. The hybrid access point will assign power to the sensor centered on its own strategy of optimizing utility (power allocation). The strategy of the sensor "*nodes*" to optimize its utility (efficiency transferring information) is therefore dependent on the strategy followed by the hybrid access point. The mathematical problem formulation is presented and analyzed in Su et al. [26], but this example gives a good indication of the relevance of applying the Stackelberg game theory to 5G and mm-Wave communications, which is especially relevant in the IoT market as well. Another method or model to allocate resources such as spectrum to buyers that is followed commonly is auctioning. Auctioning, however, typically favors the buyers that require the resources most and have the financial means to acquire these and could therefore lead to a degree of unfair practice, but the technique is still used in numerous markets. The following section, again referring to Luong et al. [19], reviews the auction processes in the telecommunication market.

5.5.3 Auction

Auctioning, essentially a procedure where the players (bidders) are bidding for a resource and the highest offer wins, has also found its way into resource allocation to wireless operators. In a survey conducted by Luong et al. [18] on the uses of commercial and valuing models for *"wireless network security"*, a number of auctioning techniques are identified that have been adopted in the telecommunications industry. Luong et al. [18] categorize and discuss the type of auctions that have specifically been used in the wireless security industry and that have an influence on spectrum purchase as well. These types of auction identified by Luong et al. [18] include

- *"conventional auction (English auctions and Dutch auctions),*
- *sealed-bid auctions,*
- *double auctions,*
- *share auctions, and*
- *ascending clock auctions (ACA)"*.

The detailed description of each type of auction is provided in Luong et al. [18, 19], among other references from the same main author, therefore only a brief summary of each auction type is provided here to identify the primary differences when considering how developed and emerging markets might differ in their approach. Firstly, conventional auctioning is also referred to as open-outcry auctioning and bids are disclosed to all buyers during the bidding process. *"The difference between English and Dutch auctions is that the English auction is an ascending-bid auction and the Dutch auction is a descending-bid auction. In a sealed-bid auction, bidders submit sealed bids to the auctioneer and bidders do not know the bidding value of others"* [18]. Sealed bids are also categorized into types, including

- first-prize (highest bid wins),
- second-prize/Vickrey (highest bid wins, pays the price of second highest bid),
- Vickrey-Clarke-Groves (VCG) (winning bid pays the social value caused by winning the commodity) [18].

Luong et al. [18] define double auctions as a process where *"buyers and sellers"* simultaneously hand in their propositions to an auctioneer which will then perform matching between the two submission categories and determine a transaction price, also referred to as the clearing price. Conversely, with a double auction dividing resources based on the calculated clearing price, sharing auctions allocate perfectly divisible resources and available resources are proportional to individual bids. This model works well if a bidder wants additional functionality during expansion of its resources, for example additional security resources through purchasing more spectrum. Moreover, this process incentivizes bidders not to deviate from the bidding strategy unilaterally. Finally, the ACA model is essentially a multi-round auction with a single seller with multiple buyers and the price of the resource is enlarged in respective rounds in anticipation that the overall demand matches the available supply.

The solutions of equilibrium of these listed auctioning models range between Nash equilibrium (English, Dutch, first-prize, Vickrey, VCG and shared auction), Walrasian equilibrium (ACA), and market equilibrium (double auction).

Another popular economic and pricing model used in 5G resource allocation is NUM-based pricing. This model is suggested to assign resources such as bandwidth and power budget to subscribers in order to take full advantage of the overall remaining utility of subscribers, given the capability constraints of the network [19]. A review of NUM-based pricing in telecommunications is presented in the following section.

5.5.4 NUM-Based Pricing

According to Lee et al. [17], and still relevant, current and earlier generational frameworks of network utility maximization for rate allocation as well as the pricing algorithms typically assume that individual links (users) offers a fixed-size broadcast bandwidth and that the utility of each user is "*a function of transmission rate only*" [17]. Even though the work presented in Lee et al. [17] was published long before work on 5G began, their arguments on NUM-based pricing can be related directly to problems that 5G have essentially solved from a technical perspective. Lee et al. [17] proposed adjusting the "*physical layer channel coding or transmission diversity to achieve variable tradeoffs between data rate and reliability*". The intrinsic tradeoff between data rate and signal quality is the basis of the work and the model is presented on methods to optimize the utility of the user based on these factors. Lee et al. [17] additionally refer to the use of multiple-input multiple-output (MIMO) multi-hop networks where the diversity and multiplexing gains of each link are controlled to achieve the optimal rate-reliability tradeoff. Lee et al. [17] raise concern that the current (in 2006) NUM framework, in the physical layer, assumed that individual links offered a constant coding and modulation pattern and the utility of each subscriber is merely a utility of native source speed. The paper then suggests a novel enhancement framework and pricing-based dispersed algorithm by taking into account a dual policy approach (identified as being non-effective in the NUM framework at the time):

- "*integrated dynamic reliability, and*
- *differentiated dynamic reliability*" [17].

Lee et al. [17] define the *integrated dynamic reliability* policy as a functionality where the communication link offers an equal error likelihood, therefore an equivalent code speed, to all of the individual sources making use of it. As a result, the link ought to offer the lowermost code speed that fulfills the condition of the source with the uppermost dependability. From this, Lee et al. [17] propose the differentiated dynamic reliability policy where multi-class sources have varied dependability needs in the system. The proposed link is thus capable of providing a different error probability (code rate) to each source.

In [20], Luong et al. defined the NUM-based pricing policy as a network model using a single source with multiple users, with the net function of individual users being the variance amid the function related with its resource distribution and the rate the users pay the source for consuming the *assets* (for example bandwidth and power). Importantly, the source iteratively brings up-to-date the cost of the specific asset prices and consumers iteratively choose the amount of assets to take full advantage of their remaining utility. The NUM result come together to a distinct and optimum result for the remaining asset distribution. In 5G, which allows for this pricing model, the NUM method can be practically used to take full advantage of the overall data rate of subscribers [19] through for example aggregated carrier allocation in heterogeneous networks. The NUM-based pricing model will also be explored as an enabling model for emerging markets to distribute resources to users through variable capacity and prices. In a 5G network, the users are diverse and have different quality of service (QoS) requirements from the network and an extension to the NUM-based pricing model that allocates resources based on user utility requirement, the PMP model [23], also adapts service prices based on user requirement. The following section briefly reviews the PMP model in 5G.

5.5.5 Paris Metro Pricing Model

PMP is a type of variance appraising that arranges variable prices for services, grounded on QoS (communication delay, reliability, and security). The source MNO is therefore able to fulfill the QoS necessities of the user and maintain a sustainable profit by varying the price of the said service as well as services required less frequently at the time. PMP therefore simply entails having several (ideally a small number of) channels that differ in price, offering different QoS, and the user selects on which channel to transmit based on the immediate needs. In addition, matching theory and contract theory can be combined with the pricing model (in fact with any pricing model) to match resource-price requirements of source MNOs and users effectively and efficiently. Ros and Tuffin [24] published a mathematical model of the PMP model for charging packet networks where stability, maximization of revenue compared to a single network (channel), and multi-applications extensions are proposed and presented.

To allow the development of 5G in its effective rollout and adoption, there are still important challenges to overcome and a fair amount of research and development to take place, not only from a technical perspective, but also in terms of research into economic and pricing models that will serve different markets based on user demand, operator supply, and third-order aspects such as geographical limitations. From a technical perspective, a combination of 5G heterogeneous networks (HetNet) with massive MIMO and mm-Wave technologies is enough to serve as a next-generation network. However, as outlined in Bogale and Le [3], additional research challenges still exist. Although these challenges were already listed in 2016, at the time of writing they are still relevant. The following section highlights the research challenges in 5G networks as identified in Bogale and Le [3].

5.6 5G Research Challenges

In Bogale and Le [3], the research difficulties that are to be looked at from the non-technical perspective of 5G include

- *"network planning and traffic management,*
- *radio resource management,*
- *cell association and mobility management,*
- *backhaul design, and*
- *low-cost beamforming*".

These research challenges are additional to the economic and pricing models reviewed in the previous section of this chapter and all issues have varied effects on emerging markets when compared to developed countries. In general, emerging markets have additional hurdles to overcome, such as financial limitations, corruption, workers with limited skills, and little to no current infrastructure (which could in certain circumstances be considered an advantage). Regardless of that, the research challenges listed in Bogale and Le [3] are valid in both circumstances and are briefly outlined in this section in order to consider their impact and relevance in emerging markets. The first challenges in the list are network planning and traffic management [3].

Network planning and management for 5G differ from earlier generations primarily owing to the types of economic and pricing models available to operators and users. Previous generations of telecommunications infrastructure considered primarily the number of BSs in order to maximize reception (therefore quality and speed) of the network within the largest area possible. In 5G, the placement and locations of macro-, micro-, and pico-BSs should be optimized not only to maximize signal strength, but also to prioritize the types of service delivery *expected* in a region. 5G mm-Wave BSs also require a larger number of femtocells owing to the high inherent free space loss experienced at mm-Wave frequencies (as well as high penetration losses). In addition, complexity in planning and management arise when considering the difficulty of estimating the network utility in indoors environments, which can vary significantly over time. Contrary to network planning for earlier generations, traditional methods for site locations, such as evenly spaced lattice structures with hexagonal grids, are not the most effective means of BS placement and could lead to lower profitability for operators and lower QoS for users [3]. Furthermore, since 5G can operate at both microwave and mm-Wave frequencies, with initial implementations likely within the microwave spectrum, the optimal solution for frequency management is complex and has not yet been tested for all types of environments and user utility. These optimizations are likely only to become near-ideal once data have been gathered in these scenarios and could therefore have a detrimental effect over the long term if initial planning is substandard. Huawei [12, 13] published a white paper on potential strategies to solve network planning for 5G, which specifically mentions challenges faced by 5G network planning. Deviating from Bogale and Le [3] at this point, the factors identified by Huawei [12, 13] are reviewed. According to Huawei [12, 13], these challenges to network planning include

- the addition of new frequency bands,
- the 5G new radio,
- 5G services and applications, and
- 5G network architectures.

The white paper presents relatively detailed reviews on these challenges in the 5G network that were raised, and certain highlights that are also applicable to emerging markets are outlined. The first challenge concerns the additional frequency bands of which 5G with mm-Wave will take advantage. Although being a large benefit in terms of uncontested bandwidth, these new frequency bands present high propagation losses and penetration losses. The signals are therefore more susceptible to architecture, vegetation, rain attenuation and oxygen attenuation. Line of sight (LoS) is therefore preferable; this has an impact on the number of BSs required and therefore the cost of the infrastructure. In emerging markets where financial backing is often lacking, thorough investigation and planning of infrastructure are required to ensure that cost-effective placement of BSs is achieved. High-precision planning, simulation and ray-tracing propagation models are needed to compensate for the limited coverage of an mm-Wave 5G network. Furthermore, spectrum planning is complex and new and the rules and constraints in these spectrum allocations are different, requiring additional research and funding to deploy.

The second challenge, according to Huawei [12, 13], is the use of the 5G new radio (NR). MIMO, a sector-level beam-based networking feature, has been identified as one of the most important technologies facilitating 5G. Traditional network planning, however, cannot meet the requirements of massive MIMO antenna arrays, as prediction of coverage, expected data rates and capacity varies from traditional methods. Again, increased engineering complexity and infrastructure planning place additional strain on developing an infrastructure with cost implications. Another factor that is typically overlooked in developing countries is the availability of skilled workers during not only the planning and deployment phases, but also for maintenance and repairs. Skilled workers need to be deployed to sites that use the 5G NR and these workers need to be trained to understand and maintain these systems, assuming a high quality of secondary education has already been obtained.

The third challenge identified by Huawei [12, 13] involves the new services and applications offered by 5G and mm-Wave networking. Modern technology has moved towards user experience-centric network construction where video coverage and high-Mbps services are needed to prioritize user experience. Key functions to achieve these services when offered to users involve evaluating the user experience in real-time, gap analysis, simulation, and careful planning. These are again high cost drivers that need to be considered not only if new infrastructure is developed, but also if traditional infrastructure is upgraded to support mm-Wave 5G services. The applications that 5G will serve as a function of the services offered include high-bandwidth mobile hotspots where LoS is typically required, new IoT applications driven by Industry 4.0, and low- or high-altitude mobile coverage using drones, for example. Again, propagation characteristics of mm-Wave is a key factor in all these applications and successful implementation over the long term depends on the quality of initial planning.

Finally, as has been touched on when discussing previous challenges, according to Huawei [12, 13] the network architecture of 5G and mm-Wave networks (network slicing, user-centric dynamic networks, and user-centric channel resource cloudification) are all relatively new terms. Even though these terms have been defined, each should now be regarded as forming an integral part of network planning. It is not ideal if these terms are considered an additional supplement after a network has been established. Taking these challenges listed in Huawei [12, 13] into consideration, some important deductions can be made when considering the implementation of 5G and mm-Wave networks in emerging markets:

- High propagation losses increase overall cost in planning, simulation, and ray-tracing requirements. Engineering and technical skills are required in this phase and if outsourced by emerging markets, the opportunity to educate the local population in emerging markets and to spearhead socioeconomic development will be lost.
- These high propagation losses drive the requirement for numerous BSs and this increases overall cost. Financially strained emerging economies might opt for using fewer BSs but the impact on network stability and efficiency could be more detrimental in the long term.
- The services and application of mm-Wave 5G are user-centric and need to established even before the network is planned and eventually deployed to ensure that the capacity and capability of the network align with the needs of the clients. In emerging markets, the needs of users are typically different from those in developed countries. Thorough research into these needs must be conducted and the local population must be consulted in each step of the planning phase.
- The operating frequency and transceiver requirements (of the 5G NR) are different from traditional networks, and emerging markets must prioritize not only spectrum allocation, but also upskilling of the local population to be able to use, maintain and fault-find the 5G NR.

The White Paper presented by Huawei [12, 13] follows the discussion of the challenges of 5G and mm-Wave with proposed solutions of network planning (with ray tracing), high-fidelity modeling of product features, coverage prediction, automatic cell planning and accurate site planning. Application cases also follow these proposed solutions. These proposed solutions are not within the scope of this chapter and consulting Huawei [12, 13] for more information on these is recommended.

Returning to the work presented by Bogale and Le [3], the second research challenge from a non-technical perspective relates to radio resource management. Again, the use of mm-Wave and MIMO technologies, which are among the enabling technologies of 5G, is mentioned and it is clear that the HetNet architecture presents significant challenges to radio resource management. The mm-Wave spectrum is notorious for its concentrated beam pattern and restricted scattering, essentially translating to LoS communications. Furthermore, the high attenuation and large bandwidth of mm-Wave signals make it especially susceptible to noise (considering the unit for noise, dBm/Hz, defines this relationship clearly). Interference from adjacent signals generated by beamforming is an additional limiting factor for microwave signals [3] and will require reliable backhaul design and implementation to minimize. Bogale and Le [3] present typical strategies for radio resource planning and management and like Huawei [12, 13], these techniques are not covered in this chapter; these sources should be consulted for an overview of the techniques.

Cell association and mobility management are further research challenges of 5G identified by Bogale and Le [3]. The challenge is recognized by the fact that in dense HetNet infrastructures with multiple tiers and multiple radio access technologies (3G, 4G, Wi-Fi, WiMAX etc.), a device can have several cell associations during an active session. Cell associations should ideally be capable of efficiently utilizing network resources across all technologies to deliver high QoS to the user and avoiding overly frequent handover between cells. Infrastructure planning must therefore take into account numerous factors such as

- signal quality,
- interference,
- loading on a channel (traffic),
- data-offloading capabilities, and
- mobility.

Traditionally, QoS could be maintained relatively easily by considering signal strength (in dBm) or the signal-to-noise ratio. However, if 5G HetNet is introduced in the cell-association strategy, these two parameters alone would not be enough to ensure high QoS, where mobility management for example is also required to treat low- and high-speed user equipment. To achieve the speed and latency requirements of a 5G network effectively, the offloading approach might need to exploit the advantages of all radio access technologies in real-time and this can become a complex task, especially in urban environments. In rural areas and in some emerging markets where the current infrastructure is lacking, the complexity of these strategies could potentially be lower and may be an advantage, considering the financial constraints and engineering skills to develop these strategies. Backhaul design is also affected by these strategies (and indirectly by the status of the current infrastructure), as identified by Bogale and Le [3].

The backhaul of 5G infrastructure must support various categories of traffic and collaboration between diverse BSs and manage and optimize the throughput and user experience. According to Bogale and Le [3], studies conducted in 2012 showed that cellular networks at the time comprised 70% copper, 10% fiber and 15% wireless backhaul. Fiber has gained a lot of traction since then, even in emerging markets, and it can be assumed that the percentage of copper use in backhaul has declined significantly. Wired infrastructure is still needed in backhaul to deliver large bandwidth, but according to Bogale and Le [3], the percentage use of wireless infrastructure should ideally increase from 15% in 2012 and complement the use of wired (fiber) technologies. 5G and mm-Wave make this possible, inherent to its high bandwidth and low latency capabilities, especially considering that the need for large aperture high-power antennas used in previous generation wireless technologies has migrated to a more economically feasible solution using numerous cells as BSs to deliver scalable internet. However, the engineering effort to deliver such a scalable, efficient, reliable and cost-effective infrastructure remains a complex challenge.

Finally, the last 5G research challenge listed in Bogale and Le [3] concerns low-cost beamforming for massive MIMO from the inherent requirement of 5G typically needing channel state information (CSI) at the transmitting and/or the receiving systems. CSI and beamforming are key design issues in 5G and mm-Wave networks and ought to have a large bearing on the real-time network performance and scalability of the network. In earlier generations of mobile networks such as 4G, the number of radio frequency chains (amplifiers, analogue-to-digital converters, digital-to-analogue converters) had to be scaled with the number of antennas used (typically only between 1 and 10). However, in a massive MIMO 5G system, this becomes impractical and channel estimation and beamforming algorithms become crucial. Bogale and Le [3] give a relatively in-depth analysis on the types of solutions to approach these complex algorithms. Two design case studies that illustrate certain challenges of the 5G MIMO HetNet are also presented in Bogale and Le [3] and can be consulted for a technical perspective of these research difficulties.

The next section gives an evaluation of the techno-economic perspective on 5G and mm-Wave with the focus on emerging markets and the potential impact of these technologies on overall wellbeing and sustainability over the long term.

5.7 Techno-economic Perspective on 5G and mm-Wave

5G is more than an incremental improvement on 4G, and as Wisely et al. [30] point out, speed increases in user-experienced data rates over the last few generations of mobile networks show this:

- 2G provided speeds of around 0.1 Mbps.
- 3G increased data rates to around 1 Mbps.
- 4G reached average speeds of around 10 Mbps.
- 5G rates are estimated to exceed 1 Gbps or even 10 Gbps.

What has not been addressed for 5G, apart from increases in capacity, latency and traffic density, among other performance metrics, is the cost of deploying such an infrastructure. It is, however, imperative to realize the cost implications of 5G and mm-Wave from the perspective of backhaul, the MNOs and the users. In Wisely et al. [30], a techno-economic perspective into the proportional cost of two proposed methodologies is given and despite widespread interest in 5G, the proposed models are still based on limited studies on the cost of 5G. The model proposed in Wisely et al. [30] is defined for a 1 km^2 urban area in London, England, and the decision was purposefully made to choose a dense urban area and adapt the model for suburban and rural areas. Furthermore, the deployment strategy for the model involves different 5G technologies (frequencies) suggested for 5G, as discussed in Chap. 1 of this book. These include 700 MHz, 3.5 GHz, 5 GHz indoor and outdoor, as well as 24–27.5 GHz implementations of 5G. A variation of operating frequency allows for in-depth analysis of propagation models and penetration losses associated typically with higher frequencies and the effect these have on cost. Wisely et al. [30] clearly state that it is difficult and complex to determine costs related to various radio access technologies (RANs) accurately and to include interpolation of 5G costs based on current operating expenses (OPEX) and capital expenditure (CAPEX) of traditional mobile communications technology. Wisely et al. [30] therefore contribute to the body of knowledge by emphasizing the complexity of giving high-capacity indoor communications and point to the requirement of cost decrease and novel income areas in 5G, without following the traditional route of simply measuring throughput and subscriber base.

Wisely et al. [30] provide a detailed breakdown of their model development and describe the method of techno-economic modeling. In this method description, the primary considerations are listed as

- clearly defining the technologies that are modelled (thus the frequency and bandwidth for the various 5G implementations),
- the capacity and coverage model, which typically include considering
  - topology,
  - link budget,
  - propagation model,
  - signal-to-noise and interference ratio (SINR),
  - *technology mapping*,
  - queuing strategy, and
  - coverage and capacity calculations,
- BS density and placement based on population density, and
- throughput from SINR mapping.

These technical modeling criteria are *relatively* easy to define and involve characterizing the signal, channel and receiver, using for example Shannon's information theory, concepts that have been established and are well known in the scientific community. However, Wisely et al. [30] also present a cost modeling section based on limited published data on RAN networking, OPEX and CAPEX cost. Wisely et al. [30] combine several works to develop the cost model and integrate the findings in these publications based on the cost model, deployment environment (mostly urban), and BS classes. Wisely et al. [30] also break down the cost of building a RAN into six key areas, namely

- the initial cost of building a site, which includes the cost of both labor and equipment,
- running costs such as site rental,
- the costs associated with the backhaul,
- ongoing operational cost and maintenance by skilled workers,
- licensing fees, as well as initially allocated spectrum cost, and
- the charges associated purely with powering the site from the local utility.

The study of Wisely et al. [30], as mentioned in the introductory paragraph of this section, was specifically done for London, England, a developed urban environment. The costs associated with building and maintaining a RAN are therefore based on the situation where services such as power utility and skilled workers for operations and maintenance are reliable and relatively easy to acquire. It is, however, important to consider that these costs will escalate dramatically in emerging markets where such assumptions cannot be made. As pointed out in Lambrechts and Sinha [16], affordability and spectrum licensing fees affect widespread technology adoption in emerging markets. Additionally, concerns such as exploitation, lack of skilled workers, absence of infrastructure, cartels and monopolistic behavior, and governing problems also escalate cost significantly. Notwithstanding this, the work in Wisely et al. [30] does not focus on these issues, and rightly so, as inclusion of these variables would make a cost model and a techno-economic perspective virtually impossible. The result of the proposed cost model and techno-economic perspective of 5G at different frequencies (including mm-Wave) is adapted and presented in Table 5.1.

As discussed and analyzed in Wisely et al. [30], there are numerous factors at play when estimating the cost model and generating the data as duplicated and presented in Table 5.1. For example, variations in annualized RAN costs based on BS class have a significant effect on the model, but Wisely et al. [30] clearly state throughout where assumptions are made and linear or logarithmic interpolation is performed. In this section, however, some observations are made based on the estimates presented in Wisely et al. [30]. Interestingly, if looking at the cell density per km^2, good coverage of mm-Wave 24.5–27 GHz 5G can be obtained with a relatively low cell density when compared to the other technologies in the list. WLAN requires the highest

Table 5.1 The results of the proposed cost model (in GBP/£) and techno-economic perspective of 5G for various frequencies (including 5G) in London, England

Technology	Cell density per km^2	Capacity (Gbps/km^2)	Cost (£M/yr/km^2)
LTE	16	1	0.53
5G 700 MHz	32	0.6	0.6
5G 3.5 GHz	256	30.6	1.8
WLAN	2000	20.9	1
5G 24.5–27 GHz	128	740	0.64

These estimates are generated for one MNO in a dense urban setting and is a subjective indoor and outdoor mean, adapted from Wisely et al. [30]

cell density to obtain good coverage and the capacity benefits of WLAN are also not necessarily superior. Looking at LTE of 700 MHz 5G to 3.5 GHz 5G the results are more or less as anticipated, and a rise in frequency leads to an surge in cell density due to propagation characteristics, also leading to an increase in capacity per unit area, where the cost per year per unit area is incrementally higher. The observation on the mm-Wave 24.5–27 GHz 5G is an outlier, where the cell density is dramatically decreased, since increasing the cell density will be economically unjustified by the inherently poor indoor coverage and little benefit to the already very high capacity, which is exponentially higher (740 Gbps/km^2 vs. 30.6 Gbps/km^2 for the second-highest contender in 3.5 GHz 5G), and more importantly, the yearly cost per unit area is about average compared to the other technologies. To achieve indoor coverage in the urban environment additional cells will obviously need to be added, but there are alternatives for routing these signals in last mile strategies, as presented in Lambrechts and Sinha [16].

In Smail and Weijia [25], mention is made of MNOs upgrading their existing 4G infrastructure to 5G in line with the major developments made in IoT and the requirements placed on these applications in terms of bandwidth and latency, considering QoS in multimedia applications. Smail and Weijia [25] stated that these upgrades should be completed by 2020, although from 2019's perspective, the rollout and commercialization of 5G have become a phased approach rather than a single event. Still relevant from the work presented in Smail and Weijia [25] is the analysis of the techno-economic predictions for the deployment of 5G and mm-Wave networks as an enabler for Industry 4.0. The work takes both techno-economic analysis and pricing models (discussed separately in this chapter) into account, proposing a pricing model consistent with the growth trajectory of mobile broadband with 5G. This proposed pricing model is briefly referred to in this section and compared to the pricing models presented in this chapter. The analysis of modeling of cost-benefit predicting in Smail and Weijia [25] is specifically tailored to a geographical area (high-density urban area in China with 3854 citizens/km^2 with 1.28 mobile devices per person), a sensible approach since each area where 5G and mm-Wave are deployed differs in the number of subscribers and the requirements of each subscriber. The analysis is based on a model that is adapted from Smail and Weijia [25] and presented in Fig. 5.4.

From Fig. 5.4, the aim is to liken the overall fee of ownership of 5G (upgrading from 4G and not implemented anew) where the revenue is projected to be collected during the period of analysis. A simple analysis is done based on the ratio of the return on investment and the total cost of ownership. In addition, a capability that 5G offers is that the MNO can take full advantage of the advantage of its asking conditions and cost and the user can get a tailored offering at a minimum price through opting out of services that are not required. In the analysis presented in Smail and Weijia [25], four primary considerations influence the cost-benefit, being

- the predicted amount of subscribers,
- the churn rate (the amount of subscribers leaving the MNO in a defined period),
- pricing strategy analysis, and

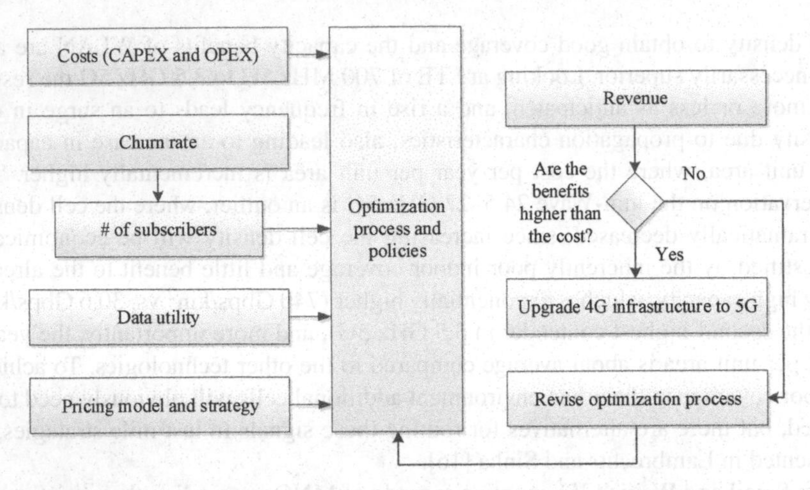

Fig. 5.4 The model presented by Smail and Weijia [25] to model cost-benefit predictions and specifically determine the number of subscribers in a particular geographical environment

- evaluation of capital expenditure and the operating expenses for various BS configurations (classes) and scenarios.

Based on these four considerations, analyzed in detail in Smail and Weijia [25], the traffic needs and network asset for the scenario are predicted. In short, the estimated generated traffic is a function of the population density, and it follows that the traffic demand $G(t)$ for a 1 km^2 area in the chosen geographical area is calculated in Smail and Weijia [25] by

$$G(t) = \rho \frac{8}{N_{dh} N_{md}} \varphi(t) D_k \tag{5.1}$$

where ρ is the population density (per km^2), N_{dh} is the number of hours of the day that are defined as busy hours, N_{md} is the number of days in a month, $\varphi(t)$ is the percentage of active users in a given time t, and D_k is the average data demand per month. The proposed model for network investment is therefore a function of the per area demand, as well as additional parameters such as discounted cost in high-density areas, overall coverage, and capacity estimates per BS class. Importantly, in indoor environments, analysis of the estimated penetration losses must also be done in order to estimate network investment effectively, a factor that was less defining in earlier generations of mobile technology. Smail and Weijia [25] continue to analyze numerous scenarios to obtain estimates of network investment for each scenario as opposed to proposing a single solution, as given in (5.1) for the traffic demand. Again, this is a sensible approach, since frequency characteristics and user density play a much larger role in the overall complexity of network planning for 5G and mm-Wave networking.

Yaghoubi et al. [32] also highlight the importance of assessing the economic viability of different transport technologies and the impact of each on both the cost and the profitability in a HetNet deployment where 5G offers high-bandwidth and low-latency communications. In Yaghoubi et al. [32], a comprehensive review is presented on a general techno-economic framework that is capable of not only assessing the total cost of ownership (TCO), but also the business viability of a 5G HetNet deployment. This general framework is then applied to a use case as defined in Yaghoubi et al. [32] for a microwave and fiber network assuming both a homogeneous and a HetNet deployment. The work in Yaghoubi et al. [32] aims to provide solutions for choosing the type of transport technology as well as defining the cost per year that should be invested to maximize profit while optimizing capacity. In defining the business feasibility assessment of a transport segment, Yaghoubi et al. [32] identify the four phases needed to deploy a communications network, these being:

- The risk of investment is reduced through planning the viability of the communications network in the identified scenario.
- A large CAPEX is then required for the initial installation phase, based on the outcome of the planning phase.
- OPEX funds are then required to maintain operation in the operational phase, which will also include maintenance and skills development to ensure future sustainability.
- A mixture of CAPEX and OPEX will contribute to the migration of customers to the new network.

Furthermore, in Yaghoubi et al. [32], the techno-economic framework for the transport segment of a new mobile network (5G in this case), is summarized and outlined in terms of modules, as shown in Fig. 5.5 and described below.

According to Fig. 5.5, adapted from Yaghoubi et al. [32], the defined models are

- time-varying *costs* associated with consumable prices, technology maturation, service expenses and support services such as human resource management,
- defining the technology used within the transport segment, as well as the components and equipment that will be installed in each location as part of the overall *architecture*,
- interconnection of the architecture to define the *topology* of the network as a function (among others) of the geographical region and existing infrastructure,
- the *market* node, which accounts for regional demand for QoS, the general behavior or users in the region, market share, service price, and availability (reach) of the network,
- a strategy and cooperation model between stakeholders, which include public, private, and governmental bodies that define market share and cross-platform investment and profitability in conducting *business*,
- planning and calculating the amount of new infrastructure as well as the volume of components needed annually in the various network locations to define the *network dimensioning* potential completely, and

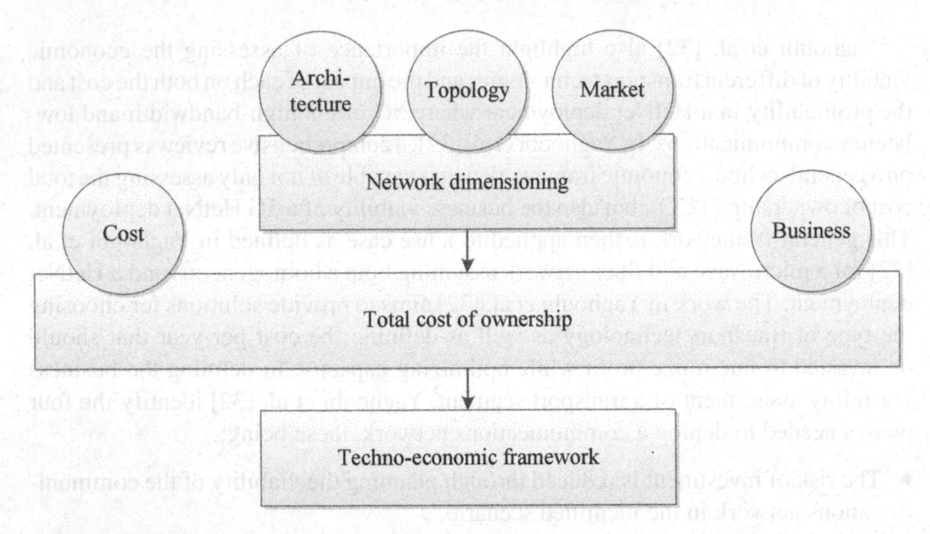

Fig. 5.5 The techno-economic framework presented by Yaghoubi et al. [32] in which the business feasibility assessment of a transport segment is categorized in terms of modules

- covering both CAPEX and OPEX as functions of the *TCO*, where equipment cost, infrastructure cost, spectrum leasing, utilities, fault management, and maintenance cost are considered.

These models all have an effect on the techno-economic model (framework) proposed by Yaghoubi et al. [32]. It is also highlighted in Yaghoubi et al. [32] that the techno-economic model must include cash flow and net present value (NPV) considerations, where

$$NPV = \sum_{j=0}^{L_n} \frac{CF_j}{(1+r)^2} \tag{5.2}$$

where L_n is the total operational number of years, r is a discount rate for estimating the present value of future cash flows through inclusion of factors such as time-value of money and uncertainty risks of future incomes, and CF_j is the cash flow for year n. Yaghoubi et al. [32] apply their proposed techno-economic model to six scenarios to compare OPEX, CAPEX, and RAN variations of a 5G HetNet and homogeneous network. The detailed results are available in the said work and in their conclusion, the model points out that there is a significant increase of the backhaul TCO in a heterogeneous network when compared to a homogenous environment. It is therefore of high importance to select the right backhaul technology to minimize cost in a heterogeneous network but also offer long-term sustainability, especially relevant in 5G networks. Yaghoubi et al. [32] conclude that fiber is a more cost-effective backhaul solution compared to heterogeneous wireless solutions in areas with a high density of users. The general conclusions made by Yaghoubi et al. [32] are identified as:

- Low TCO does not always result in high profits.
- A technology with low upfront cost that is capable of generating income as early as possible is recommended.

The techno-economic model, use cases, and conclusions presented in Yaghoubi et al. [32] contribute to the current body of knowledge in terms of analyzing a specific scenario (geographically and in terms of population density) and gives pointers towards planning for network deployment and distribution of technologies such as 5G and mm-Wave. Like most other proposed techno-economic models and frameworks, these remain extremely location-specific and might be completely different for rural areas and urban environments in emerging markets.

On a smaller scale when compared to national MNOs, Walia et al. [28] propose a deployment framework for future indoor small-cell 5G networks not only for smaller venues, university campuses, or micro-operators, but also for adaptation on scale by MNOs. The research presented by Walia et al. [28] is motivated by improving indoor network coverage of mm-Wave 5G through leveraging network slicing (see Chap. 1 of this book) to provide (in this case, campuses) with tailored services. The study focuses on

- reducing overall cost (TCO),
- generating revenue for the micro-operators,
- providing dedicated services,
- improving (indoor) coverage, and
- efficiently utilizing spectrum.

From these focus points, it is clear that although the proposed framework is on a small scale, MNOs could benefit by adapting some of the work presented in Walia et al. [28]. Walia et al. [28] define and describe the development of their framework in detail, specifically outlining the method to calculate the TCO. The framework is very area-specific in this case, with little generalization through mathematical analysis. However, the framework is primarily deployed based on

- incorporating prior considerations such as user requirements and QoS, geometry of the building, and analyzing existing infrastructure,
- planning for deployment based on coverage, capacity, network slicing, type of equipment (femtocells and/or picocells), and strategic placement of equipment,
- a deployment strategy to optimize all access points and integrate services through network slicing, and finally
- considerations of operation, administration, and managing services in real-time.

Walia et al. [28] draw some interesting conclusions based on use cases in a controlled environment; in short, these conclusions include:

- The techno-economic analysis on the framework provided in Walia et al. [28] proves to be viable for university campus integration.

- Shared indoor small-cell infrastructure improves indoor capacity and coverage of a 5G network.
- Femtocells can in addition be deployed to distribute connectivity cost-effectively.
- To maintain capacity, fewer femtocells are needed compared to traditional Wi-Fi access points, which effectively leads to a lower TCO since the prices are comparable.

The framework in Walia et al. [28] is therefore useful, since it is on a small scale and limited studies and resources are available on optimizing indoor coverage of 5G and mm-Wave. The proposed framework of Walia et al. [28], in conjunction with, for example, a technology-specific techno-analysis (of MIMO and distributed antenna systems) as presented in Bouras et al. [4], could present a viable starting point in generating area-specific small-scale techno-economic analysis of 5G and mm-Wave for more environments. Based on the discussion presented in this chapter, there is, however, a factor that still limits the successful planning, deployment, and maintainability of 5G networks and will have a detrimental effect in the long term on techno-economic as well as socioeconomic growth—especially in emerging, unequal markets. This is the affordability of current- and future-generation broadband and a benchmark for policy makers and supervisory bodies to endorse competitive and varied broadband commercial sectors and make internet access affordable. However, it is known that many emerging markets are unequal in their income distribution, and future generation technologies that do allow a high level of customization, such as 5G through network slicing, should be investigated to potentially dilute unequal access to broadband internet. The following section investigates potential policies and strategies to adapt costing of 5G and mm-Wave networks to provide internet access in unequal markets.

5.8 Internet Affordability

To investigate the potential of costing strategies catering for unequal markets, a relevant starting point is the affordability of broadband globally and highlighting the issues affecting emerging markets that are typically also unequal in wealth distribution. In the Alliance for Affordable Internet 2019 Affordability Report, the annual list of countries sorted by its broadband affordability drivers index (ADI) is published. The 2019 Affordability Report determines the ADI score of a country not by the price of broadband internet or an indication of how affordable broadband internet is in the country, but rather scores countries in terms of two primary policy groups, namely

- the degree that current internet infrastructure is installed and the policies regarding future expansion, and
- the adoption rates of current broadband as well as policies that will enable future equitable access.

The results are available publicly and without delving too much into the affordability list, it is evident, similar to previous reports, that countries in the lowest World Bank Income Groups, for example Yemen, Ethiopia, Haiti, the Democratic Republic of Congo, Sierra Leone, and Liberia, are among the countries with the lowest ADI. The upper-mid World Bank Income Groups countries, such as Malaysia, Colombia, Costa Rica, Peru, and Mexico, are among the countries with the highest ADI. Apart from the results, the report thus gives insight into numerous factors that influence the ADI of a country and the observations from countries that perform satisfactorily. Some of the points are highlighted in this section, as these influence proposed policies and strategies for costing in unequal markets.

In the executive summary of the 2019 Affordability Report, mention is made of the fact that the ADI investigates the policy development that low- and middle-income nations are creating to sustain reasonably priced internet. According to the Affordability Report, a *competitive* market effectively leads to affordable internet access, pointing to the fact that in an anticompetitive market with only a few key players distributing internet, the affordability of internet access will suffer. The analysis specifically shows that healthy market competition gives consumers choices and adds pressure to lower prices. This is obviously not only true in the telecommunications market but has also been identified as an issue in specifically emerging markets. The result of this finding is that regulators and policy makers, especially in low-income emerging markets, must encourage a healthy market through competition.

In some emerging markets, this is not easy, since the primary stakeholders in the regulating bodies and policy makers could be a corrupt government with anticompetitive market gains as their primary goal. Financial instability can also lead to inability to fund new competition if international backing cannot be obtained, another reality in many emerging markets. Another key finding from the executive summary in the 2019 Affordability Report is that *public access* to broadband internet (for example wireless open access networks) for last mile internet connectivity in urban and rural regions is vital to strengthen markets. This technique should complement the policies on competition and free Wi-Fi, or even 5G in public areas, which will add pressure to the market to adapt prices as consumers migrate to these zones. Unfortunately, the 2019 Affordability Report also found that progress with market competition had stalled over the course of a year and in certain circumstances, markets (globally and not only emerging markets) were becoming amalgamated (having exactly the opposite effect as competition). There are numerous reasons for these occurrences, but according to the 2019 Affordability Report, governments must play an even larger role to support healthy broadband markets. The three core areas on which local governments should focus to achieve this goal are

- setting fair and clear market rules and clear spectrum licensing rules for traditional providers as well as community networks to shape a competitive market,
- ensuring that backhaul access is affordable for new service providers to enter the market, and
- prioritizing public access and community networks that supplement the marketplace.

For a local government to be fully invested in pushing broadband internet through promoting competition, the 2019 Affordability Report proposes dynamic regulation to build a pro-competition regulatory framework. Such a proposed policy includes

- defining the number as well as the diversity of service providers through fair and efficient licensing,
- regulating the number of service providers through spectrum allocation,
- reducing retail prices for the consumers by actively facilitating infrastructure sharing, and
- ensuring market stability through evidence-based policies and consultative processes that are inclusive and build trust with operators.

The core focus areas for local governments to achieve the goal of supporting healthy broadband markets seem relatively simple to achieve; however, the local government should be fully invested in achieving these goals without monopolizing the industry. Furthermore, as in many emerging markets, broadband can only be prioritized once more pressing issues such as poverty alleviation and access to basic services have become sustainable. In Chap. 1 of this book, reference is made to the fact that access to cost-effective or free broadband internet should improve all sectors within a country and boost socioeconomic growth. It is still difficult to prioritize while people are dying of hunger. For example, the 2019 Venezuelan power blackouts resulting from lack of maintenance and lack of technical expertise will, and this is fair, receive the highest priority to mitigate and solve and resources and funding from the government are likely to be invested in that, as opposed to, for example, commercial rollout of 5G. This reality in emerging markets makes it difficult for broadband internet to become accessible to all and have the techno-economic benefits relished in other countries. South Africa is in a similar situation, where a large percentage of the GDP is pumped back to save its (only) power utility from collapse, leaving fewer resources for developing the infrastructure needed to supply rural areas and poverty-stricken areas with broadband internet access.

Regardless of local issues and challenges in emerging markets, one strategy to encourage 5G and mm-Wave networking as technological enablers of Industry 4.0 and a baseline for techno-economic growth and socioeconomic wellbeing, is effective, sustainable, and dynamic costing. Unequal markets in emerging countries are common, and strategizing around this fact could be beneficial for such countries. The following section proposes the idea of costing in unequal markets to achieve phased and dynamic deployment and subscription to next generation communications technology.

5.9 Competition and Productivity in Unequal Markets

A degree of market inequality is healthy and necessary for growth, by producing compensations for "*effort, achievement, and innovation*" [10]. However, stagnation in median wages and bigger gaps between various groups of a population typically have adverse effects. For policy makers to reverse the tendencies of large inequality, the contributing factors should be identified and understood. According to Ennis et al. [10], although the origins of these tendencies are debatable, some traditional explanations are agreed on, including

- "*reduced fiscal transfers from rich to poor,*
- *differences in human capital value, or*
- *differences in the demand for types of workers*".

5.9.1 Competition and Market Power

Ennis et al. [10] investigate the bearing of opposition on inequity and develop a model that illustrates how greater revenues from market power, and accompanying inflated prices, could affect the dispersal of prosperity and power. Market power in this context is defined as "*the ability to drive prices and returns to competitive levels*" [10]. The work presented in Ennis et al. [10] can therefore be used to grasp inequality and its effects and relate these issues (and the proposed solutions) to the telecommunications industry and specifically 5G and mm-Wave networking. The model presented by Ennis et al. [10] is defined as a steady-state model intended to capture the most essential factors, such as real profits, affluence distribution and spending distribution to describe the prospective consequence that inflated fees would have on diverse populations. A comparative analysis is therefore presented between two different steady-states:

- Business proprietors hold market power and are categorized through observing the three essential factors.
- Markets are competitive and the three essential factors are implied by the model.

For this model to be developed and for the redistributed effect of market power to be analyzed, certain assumptions are also made in Ennis et al. [10], including:

- The market power of the countries represented in the study is estimated by the dissimilarity amid the mean markup over all segments in the specific nation and the smallest markup that mirrors the most notable practices of some of the most economical markets.
- The marginal tendency to save that results from increased income as a function of lower prices is not adapted over affluence categories but kept constant.
- Power gains in the market are dispersed in fractions of the present net affluence.
- The price of goods across sectors is exaggerated by market power with equivalent proportion.

Implementing the assumptions listed above, Ennis et al. [10] derive the monopolistic and competitive steady-state models. The redistributive consequence of market power is isolated and the mere influence of monopolies is to elevate fees; all outputs remain constant. Therefore, price increase drives aggregate income, consumption expenditure, as well as wealth in both monopolistic and competitive markets. Ennis et al. [10] define the following relationships:

$$F^m = \mu F^c \tag{5.3}$$

$$Y^m = \mu Y^c \tag{5.4}$$

and

$$C^m = \mu C^c \tag{5.5}$$

where F is the net wealth assets, Y is the total income in an economy, C is the aggregate consumption expenditure and m and c represent the monopolistic and competitive markets, respectively. It follows [10] that the national income identity (total income should be equal to the sum of the various components that make up a country's GDP), defining the aggregate output, is determined by

$$Y^j = W + R^j \tag{5.6}$$

where W is the labor income and R is the capital income, and j represents both m and c. Noticeable from (5.6) is that a change in price moving to a competitive steady-state does not affect labor income and is assumed constant in both steady-states. Since it is unknown how capital income R affects market power, Ennis et al. [10] obtain a relationship between R^c and R^m as

$$Y^c = W + \frac{1-\mu}{\mu} W + \frac{R^m}{\mu} \tag{5.7}$$

where

$$R^c = \frac{1-\mu}{\mu} W + \frac{R^m}{\mu} \tag{5.8}$$

and to define the segments of prosperity and salary of a specific populace (using lowercase letters), the salary cleared by a populace i in every steady-state, is defined in Ennis et al. [10] as

$$y_i^m Y^m = W_i + f_i^m R^m \tag{5.9}$$

and

$$y_i^c Y^c = W_i + f_i^c \left[\frac{1-\mu}{\mu} W + \frac{R^m}{\mu} \right] \tag{5.10}$$

where W_i is the labor salary cleared by every populace group. Therefore, to define salary and prosperity portions differences of market power implicitly as functions of apparent revenue and affluence dividends, as well as the markup and the revenue portion of labor, Ennis et al. [10] propose

$$y_i^m - y_i^c = (\mu - 1)\left(f_i^m - y_i^c\right) + (1 - \mu\alpha_L)\left(f_i^m - f_i^c\right) \tag{5.11}$$

where α_L is the shares of labor income in the monopolistic steady-state. Through similar analysis, Ennis et al. [10] derive the equations for the consumption function and wealth dynamics, given by

$$c_i^m - c_i^c = \frac{1 - s'}{1 - \bar{s}}\left(y_i^m - y_i^c\right) \tag{5.12}$$

where s' is the fringe inclination to save, s—is the typical saving percentage of the market, and c_i is the expenditure part of a particular populace group and

$$f_i^m - f_i^c = \frac{1}{\bar{s}}\left(y_i^m - y_i^c\right) - \frac{1 - s'}{\bar{s}}\left(c_i^m - c_i^c\right). \tag{5.13}$$

Ennis et al. [10] characterize the stability dynamic forces of affluence, salary as well as expenditure shares regarding f_i^c and y_i^c to obtain the model of redistributive consequence of market power on affluence and salary, defined by

$$f_i^c = f_i^m + \frac{\frac{s'}{\bar{s}}(\mu - 1)}{1 - \frac{s'}{\bar{s}}(1 - \mu\alpha_L)}\left(y_i^m - f_i^m\right) \tag{14}$$

and

$$y_i^c = y_i^m + \frac{(\mu - 1)}{1 - \frac{s'}{\bar{s}}(1 - \mu\alpha_L)}\left(y_i^m - f_i^m\right) \tag{15}$$

and this model has the benefit of stemming processes of the redistributive consequences of market power from perceived data that can differ between nations. The variables and data used to calibrate the model are also presented in Ennis et al. [10], considering eight countries (Canada, France, Germany, Korea, Japan, Spain, the United Kingdom, and the USA). Noticeable from the list of nations is that these are developed countries and adding emerging markets to calibrate the model could be useful. Ennis et al. [10] present the results section graphically, showing the effect of market power on affluence distribution in each country, as well as the disparity in affluence proportions from a 1% markup reduction. The full set of results is available in Ennis et al. [10]. Some highlights of these results that are directly based on the proposed model are summarized below.

- *"Redistribution from eliminating market power (competitive market) would occur primarily from the wealthiest 10% of a population to the bottom 80%.*
- *On average, between 12% and 21% of the wealth of the richest 10% reflects the presence of market power.*
- *A 1% reduction in markup increases the wealth of the poorest 20% in a specific country by 0.19 percentage points.*
- *The income of the poorest 20% of households is expected to rise between 14% and 19% in the absence of market power (competitive market)."*

Therefore, from the work published by Ennis et al. [10], market power and therefore anticompetitive behavior could subsidize considerably to affluence disparity. Only the wealthiest 10% in a country benefit from market power and the poorest 80% of the population are adversely affected by lack of competition. In the telecommunications industry, the trends shown in Ennis et al. [10] could be similar and for an emerging market to gain wealth and prosperity from technology, a competitive market is encouraged. It is however known that many emerging markets have high inequality in various market sectors and government policy makers should address these issues related to Industry 4.0 and its potential benefits. Furthermore, as Ennis et al. [10] show that a small reduction in markup (1%) have a significant effect on the wealth and wellbeing of the poorest people in a country, this result should also be applied in pricing of mobile data (and voice). In the modern age and especially Industry 4.0, data and voice are enablers in doing business remotely, education, and offering services that are not geographically limited. The price of data and voice can, however, be a limiting factor and spectrum allocation by governmental bodies should be designed to benefit all users. 5G and mm-Wave networks allow for adaptive and tailored pricing and this can have major effects on socioeconomic growth through internet access if applied strategically and fairly.

5.9.2 Inequality and Productivity

The United Nations (UN) Economic Commission for Latin America and the Caribbean [9] provides concise proof of just how inefficient a population is an unequal market. The work is not specifically aimed at the telecommunications market, but rather takes into account all markets in specific countries. The results are therefore relevant and can be adapted to the telecommunications market. Numerous references to broadband availability as an enabler for mitigating inequality in mostly emerging markets are made and add to the relevance of this section. ECLAC [9] proposed that *equality* (when discrimination is not factored in, although it will become apparent in this section that discrimination is an inherent characteristic of human nature) is at the center of techno-economic development, defending this statement by providing two motives:

- firstly, equality imparts policies with a rights-based approach and not the converse, a needs-based approach.

- Equality is a requirement for progress and hence for development (of productivity, socioeconomics, sustainability, knowledge transfer, and skills development).

ECLAC [9] demonstrates a negative relationship amid disparity and efficiency in a wide range of nations and markets. In a further argument it is shown that the relationship between inequality and productivity is complex and runs in both directions. Equality is also shown to have a vital role in efficiency on the supply side and to have a positive impact on demand. Furthermore, it is argued that revenue dispersal is more probable to drive the increase of demand in a nation that has a larger diversity and *competitive* manufacture configuration. Importantly, ECLAC [9] also maintains that a lag in technology development in one country might be a result of global uncertainties, among others highlighting the importance of prominent countries to take the lead and distribute knowledge, skills, and technology to the rest of the world. The identified global uncertainties in ECLAC [9] that could lead to especially emerging markets becoming uncertain about the market are listed as the following:

- Globalization from the move towards a digital world is creating international tension; since the global financial crisis in 2008, many developed countries have effectively started over in their economic growth and there is competition for control in Industry 4.0. Emerging markets that faced the same financial crisis are typically in a worse position than developing countries and regaining sustainable economic growth is more difficult. Investment in a digital world might be an aspiration, but is not always practical, as economic recovery (in numerous other markets) is still under way.
- Technology regarding performance and impact is also causing uncertainty in the market, especially considering technologies such as 5G and mm-Wave communications that are still a long way from maturation. Markets are hesitant to overcapitalize in these technologies and are effectively waiting for other countries to make the first mistakes. It is less risky for emerging markets to enter a mature market, but they typically do not have the monetary assets and human capital to spearhead these expertise. As a result, market uncertainty during the maturation of a technology is high in emerging markets. Technology development also moves much more quickly compared to historical development and even a year or two of not participating in a new technology could have severe long-term effects.

Apart from these global uncertainties that respond primarily to changes and the maturation of technology, ECLAC [9] also identify external vulnerabilities that could widen the gap of inequality in many countries, especially in emerging markets. The two primary external vulnerabilities, according to ECLAC [9], are:

- Economic growth has fluctuated since the financial crisis and global economic growth have been slower than in the period leading up to the crisis. External vulnerability also tends to be intensified by technological revolution (for example digitization and automation) and the corresponding drop in labor demand have effects on unemployment.

- Debt-to-equity ratios have increased since 2008, especially in emerging markets, and this financialization (financial capitalism) has led to decreased interdependence between monetary policies and risk aversion, resulting in greater financial openness. Emerging markets have increased their indebtedness (whereas developed countries have decreased it), which leads to financial fragility and funds are not implemented to financially back reserves in manufacture. Future rates of investments are therefore uncertain and this leads to lower growth in these markets.

ECLAC [9] also presents a detailed analysis of how inequality creates a barrier around productivity and compromises innovation. Education is identified as one of the major causes of inequality, since people with no formal education will typically receive a lower income compared to people with a formal education. The inequality gap between these population groups will also increase over a long period during which lower growth is expected for the non-educated population. Over this period the benefits of educating a population normally outweighs the private benefits of paying lower wages for work where educated workers are not essential. This effect will spread through an economy and is not localized, according to ECLAC [9].

ECLAC [9] also simulates the incomes of people that are aged between 25 and 55 who have concluded the initial cycle of tertiary schooling, and the results indicate that the income levels are higher (given as percentage change) than the current household income in the selected countries. Countries with larger inequality in terms of education (Guatemala and Honduras) show the largest percentage change (increase) of up to 25% based on this simulation, therefore indicating the biggest effect in terms of increasing income levels through education. The immediate effects of lower income and fewer opportunities also have a macroeconomic effect on the socioeconomic status of these countries and limits productivity and growth over the long term. ECLAC [9] also highlights that even if every person had alike opportunities and access to aptitudes at the beginning of their lives, external factors such as discrimination (race, sexual orientation, ethnicity and religion) would (again) lead to inequality.

Unfortunately, in emerging markets discrimination is common. (It is for example illegal to be bisexual in Brunei and this offence carries a death sentence in that country.) These issues are therefore likely to spread inequality until they are faced. Finally, in ECLAC [9], apart from discrimination, several reasons for inequality that cause segregation and deterioration are listed, with specific reference to territory and the environment. These reasons include:

- Spatial focus of salary and aptitudes are vital in the changing aspects of a local economy. Territorial inequalities and disparities are a direct consequence.
- Historical infrastructure from traditional mining and agricultural exports (and other natural resources) is indicative of geographical fragmentation if already insufficient.
- Urban segregation in an asymmetric rural-urban dynamic causes disproportions in sanitation, energy supply and basic drinking water and can also lead to inequality

(especially when comparing the poorest and the richest population groups in an emerging market).

- Metropolitan mobility, which is an amalgamation of disparity in manufacture, energy efficiency and ecological decline, contributes to inequality between for example urban and rural populations in a country.
- Environmental sustainability is associated with costs in future productivity as production services decrease and inequality between generations increases.

These dimensions of inequality are closely related to education, discrimination and the response to the financial crisis and vary between emerging markets and developing countries. Poverty levels are typically concentrated in areas, mostly in rural areas, and furthermore, compounded environmental degradation in urban areas has negative consequences for the already poorer rural areas as farming ecosystems suffer, leading to uncertainty about the future. Analyzing these effects individually is convenient to identify and focus on particular uncertainties, but in reality, the complex ecosystem of uncertainty and inequality is difficult to analyze in terms of its economic impact. Identifying these issues and highlighting the cause and effect of each is useful, but to achieve sustainability in unequal markets, policies and regulations must be addressed too. The following section briefly highlights key points when considering sustainable policies in unequal markets and acts as a precursor to potential solutions (policies) that take advantage of unequal markets and distribute wealth, in this case through connectivity for Industry 4.0, to all population groups.

5.9.3 Sustainable Policies in an Unequal Market

Sustainable development needs to be encouraged to mitigate inequality and provide services and commodities to entire populations with few to no limitations to access. In the modern world, a large degree of sustainability and economic growth involves broadband internet and Industry 4.0. It is therefore quite common to compare sustainable policies in terms of their consideration of technology and the roadmap for each country. In this section, reference is made to sustainable policies from ECLAC [9] and a similar trend is observed, with some reference to 5G and mm-Wave as enabling technologies in creating sustainable and tailored solutions to connectivity. Three primary pillars based on innovation in technology are defined in ECLAC [9] to achieve a roadmap to sustainable policies, namely

- digitalization,
- sustainable cities, and
- renewable energies.

Without delving into these sustainability goals independently (and relating them to the UN Sustainable Development Goals), it can be noted that internet connectivity, particularly 5G and mm-Wave services, is a backbone of digitization and sustainable cities that can operate completely ubiquitously and with adequate bandwidth for sustainable expansion. Incorporation among technologies can additionally decrease the ecological imprint of digitization and advance sources of renewable energy. Digital technologies need energy supply and depending on the technology, the amount of energy supply varies. However, the sheer amount of digital equipment, including low-power IoT devices, adds up and energy supplies are under severe strain to provide efficient and uninterrupted power to these devices (for example data centers). The development of renewable energy sources is interdependent with advances in digitization and necessary for sustainability and decreasing environmental degradation.

Transmission networks and infrastructure of digitization and sustainable cities in a digital industry are dependent on energy, but also on policies that prioritize areas of development. Governmental alignment with these areas is crucial and should be identified as soon as possible if not already addressed. According to ECLAC [9], three key areas should be prioritized, namely

- immovable and mobile broadband infrastructure to reach infiltration intensities comparable to that of the Organization for Economic Co-operation and Development *average-income* nations and comply with worldwide performance criteria based on speed, latency and pricing,
- increasing the quality stream of goods and services in accordance with modern digitization within the information technology industry, and
- aggregating business capabilities towards a digital revolution of commercial models, goods, and services.

These policies must, however, achieve collaboration between governmental entities and public-private partnerships and a high level of transparency and accountability must be ensured. Such policies encourage productivity and innovation. Regulation and institutional challenges related to these policies, as outlined in ECLAC [9], include

- the extension of digital development strategies with policies that place strong focus on production restructuring as well as specialization, skills training, and innovation in technology, which includes micro-, small, and medium enterprises,
- a shift towards incentive-based regulatory models to incorporate novel requirements for investments in networks and incorporate the characteristics of the industrial internet and its modern policies of placement, interoperability, digital safety, data safety, and confidentiality, and
- especially relevant in emerging markets, the implementation of regional digital markets that are *scalable* through harmonization of spectrum and national regulations.

These proposals for the structure of policies are specifically in line with positioning a region internationally in terms of its infrastructure, regulations and actions to support local supply and demand. These regions need to be connected to global technology networks and looking after the handover of skills and knowledge of technical capabilities (managing networks and implementing digital platforms) and in market integration. As a result, new business and technology skills can be attracted and a sustainable infrastructure is possible if successful skills and knowledge transfer are achieved. Delays in investments in modern networks, such as 5G and mm-Wave technology in these regions, could have immediate effects such as network congestion, but more importantly, stifle socioeconomic growth in these regions. It has been shown [15] in many modern smart cities (and in general, urban areas) that digital technologies play a crucial role in regulating resource use and improving QoS. Strengthening regulation on the ecological wellbeing of civic amenities (emissions from the transportation sector), consolidating local finances towards policies, developing countrywide metropolitan guidelines and coordinating these with industrialized policies and strategies on technologies all complement the shift towards digitalization. In emerging markets, however, there are still numerous pitfalls related to the internet, not typically experienced in developed countries, which must be addressed before structuring potential costing in unequal markets. These pitfalls are discussed in the following section.

5.10 Internet Pitfalls in Emerging Markets

The internet, initially a government-funded research project for the military, has become an integral part of society and most work and entertainment have an internet-related service associated with it. It has also become a necessity for many businesses and towards Industry 4.0, it will become compulsory. In emerging markets, however, internet use is still low; in Pakistan and Bangladesh respectively only 8% and 11% of people occasionally access the internet from a mobile device [6]. Coronel [6] clearly links high penetration rates of the internet in emerging markets with economic growth, two factors that go hand in hand and operate both ways (increased economic growth leads to higher internet penetration). Coronel [6] additionally mentions that although broadband internet penetration in emerging markets has increased, the lower prices paid in these areas when compared to the rest of the developed world are unfortunately also indicative of lower quality. 5G and mm-Wave therefore have a definite role to play in these markets, not only in preparing these economies for Industry 4.0, but also in improving network quality and tailoring the solutions to maintain a price that is in line with what the local population can afford. Coronel [6] identified some issues specific to emerging markets that must first be addressed to spearhead new infrastructure in these regions. An interesting comment from Coronel [6] is that people in emerging markets, such as countries in Africa, South America and Southeast Asia, are more receptive to trusting the internet than those in developed countries such as the USA, Japan, and Singapore. As a result, these markets are very susceptible and vulnerable to untrustworthy and illegal internet activity, which should be addressed through awareness campaigns. Apart from this, the primary challenges identified in Coronel [6] are discussed in this section.

5.10.1 *Divergence Between Consumer Needs and Business Opportunity*

In general, the needs of the market have shifted more towards mobile broadband as opposed to fixed line connectivity. Technology has also advanced to a point where these solutions (5G and mm-Wave networking) are capable of delivering services at the speeds and latency required for accessing mobile content. MNOs, however, must be able to provide the infrastructure to satisfy consumer needs, and maintain and expand these networks to avoid congestion as user subscriptions increase. Networks that do not have sufficient capacity and are prone to outages or sub-standard reception will cause user dissatisfaction and will not be sustainable as a viable business opportunity. Infrastructure development and initial capital investment are also high and in emerging markets with limited financial resources, this becomes a challenge (especially doing it right the first time and not using cheap equipment that could later result in even higher costs). There is therefore a divergence between consumer needs (high-speed and dependable internet connections) and business opportunities (high initial cost combined with regular maintenance and expansion).

5.10.2 *Data Monetization*

The concept of data monetization, a technique to generate revenue for gathered (personal) data, is not new, but it is related to certain ethical and regulatory considerations. Furthermore, in recent times data monetization has become a contentious issue and there are concerns about the buying and selling of personal data between large corporations [29]. Data monetization has become a relatively inexpensive and quick way for businesses to meet their financial objectives, although in fairness, certain companies also use the data for social good and research [6]. In emerging markets issues such as unethical reselling of personal data with compensation to individuals, data aggregation and data ownership are likely to have to be dealt with through policies and strict regulations and governments and governing bodies should be aware of these issues.

5.10.3 *Personal Data in Exchange for Free Digital Services*

In line with the discussion on data monetization, another method of extracting personal data from users that would otherwise not be willing or aware of its dangers, is for businesses to offer free (mostly digital) services in exchange for personal data. Companies have been found guilty of this practice, where for example data were gathered during participation in digital surveys and used to deliver specific content during a presidential campaign. Users willingly provide their personal data; however, in many of these cases the terms and conditions of participating in the digital service is unclear or purposefully complex and lengthy. Chakravorti [5] reports that users in emerging markets are more willing to share personal information with companies and agree to their requests to aggregate their data. The offering of free digital services usually encourages these users, but it has been found [5] that users will supply their data for no immediate benefits too. Again, emerging markets need to be made aware of the dangers and pitfalls of sharing their personal information and businesses should not be incentivized to collect and sell personal data. This is, however, virtually impossible and according to Coronel [6] is likely to be a serious factor to deal with as more people in emerging markets go online.

5.10.4 Net Neutrality

The purchasing decision of users in terms of telecommunications service provider should not be influenced by offerings of digital services and subscription fees to websites and/or applications that may create an unfair advantage for individual MNOs. There are, however, no specific rules that prevent MNOs and telecommunications providers from offering such deals and they are thus able to influence potential clients and create an anti-competitive environment. Net neutrality has also remained severely argued as a possible strategy to regulate the internet [27]. There are two schools of thought regarding this practice:

- Advocates of net neutrality argue that there are concerns about ISPs' pricing power and this could be implemented to victimize against content providers (CPs), hurt innovation and find its way down to consumers.
- Arguably, without service differentiation there is little incentive for ISPs to expand services and technology infrastructure and this will have an impact on future internet capabilities.

Both these arguments are valid in devising costing strategies in unequal markets and Tang and Ma [27] and Ma et al. [21] proposed models to identify the shortcomings in internet deployment in areas where the incentive for the service providers might be low in terms of business opportunity. These models, among others, are reviewed in the following section and form part of the discussion on costing approaches in unequal markets. Couchenery et al. [7] provide an in-depth model that analyzes the relationship between two competing ISPs and one CP and model the interactions based on a game (therefore the goal is to *win*). This model is depicted in Couchenery et al. [7] and adapted and presented in Fig. 5.6.

From Fig. 5.6, to study non-neutrality, the per unit prices are also introduced, as these would form part of an incentive-based approach. For a neutral environment, access fees charged to subscribers are fixed amount payment subscriptions for respectively the CP and the ISPs. The work presented by Couchenery et al. [7] not only serves as a method to analyse various implementations of net neutrality, but also defines several pricing structures that are discussed in the following section,

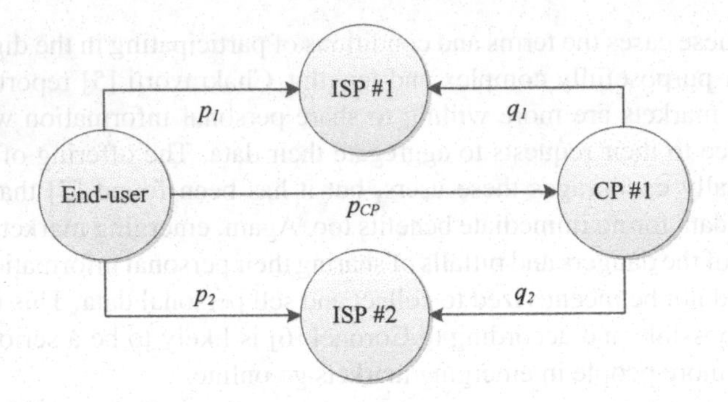

Fig. 5.6 The proposed game-theory model in Couchenery et al. [7] where the relationship between two competitive ISPs and one content provider is analyzed. Positive flat rate pricing is defined by the variables p and positive per volume unit prices are denoted by q

forming part of internet cost strategies in unequal markets. Couchenery et al. [7] therefore specifically research the influence of horizontal costs on the revenue of suppliers to determine if these assist in mitigating unfairness in the current revenue sharing between ISPs, as is claimed in the present net neutrality deliberation. The game played is through regressive stimulation, implying that companies can forestall the results of future games if selecting a strategy. Couchenery et al. [7] highlight that "*side payments, unless decided by ISPs, have little chance to address the concern from ISPs regarding fairness of the revenue sharing associated with users accessing content through their infrastructures*". This is largely because of the competition among the ISPs on admission fees that drive their revenue lower. Conversely, with the CP being in the monopolist position, these entities always obtain significant revenues and horizontal costs are beneficial to them. If horizontal costs are chosen non-cooperatively by the providers, the ISPs can benefit but at the cost of the CP and the users. Interestingly, according to Couchenery et al. [7], horizontal costs aimed to take full advantage of social and subscriber prosperity also maximize CP profits. If horizontal costs are chosen non-cooperatively by providers, only one of the ISPs wins a big share and the other ISPs gain a small portion more than compared to the neutral case and this scenario could potentially create tension and negotiations between the ISPs.

Dai et al. [8] additionally consider if a vertically unified broadband and content supplier could unjustly create an advantageous scenario for itself by selling QoS to CPs for inflated fees or by declining to supply access to QoS to rival content, therefore effectively limiting QoS to the competition CPs. The model presented by Dai et al. [8] is therefore designed as an analytical competition where "*one vertically integrated provider and one over-the-top (OTT) CP compete for revenue*". The broadband provider determines if it will use QoS, and if this is used, sets the worth either to the OTT CP or straight to the user. The mathematical market models defined in Dai et al. [8] are presented in detail, and the scenarios that are considered include the scenario where

- the ISP opt not to use QoS,
- sells QoS to the CP, and
- sells QoS directly to users.

Furthermore, each of these scenarios is analyzed based on variations in the prices of QoS sold either to CPs or directly to the users. Dai et al. [8] provide the numerical results for further reference and conclude that in this model the market share of the ISP rises by the change in the worth of services (film) and drops by the change in cost. It is shown numerically that the provider is allowed to retail QoS to subscribers at a lesser worth compared to when it is retailed toward the OTT CP. The provider is also further expected to implement QoS exclusively for their included services if it is supplying analogous services to the OTT CP and if the apparent efficacy of QoS is great, therefore retaining QoS for its own benefit. Furthermore, it is typically in the interest of the OTT CP that the ISP sells QoS to the user directly, at a lower price, to increase revenue. Finally, if the ISP does not adopt QoS the CP and the user will earn less profit or the user's experience will deteriorate. Even if the price of QoS is not regulated, all parties are likely to derive some benefit. These results will also be considered in the costing in unequal markets section in this chapter.

The final internet pitfall that is identified by Coronel [6] is that of free internet, briefly discussed in the following section and relevant in the costing in unequal markets section in this chapter.

5.10.5 Free Internet

Increasing economic growth, as mentioned multiple times throughout this book, is one of the arguments for better internet in emerging markets, according to the World Economic Forum [31]. Within the same argument, this research also shows that citizens in emerging markets are more likely to use broadband services that influence their welfare (social, health, and education), such as online government services, when compared to developed countries. Local application (app) development is also starting to prosper in numerous countries (apart from the USA and the UK) and the World Economic Forum [31] reports that these countries spend more time on the locally developed apps as opposed to the ones developed in other larger economies. Local digital ecosystems are crucial to develop services tailored to local needs and boosting competition in an increasingly international digital service market, according to the World Economic Forum [31]. Coronel [6] highlights that decentralized data marketplaces and modern frameworks, such as open data exchange (ODX[1]), are aiming to provide free internet in emerging markets to resolve the lack of internet penetration in these areas. Currently, technical and commercial barriers hinder ISPs from providing free internet and to some extent violating net neutrality. ODX specifically argues that through the net neutrality principle, ISPs should be

[1]Open data exchange—Free internet for emerging markets—http://odx.network.

allowed the opportunity to provide free internet in emerging markets. ODX aims to provide these services by democratizing internet access through the blockchain and CP and ISPs can transact in the global marketplace at scale and through a trustworthy network. At the time of writing, and possibly under correction, research into progress and milestones of ODX and the goal to provide free internet in emerging markets show no new milestones posted since June 2018.

5.11 Conclusion

In the introductory section of this chapter, several challenges and opportunities in emerging markets with respect to internet connectivity and participation in a digital era are highlighted. Arguably the most important to take away from the section is that even though emerging markets and especially rural areas may have underdeveloped infrastructure that is possibly not maintained and is obsolete, an innovative group of technologies like 5G and mm-Wave enables these areas essentially to start over in investing in their digital future. Numerous references have been made throughout this chapter and in literature to technology (especially the internet) and economic growth. The techno-economic prospects of stable and accessible internet are clear; it is getting these services to all areas globally that will make the difference to socioeconomic sustainability. This chapter is therefore structured to encourage and prioritize specifically 5G and mm-Wave networking not only to deploy capable infrastructure in barren areas, but also to future-proof these areas to handle increased utility and network expansion. 5G has certain characteristics that previous generations of mobile networks did not have, as discussed in Chaps. 1 and 2 of this book, and these should be taken advantage of. To defend these statements, a section on the role of 5G on productivity and economic growth is presented in this chapter. A report generated by BCAR [1] summarizes and emphasizes numerous valid and crucial points on productivity as a function of economic growth. Transformation of traditional models of productivity growth that were relevant in previous industrial revolutions must be updated to represent a digital economy and expand wealth to emerging markets through the internet. The benefits of productivity and economic growth are also not just short-term but have long-lasting effects in any region that successfully adopts these technologies and participates in global trade or services or even products. Furthermore, this section reports a portion of the results in BCAR [1] showing the effects of rollout time of 5G and mm-Wave compared to the short- and long-term benefits that could be expected. The information could be used by MNOs to plan infrastructure development in rural areas and determine the viability of their proposed rollout period to maximize return on investment.

Potential strategies for resource management and pricing models of 5G and mm-Wave networks are also reviewed in this chapter (along with a separate section on economic/pricing models) and this review forms part of the following chapter in this book, related specifically to internet costing in unequal markets. These sections highlight a number of advantageous characteristics of 5G (and mm-Wave) that enable operators to structure resource management policies to benefit themselves and the users, especially since in emerging markets the cost of broadband internet could be high (or low, but proportional to the service quality). Dynamic demands and resource sharing as well as tiered pricing are investigated in this segment and the benefits and weaknesses are listed. The economic and pricing models section focuses on models that play out as games, where there are players that compete to win. These models have been used in numerous applications and can be adapted relatively easily to the telecommunications market where the players are typically spectrum regulators, MNOs and ISPs and CPs.

The remainder of the chapter is structured to discuss research challenges of 5G and mm-Wave that are to be addressed to achieve the maturation of the technology, the accompanying techno-economic perspectives, issues and potential solutions for internet affordability in emerging markets, and the effect that a healthy, competitive market has on inherently unequal markets. The closing section in this chapter discusses several internet pitfalls in emerging markets that will possibly always be applicable when considering how many emerging markets operate (in all sectors). Essentially, these issues need to be highlighted and planned around, as opposed to providing solutions to mitigate the issues themselves. The following chapter is dedicated to internet costing in unequal markets.

References

1. BCAR (2018) Impacts of 5G on productivity and economic growth: working paper. Australian Government Department of Communications and the Arts. Retrieved September 9, 2019 from http://www.communications.gov.au
2. Basar T (2010) Lecture notes on non-cooperative game theory. Coordinated Science Laboratory, University of Illinois, Urbana, IL
3. Bogale TE, Le LB (2016) Massive MIMO and mmWave for 5G wireless HetNet: potential benefits and challenges. IEEE Veh Technol Mag 11(1):64–75
4. Bouras C, Kokkalis S, Kollia A, Papazois A (2018) Techno-economic analysis of MIMO & DAS in 5G. In: 11th IFIP wireless and mobile networking conference (WMNC), pp 1–8
5. Chakravorti B (2018) As emerging economies bring their citizens online, global trust in internet media is changing. Retrieved September 28, 2019 from http://www.theconversation.com
6. Coronel M (2018) The problem with internet in emerging markets. Retrieved Oct 21, 2019 from http://www.medium.com
7. Couchenery P, Maillé P, Tuffin B (2013) Impact of competition between ISPs on the net neutrality debate. IEEE Trans Netw Serv Manage 10(4):425–433
8. Dai W, Baek JW, Jordan S (2016) Neutrality between vertically integrated cable provider and over-the-top video provider. J Commun Netw 18(6):962–974
9. ECLAC (2018) The inefficiency of inequality. Retrieved October 5, 2019 from http://www.repositorio.cepal.org
10. Ennis S, Gonzaga P, Pike C (2017) Inequality: a hidden cost of market power. Retrieved Oct 22, 2019 from http://www.oecd.org
11. GSMA (2018) The state of mobile internet connectivity 2018. Retrieved October 3, 2019 from http://www.gsma.com
12. Huawei (2018). Capturing growth opportunities in emerging markets to accelerate evolution towards 5G. Retrieved Oct 21, 2019 from http://carrier.huawei.com

13. Huawei (2018) Huawei 5G wireless network planning solution white paper. Retrieved Oct 9, 2019 from http://www-file.huawei.com
14. IHS (2017) The 5G economy: how 5G technology will contribute to the global economy. Retrieved Oct 24, 2019 from http://cdn.ihs.com
15. Lambrechts JW, Sinha S (2016) Microsensing networks for sustainable cities. Springer International Publishing Switzerland. ISBN 978-3-319-28358-6
16. Lambrechts JW, Sinha S (2019) Last mile internet access for emerging economies. Springer International Publishing Switzerland. ISBN 978-3-030-20956-8
17. Lee JW, Chiang M, Calderbank RA (2006) Network utility maximization and pricing-based distributed algorithms for rate reliability tradeoff. In: IEEE INFOCOM, pp 1–13
18. Luong NC, Hoang DT, Wang P, Niyato D, Han Z (2017) Applications of economic and pricing models for wireless network security: a survey. IEEE Commun Surv Tutor 19(4):2735–2767
19. Luong NC, Wang P, Niyato D, Liang YC, Han Z, Hou F (2018) Applications of economic and pricing models for resource management in 5G wireless networks: a survey. IEEE Early Access Article 1–42
20. Luong NC, Wang P, Niyato D, Liang Y, Han Z, Hou F (2018) Applications of economic and pricing models for resource management in 5G wireless networks: a survey. IEEE Commun Surv Tutor 21(4):3298–3339
21. Ma RTB, Wang J, Chiu DM (2017) Paid prioritization and its impact on net neutrality. IEEE J Sel Areas Commun 35(2):1666–1680
22. McKane, J. (2019). Rain's big plans to expand its 5G network. Retrieved Oct 8, 2019 from http://www.mybroadband.co.za
23. Odlyzko A (1999) Paris metro pricing for the internet. In: Proceedings of the 1st ACM conference on electronic commerce, pp 140–147
24. Ros D, Tuffin B (2004) A mathematical model of the Paris metro pricing scheme for charging packet networks. Comput Netw 46:73–85
25. Smail G, Weijia J (2017) Techno-economic analysis and prediction for the deployment of 5G mobile network. In: 20th conference on innovations in clouds, internet and networks (ICIN), pp 9–16
26. Su J, Xu H, Xin N, Cao G, Zhou X (2018) Resource allocation in wireless powered IoT system: a mean field Stackelberg game-based approach. Sensors 18(10):3173
27. Tang J, Ma RTB (2019) Regulating monopolistic ISPs without neutrality. IEEE J Sel Areas Commun 37(7):367–379
28. Walia JS, Hämmäinen H, Matinmikko M (2017) 5G micro-operators for the future campus: a techno-economic study. In: Internet of things business models, users, and networks. IEEE, pp 1–8
29. Wessling O (2018) How Facebook will monetize your data in the future. Retrieved Oct 25, 2019 from http://cpomagazine.com
30. Wisely D, Wang N, Tafazolli R (2018) Capacity and costs for 5G networks in dense urban areas. IET Commun 12(19):2502–2510
31. World Economic Forum (2019) Emerging markets: big challenges, big opportunities. Retrieved Oct 27, 2019 from http://reports.weforum.org
32. Yaghoubi F, Mahloo M, Wosinska L, Monti P, de Souza F, Crisstomot J, Costa WA, Chen J (2018) A techno-economic framework for 5G transport networks. IEEE Wirel Commun 25(5):56–63

Chapter 6
5G and Millimeter-Wave Broadband Internet Costing in Unequal Markets

Abstract Historically, the pricing of critical services and products required some forms of financial intervention in order to effect distribution to entire populations. In emerging markets, the complexity and challenges of distributing critical products and services in significantly unequal markets are much higher. Governments have been forced to innovate policies and strategies to achieve successful distribution of products and services to poor households and rural areas. This chapter reviews such policies in various market sectors. The sectors are specifically chosen as capital-intensive, critical services, since broadband internet, in line with the fourth industrial revolution (Industry 4.0), is fast becoming critical to bridge the digital divide and boost economic growth in emerging markets. The underlying focus is on fifth-generation (5G) and millimeter-wave technologies, emerging technologies that could have a major impact on unserved markets.

6.1 Introduction

In the dawn of the fourth industrial revolution (Industry 4.0), millimeter-wave (mm-Wave) and fifth-generation (5G) internet connectivity have the potential to involve emerging markets and rural areas (also in developed countries) to participate in a modern era of doing business. The decentralized approach of doing business through the internet has many advantages and can give *everyone* an equal opportunity to sell products or services, globally. A major hurdle that challenges the ubiquity of broadband internet in emerging markets is its price offered to consumers. Although the price that a consumer pays for connectivity is affected by numerous factors, the bottom line is that if the internet is not affordable, populations in poor nations will continue to be associated with the digital divide. Products and services in unequal markets are often priced in ways to bring affordability to the masses and use some

W. Lambrechts and S. Sinha, *Millimeter-wave Integrated Technologies in the Era of the Fourth Industrial Revolution*, Lecture Notes in Electrical Engineering 679,
https://doi.org/10.1007/978-3-030-50472-4_6

method of discrimination between the wealthy and the poor and are often subsidized to achieve this. In this chapter, a review of the pricing of various critical services is presented to determine how these sectors distribute themselves across unequal markets. The reviews are based on the pricing methodologies in the following sectors:

- medicine,
- energy,
- petroleum, and
- water.

By reviewing the pricing strategies of each of these sectors, in emerging markets, this chapter aims to draw similarities to the broadband internet sector, specifically 5G and mm-Wave networking. Broadband internet has in recent years been proposed as a basic human right and its global importance for economic growth has therefore been realized. The price of broadband internet, specifically mobile service, remains high in many emerging markets (also in some developed countries) and can be a limiting factor to providing internet access to emerging populations. The long-term benefits and economic perspective of internet connectivity have been established in Chap. 5 of this book, and this chapter takes the argument forward to determine strategies to achieve internet-for-all and bridge the digital divide towards Industry 4.0.

The following section reviews the pricing of medicine in emerging markets and critically reviews a publication by Everard [6] that investigates methods to distribute and price medicine in these countries. The importance and impact of providing medicine in any country is high and therefore the priority of developing distribution and pricing strategies is typically very high on a government's list. As a result, innovative and large-scale strategies have been implemented in the health sector to ensure medicine is distributed to the population, especially considering that many people in these countries cannot afford basic health care insurance. The identified strategies could be adapted for the mobile broadband sector, which is reviewed in the final section of this chapter.

6.2 Pricing of Medicine in Emerging Markets

The distinction between the three classes of nations, low-, middle-, and high-income, has also led to studies on access to medicines based on the wealth of a country. This research, as proposed in Everard [6], follows a strategy that investigates variable methods to distribute and price medicines to countries that are developing, and explains the differences when compared to the strategies of developed countries.

Everard [6] acknowledges that low-income countries are commonly considered by extraordinary numbers of mortality, illness and debility, and these countries have limited or no access to health care and other health-related services. The study and results published in Everard [6] are analyzed in this section primarily to determine if there are similarities and policies that could be adapted in the telecommunications industry to cost internet distribution (mobile or fixed) in uneven markets. Across many industries in emerging markets, the availability and affordability of crucial products and services are not aligned with the average income of the population. There are numerous reasons for this, and these reasons can be complex. Not only are financial constraints in these countries a major concern; the priorities of government, public, and private organizations also have to be aligned to work together in providing the population with critical services.

6.2.1 Access to Essential Medicines

Everard [6] reviews access to essential medicines in emerging markets through four components, originally developed by the World Health Organization (WHO), namely

- *"rational selection and use,*
- *affordable prices,*
- *sustainable financing, and*
- *reliable health and supply systems".*

These four components are specifically subjected differently in each country, whether emerging or developed, by challenges in

- *"political,*
- *social,*
- *ethical,*
- *economic, and*
- *industry-related difficulties and developments".*

The four components and the issues that affect each could sensibly be compared to broadband internet distribution in emerging markets and developed countries.

6.2.2 Rational Selection and Use

Rational selection and use would be based on the differences in demand for mobile broadband, be it basic internet access to run a website that markets products or services, or high-capacity video-on-demand streaming services. The demands of consumers vary greatly and identifying this component is essential to ensure that demands are met without over- or underspending on infrastructure development. Affordable prices are also relevant in the broadband internet sector and consumers would not be able to use internet services to conduct business if these are not affordable. In terms of sustainable financing, mobile operators often do not enter unserved markets, since these areas have already been identified as unsustainable. Recovering the cost of infrastructure development is difficult in these areas and a form of sustainable finance is necessary to incentivize mobile operators to enter these markets. Finally, reliability is crucial in the broadband internet sector. If a service is introduced into a market or area, consumers will inevitably start relying on it and its reliability will determine the longevity of the service and the impact on socioeconomic growth.

In terms of medicine, all stakeholders in a country have an obligation and a role to play in developing the four components identified by the WHO. These stakeholders should aim to remove obstacles and artifacts that hinder development and evaluate policies that enable opportunities to implement these components. In terms of rational selection and use, specifically in terms of essential medicines, it is noted in Everard [6] that treatments for major infectious diseases and related conditions can be extremely complex and costly, as are the medicines that treat these conditions effectively. Through rational selection and use, the aim is to make these medicines available at low cost by defining which medicines are most needed in a specific area (demand), identifying the most cost-effective treatment of the condition, as well as taking responsibility for quality and safety, and guaranteeing that the treatment and medicine are administered and taken effectively. Everard [6] argues that national-level treatment guidelines and protocols are to be introduced and that key pharmaceuticals must be prepared and disseminated to administer services and distribute medicines, accompanied by in-service training for the medical workers. Stepping away from essential medicine rational selection and use for a moment and focusing again on internet distribution in emerging markets, an analogy can therefore be drawn. 5G and mm-Wave networks are designed, primarily through network slicing as reviewed in Chap. 2 of this book, to enable "*multiplexing of virtualized and independent logical networks on one physical network infrastructure*" [6]. Through optimizing resources by implementing network slicing, a network topology can

- maintain specific factors such as connectivity, speed, and capacity, and
- distribute these resources based on the requirements of the application (user).

This enables opportunities to create agile networks that adapt, and more importantly, ration resources for the needs of specific industries, commercial users, and the economy. Emerging markets therefore have the opportunity to select which services are required from a 5G and mm-Wave network and implement cost-effective solutions tailored to the economy, ideally benefiting all stakeholders involved, down to the consumer. If effective research into the needs of consumers is prepared, as Everard [6] refers to, in terms of essential medicine, the implementation and distribution of essential internet-related services can be achieved. These services will inevitably differ vastly between countries, economies, and emerging markets (when compared to developed countries) and can be upscaled (or downscaled) if necessary and over a period.

6.2.3 Affordable Pricing

The second component identified by the WHO to distribute essential medicines to emerging markets regards affordable pricing. Although the idea of affordable pricing seems relatively obvious, challenges in emerging markets typically obstruct its effective implementation. According to Everard [6], the unaffordability of medicines to individual patients in low-income countries influences access to care and treatment. According to Everard [6], looking for personal is more frequently a request to buy medications rather than a consultation with a competent health employee. The price of medicine, which is sometimes higher in emerging markets when compared to developed countries, is most commonly paid out of pocket and in low-income families these purchases can make up a large percentage of their monthly income. There are however, according to Everard [6], various means to lessen the burden on these families, but primarily and related to essential medicine, the cost of drugs paid out of pocket should be brought to a reasonable level. Practices to ensure that medicine prices are kept low that Everard [6] lists include

- endorsing *competition* amid good general medications that concern off-patent matters,
- pure *negotiation* of fees between government, public, private sector and pharmaceutical companies that distribute medicines in these markets,
- promoting competition of on-patent medicines and distributors,
- reducing duties and *taxes*,
- reducing wholesale and retail *margins*, and
- ensuring *transparency* of cost facts for health care suppliers and users such that the local public is informed about the actual cost of medicine and how/where to find the most affordable products.

Referring to 5G and mm-Wave internet distribution in emerging markets, the techniques presented by Everard [6] to maintain low prices for essential medicines have similarities that can also be implemented in the telecommunications industry. Among the most important factors, and sometimes lacking in emerging markets, is promoting competition. In economics, price competition [5] is one of numerous ways (and one of the most effective ways) to enable a product or service to exist and compete in a fair marketplace. Prospective consumers typically judge two products or services that are substantially similar based only on their respective pricing. The cheaper of the competing and similar products is then usually purchased.

Competition has been stimulated in emerging markets and there are numerous new opportunities for growth and larger interest in global competition, according to Madgavkar [14]. As reviewed in Chap. 5 of this book, several pricing strategies affecting competition can be developed to promote competition equally and fairly and provide the best quality of service (QoS) to consumers. With 5G and mm-Wave technology having the ability to scale network resources (see Chap. 2 of this book), new service providers are encouraged to offer tailored services and do not necessarily require large initial capital investments. Furthermore, as described in Chap. 5 of this book, these service providers do not necessarily require existing (4G) infrastructure to roll out 5G networks. In this sense, competition in the telecommunications industry and between service providers can be promoted through the right governance and fair pricing of spectrum, which, inevitably in the mm-Wave bands, should be enough for all service providers to share. Factors listed by Everard [6], such as price negotiations, are also relevant in the 5G and mm-Wave market, since service providers are able to negotiate prices on spectrum and licensing based only on the services and applications they aim to provide, thus not having to pay for spectrum that they would not require initially.

Duties and taxes on products and services can also be negotiated with government and retail margins with suppliers, to promote fair distribution of resources that would enable small and medium-sized enterprises (SMEs) to compete in a naturally monopolized [18] industry. Finally, related to affordable pricing, transparency of pricing is crucial and potential consumers should be informed exactly what the services offered entail and how the pricing was established, allowing them to choose a best-fit solution for their requirements. In the final section of this chapter that reviews the pricing of broadband in emerging markets, the concept of transparency in broadband pricing is highlighted.

6.2.4 Sustainable Financing

The third component identified by the WHO is that of sustainable financing and according to Everard [6], this factor relates to the context of *overall* funding in health care. Therefore, overall sustainable funding also includes the avoidance and treating urgent transferable illnesses that have a large impact on civic health (this could perhaps be related to a holistic view on financing broadband internet that benefits all stakeholders all the way down to the consumer). According to Everard [6], governments have usually been the sole financing entity of the health sector in emerging markets and medicine has been distributed to patients free of charge. However, as funds have diminished in these countries, shortages of essential medicines have occurred. The problem is exacerbated in rural areas where the supply has essentially collapsed and the demand has risen to alarming levels. As a result, health care in numerous emerging markets has become privately funded, or at least only partially funded by government. Furthermore, since low-income countries typically cannot afford private health financing, such as medical aid schemes, up to 90% [6] of treatments and medicines are funded out of pocket, whereas this figure drops below 20% in developed countries. The burden inevitably also falls more heavily on poorer families, which are typically prone to a higher incidence of diseases and illness, accompanied by challenges such as a general lack of clean drinking water, sanitation, or just generally inadequate and unhealthy living conditions. According to Everard [6], a large number of Sub-Saharan African countries have re-introduced public health services subsidized by domestic tax revenue as fee structures or co-payments. To achieve sustainable financing of the health care systems in emerging markets, combinations of several feasible financing mechanisms can also be introduced to distribute contributions between low-, medium-, and high-income families. Perhaps a similar approach could be followed in telecommunications industries that aim to provide products and services based specifically on 5G and mm-Wave technology (the reasoning being the long-term benefits of introducing next-generation technology now).

To move away from the typical scenario of communications authorities selling spectrum to privately owned service providers, and consumers all paying an equal price for these services, low-income families could be offered these products and services at reduced prices. Prices would *have to be* subsidized by governments, essentially through tax revenue from the wealthier families. Since 5G and mm-Wave technologies offer tailored services for each user (see Chap. 2 of this book), the speed and latency of lower-cost services can be adjusted for each user, and essentially scaled based on the price paid for the service. Therefore, the subsidized prices would also be lower compared to the service offerings for wealthier families.

Current broadband services (2G, 3G, and 4G) typically only scale pricing based on the amount of data purchased or by limiting the time that the data are valid. For example, buying 1 GB of data that expires within 24 h is cheaper than buying a 30-day 1 GB data bundle, but the access speeds of both options are equal, dependent on the speed and latency offered in the reception area. Scaling the speed and latency of a 5G mm-Wave connection could essentially lead to scenarios where a 1 GB data bundle would have differing prices, categorized into several options, from the highest possible data speed down to for example 1–2 Mbps. The service configuration could exist on the 5G handset of the user (within the SIM card) and give the service provider the ability to change these configurations based on the purchase of the user. A form of sustainable financing could result from such new techniques, since infrastructure capital investment will also differ if consumers are provided with lower QoS at lower cost. Subsidies could assist initial investment, which will be lower in, for example, rural areas where internet demand is relatively basic.

6.2.5 *Reliability*

The fourth and final component identified by the WHO is ensuring reliability, specifically in the context of health and supply systems, but with suggestions that could be adapted by the telecommunications industry. The main argument brought forward by Everard [6] in terms of reliable health care is to improve existing essential medicine resource schemes that are vital to health care expansion in emerging markets. The primary challenge is again related to the overall cost of health care resulting from political agendas of governments owing to the increasing costs of products and services from the key pharmaceutical industries. Governments, especially in emerging markets, have been forced to investigate new methods of distributing health care services to low-income families, since their subsidy contributions are diminishing with low economic growth and other issues such as corruption, war and famine [11]. Mechanisms to mitigate these high costs that have been put forward by some developed countries include

- "*national drug formularies,*
- *non-reimbursable drug lists,*
- *restricted reimbursable schemes,*
- *national price regulation,*
- *promotion of generic drugs, and*
- *monitoring of prescribing costs*".

Governments in some developed countries have been encouraged to manage the priorities of health services and ensure universal access to various national health packages. These governments aim to assign a dedicated sum of assets to fulfill the requirements of low-income families that require interventions by these governments to offer cost-effective health services. Again, emerging markets often do not have the resources to introduce such techniques, but the WHO encourages developed countries to assist. Again, to boost economic growth in emerging markets through distributing 5G and mm-Wave-based internet services to all citizens, intervention and adaptable pricing are required, with a fair subsidy from government and from developed countries, which can also benefit from skills development and investment. A similar argument to that supporting sustainability can be offered, i.e. providing scaled internet services to low-income families, which would lessen the burden on governments and reduce the overall budget to realize an internet-for-all approach. This would involve a relatively similar approach as listed above, where national formularies could be implemented, restricted service offerings made available, national price regulation incorporated to encourage competition, and the pricing of service being monitored as an active assignment by these governments.

Returning to the second component identified by the WHO to distribute essential medicines to emerging markets, techniques listed by Everard [6] to ensure affordable prices are briefly reviewed in a subsection, since these techniques are uniquely adaptable to other sectors. A more detailed investigation of some of these suggestions, and newly introduced ones, are also discussed in Everard [6] and briefly summarized in this section. These suggestions include:

- The significance of competition. Essentially, almost any sector can be stimulated through competition. As 5G and mm-Wave networks offer large spectrum and tailored services, and service providers do not need existing infrastructure, competition can be encouraged to offer consumers options and alternatives when choosing a cost-effective service provider. New service providers will have the ability to purchase spectrum that only satisfies their requirements based on the current demand on their networks and will be allowed to purchase more spectrum as their consumer base increases.

- Reliable pricing information. If the pricing information on products and services in the 5G and mm-Wave spectrum is transparent and reliable, consumers will be able to compare the offers from various service providers and choose the most cost-effective offers. Reliable pricing information will also stimulate competition among service providers and essentially drive down the cost of internet access. In the information age, consumers have become more price-sensitive and can demand transparent breakdowns on the cost of products they purchase and services to which they subscribe.
- Differential pricing. As mentioned in this section, 5G and mm-Wave networks can be scaled up or down for each user, which would determine the QoS of the products and services as a function of the price paid by consumers, essentially leading to innovative differential pricing strategies. Smaller service providers will be able to buy less spectrum from communications authorities at a reduced cost and only service their current client base, with the option to purchase additional spectrum as the number of subscribers increases.
- Large-scale competitive procurement. 5G and mm-Wave have the important advantage over previous generations of extremely large amounts of bandwidth available within their spectrum. Service providers could therefore essentially buy spectrum on large scales at discounted rates and use the additional spectrum for service upgrades as the number of subscribers increases, or offer completely different services that complement Industry 4.0 innovations such as the internet-of things (IoT) and wireless sensor networks.
- Price controls. Governmental monitoring of the prices set by service providers will ensure that fair pricing for services is offered to consumers. This intervention could mitigate the opportunities for large corporations to monopolize the industry.
- Lowering or eliminating tariffs, duties and taxes. Governments as well as public and private sectors should be encouraged, at least at first, to lower the monetary incentives from broadband offerings and allow the sector to grow naturally until some form of maturity has been established. At this time, prices could be adjusted based on supply and demand and cost recovery realized.
- Investing in local skills and production. Although encouraging international investors to develop and mature 5G and mm-Wave networks has its advantages, it is crucial to develop local skills to maintain the solutions and therefore invest in socioeconomic development of the local population. Job creation and human capital investment are among the most important activities needed in emerging markets and Industry 4.0 has the ability to develop skills through technologies that offer ubiquitous internet connections in urban, rural, and suburban areas.

Relating the strategies and theories presented in Everard [6] to cost effective access to essential medicines in emerging markets has revealed several comparisons that can be adopted in the telecommunications industry. Cost-effective and ubiquitous internet access is a crucial factor for emerging countries to partake in Industry 4.0 and learning from other sectors (and markets) to obtain this can be beneficial.

Techniques to adapt costing in unequal markets are therefore shared between sectors and industries and potential solutions should not be looked at in isolation, but shared. The health sector, specifically, is an interesting sector to analyze because of its high priority in government and its global relevance.

In the following section, a report by the World Energy Council (WEC) that was published in 2001 is analyzed and the pricing structures of energy in emerging markets are studied. Although this report is relatively old (published in 2001), the reported statistics will not be quoted directly in this analysis; the focus is placed on the general policies and strategies that were presented. Furthermore, the more recent energy milieu according to WEC [22] is also considered if any significant changes occurred in the time between the reports. These policies and strategies will be specifically analyzed in terms of their potential similarities and application to the telecommunications sector, as is done in this section referring to access to essential medicines in emerging markets. The policies that could be adapted in the telecommunications industry are outlined in the following section.

6.3 Pricing of Energy in Emerging Markets

In the report published by the WEC [21], the premise of the research is drawn from three identified energy goals, developed in 2000 by the WEC millennium statement (Energy for Tomorrow's World—Acting Now), namely

- accessibility,
- availability, and
- acceptability.

These three goals can easily be adapted for various other sectors and industries; however, in terms of broadband internet distribution in emerging markets, the goals are remarkably similar. These three goals have been identified as three pillars of energy pricing policies by the WEC [21] and developed to be adapted for developed countries as well.

6.3.1 Accessibility

According to the WEC [21], *accessibility* pertains specifically to the endowment of dependable and inexpensive up-to-date (energy) services to all families in a specific market, for which an imbursement is prepared. In our comparison with broadband internet, the important factors to relate are the reliability and affordability of the given service (or product). To achieve the goal of accessibility, policies that are modernized and scrutinized by all stakeholders should be adapted to meet the needs of specifically low-income families, according to the WEC [21]. Such policies must

- speed up commercial development,
- encourage trading with neighboring countries, and
- promote further unbiased revenue dispersal.

The WEC [21] acknowledges that to succeed at accessibility, all costs that relate to the service (in this case energy) must be reflected in potential policies (for example emission costs and waste management for energy generators); however, it has been found that the associated costs typically increase the cost of the *basic service* above what the local population in emerging markets can afford. Critical costs that should be included in the price or the service are

- variability in the market,
- maintenance, and
- extension costs,

since these are crucial for long-term sustainability. Parts of these costs could, however, be subsidized by local governments (at least for a period) to offset the price paid by consumers to a minimum and boost economic growth. A deeper investigation on such policies is given by the WEC [21] and adapted and reported later in this section. The importance of outlining market variability, maintenance and extension cost is, however, also applicable to the broadband internet sector, since these factors influence the cost of the basic service (internet connectivity) as well.

6.3.2 Availability

To define the WEC's second goal, *availability*, the WEC [21] states that this goal includes both the quality and reliability of the service (in the specific case of energy); this is again an important factor in broadband internet distribution. Although the continuity of energy supply is arguably more essential when compared to that of broadband internet connectivity (considering for example the importance of reliable energy in hospitals), a high level of availability and reliability is also required to stimulate economic growth in emerging markets. The WEC [21] argues that to achieve high levels of availability of a service in an emerging market, good practice is to found guidelines and procedures that permit service providers to recuperate their reserves in services that excerpt, manufacture and bring services to the consumer. These policies should also incentivize maintenance and service expansion and allow reasonable return on investments. If such policies in the energy sector can be adapted for broadband internet access, a large number of smaller service providers would be encouraged to invest in 5G and mm-Wave networks, knowing that the policies are designed to ensure good return on investment with sufficient government backing over the long term. The process will therefore drive competition, lower prices for consumers and expand services to rural areas and low-income families.

6.3.3 Acceptability

In terms of *acceptability*, the WEC [21] specifically addresses

- environmental goals, and
- public attitudes.

In the energy sector, pollution from traditional coal-based energy production can harm large populations, especially in emerging markets where regulations might not have been put in place to safeguard communities from harmful particulate matter and toxins. Therefore, services such as energy generation must be rendered while maintaining a low environmental impact and be driven by clean and energy-efficient solutions. In terms of broadband internet infrastructure, acceptability can also serve to realize solutions that have as little environmental impact as possible through energy efficiency and ideally renewable energy. As reviewed in Chap. 2 of this book, 5G and mm-Wave networking, specifically through small-cell deployments, offer new levels of energy efficiency that can be exploited in urban, suburban, and rural areas. Policies that encourage renewable energy for base stations (BSs), for example, should be prioritized, not only to protect the environment, but also to ensure reliable connectivity in areas where energy supply is unpredictable.

In working towards these three goals set out by the WEC (accessibility, availability, and acceptability), the WEC [21] presents a more in-depth review on the costing strategies that are typically implemented in emerging markets. These strategies do not necessarily follow the idealized approach set forth by the WEC [21], but show the advantages and shortcomings of energy costing in unequal markets.

6.3.4 Energy Costing Strategies in Emerging Markets

The WEC [21] analyzes the costing strategies of energy in emerging markets based on two broad outlines,

- the policies that are in place and how these align with what the population needs (with respect to accessibility, affordability, and acceptability), and
- the cost of the service delivered to consumers.

Firstly, in terms of the policies, these include economic, social, and service (energy in the case of the WEC [21]) policies that a local government adopts during its planning and decision-making phases. The government therefore plays a large role in implementing these policies; although typically developed by some local regulator, the government essentially has the last say in the process. From the *economic* perspective, a government can implement broader policies, for example in the energy sector, to

- encourage,
- subsidize, and
- incentivize

public or private energy consumption from renewable sources. 5G and mm-Wave internet infrastructure, for example, requires relatively little existing infrastructure of previous generation networks, which effectively levels the playing field for SMEs to enter the market and assist in spearheading Industry 4.0 towards ubiquitous broadband internet in emerging markets (especially in rural areas). Inevitably, these policies have a major *social* aspect that is driven by the economic impact. Policies that enable internet connectivity for an entire local population, focusing on rural and poor areas, equip the population to participate in economic growth. There are, however, numerous challenges that make implementing such policies difficult in emerging markets (financial constraints, unskilled workers, corruption, political misalignment, civil unrest), as reviewed in Chap. 5 of this book, as well as in Lambrechts and Sinha [11], but essentially these challenges make it difficult to recover the cost of investments fully.

Secondly, the total/overall cost of the service plays an important role, since this directly determines what the consumers will pay, regardless of policies to lower prices through external strategies such as lowering profit margins (through for example subsidized government intervention). Another analogy that can be drawn between energy and the broadband internet distribution sectors, set out by the WEC [21], is that the cost of electricity supply is divided into four primary categories (or *phases*), these being the

- generation (generating energy at the power plant),
- transmission (stepping up the voltage using a transformer for transmission across transmission lines over long distances),
- distribution (stepping down the voltage at the substations for distribution through distribution lines that transport power to buildings), and
- supply (street poles have transformers that phase down the electrical voltage used by the community).

It is possible to relate these four categories to the generation and supply of broadband internet services to populations in areas where there are currently no such solutions (the unserved markets). Essentially, the cost of a service, whether energy supply or broadband internet, should be minimized through each of these *phases* of delivering the service to the consumer. Important to consider, as described by the WEC [21], specifically in the energy generation industry, generation costs include

- "*fixed capital cost of the generating plant, and*
- *all variable costs such as operations, maintenance, and fuel*".

Transmission of energy supply according to the WEC [21] is predominantly a static fee, whereas delivery is a combination of static and adjustable fees. Broadband internet generation is also dependent on transmission between the first mile (where the internet enters a country), through the internet backbone and finally distributed in the last mile and supplied to the consumers. Any remaining costs to provide the service are captured under supply costs and include overheads to maintain service delivery (salaries, metering, customer service, etc.). The supply costs are relatively static per user; however, the supply fee per user per unit (energy for example) can vary significantly, for example between users in urban and rural areas. According to the WEC [21], supply costs (albeit highly variable) make up a relatively small percentage of the total cost for a service such as energy or internet distribution, and these variations can be absorbed within the industry. Of the four listed components, distribution has been identified by the WEC [21] as the largest varying factor in service delivery. In terms of energy supply, distribution costs vary with the

- type of customer (business, residential),
- quality of the service,
- intensity of consumption, and
- distance from the grid (transmission network).

In supplying broadband internet connectivity in the last mile, similar variations may occur, where large corporations require high-bandwidth and ultra-reliable internet connections; however, a single entry point into the corporation mitigates some of the infrastructure challenges to provide such a service (it is distributed to one building and serves multiple consumers). For rural areas, infrastructure cost is typically high and the return on investment low, with fewer consumers also using fewer services. Similar to energy distribution to rural areas and in many suburban parts in emerging markets, service providers are hesitant to extend their infrastructure to these areas. Distribution is required to numerous entry points with typically a low number of consumers at each entry point. It is therefore again useful to refer to the WEC [21], analyze the potential solutions in costing energy in unequal markets and relate these techniques to the broadband internet sector.

The WEC [21] investigates four approaches to pricing energy in emerging markets in order to expand specifically transmission and distribution networks to low-income areas. These four approaches include:

- "*cost-of-service ratemaking (price setting),*
- *marginal cost pricing,*
- *opportunity cost pricing, and*
- *market pricing*".

A brief analysis of each of these pricing strategies is presented in this section and related to the broadband internet sector. The WEC [21] additionally reviews the issues and challenges in the various pricing strategies and these are also cited this section. The first approach to pricing reviewed by the WEC [21] is cost regaining pricing/cost-of-service ratemaking.

6.3.5 Cost-of-Service Ratemaking

Among one of the oldest practices, or tools, of ratemaking, cost-of-service ratemaking can be employed to price services dynamically. The technique and regulation in general are driven by three issues that typically emerge in any marketplace, namely

- natural monopolies,
- undue price discrimination, and
- destructive competition.

In simple terms, cost-of-service ratemaking involves determining rates with reference to a revenue requirement. In turn, this requirement is determined by summing two aggregated components of a firm or industry's costs, namely

- the expenses incurred to provide the service and
- the return on capital investments.

The profit on asset investment is calculated by applying an allowed rate of return to the rate base, the net book value of the plant venture. Its value is determined by subtracting accumulated depreciation from the original cost of the investment. The cost-of-service ratemaking principle has been adopted to permit a supplier to recuperate all operational expenditures as well as devaluation and receive a static rate of profit on the asset in the rate base [21]. This technique is frequently implemented where there is little or no likelihood of introducing economical market pricing and where the assets are big related to the supplementary annual monies. It therefore lends itself well to service deliveries such as energy utilities and broadband infrastructure (which inevitably also lends itself well to forming natural monopolies). To determine a fair return on investment, an amount of aspects need to be deliberated, such as

- "*price stability,*
- *price predictability,*
- *levels of risk,*
- *the demand to attract capital,*
- *income tax obligations, and*
- *social policies*".

Cost-of-service ratemaking also has many benefits, since as soon as the rate of return is established, the price of manufacture can easily be deliberated. To its detriment, the supplier is not rewarded for efficient capital investment management, primarily because of the Averch-Johnson effect (the return is greater in outright facets the broader the principal venture is). To address the dilemma of natural monopolies developing and low rewards for the suppliers, a move towards performance-based ratemaking, as opposed to focusing purely on cost, has received attention over the past few decades [21]. An incentive is therefore created for the controlled bodies to increase efficacy and performance if the reserves from such enhancements are mutual for the consumers.

Another pricing approach proposed by the WEC [21] to distribute energy fairly in emerging markets is that of *marginal cost pricing*. This technique is analyzed and related to the broadband internet servicing market in the following section.

6.3.6 Marginal Cost Pricing

The marginal cost pricing principle entails "*setting the price of a product at (or slightly above) the variable cost to produce it and is typically related to short-term price-setting situations*" [3]. The technique is either applied if a firm is financially healthy and aims to maximize profitability by selling 'a few more' units, or in desperation if the firm is unable to sell units otherwise (at the price it initially aimed for). In either event, marginal cost pricing is a short-term variable pricing strategy and can be implemented for numerous reasons, depending on the product or service, market size, market penetration and long-term goals. Marginal cost pricing typically relates to price variations of physical materials as opposed to labor. Economically, the ideal distribution of resources is achieved if the marginal price is equivalent to the marginal cost [21]. Short-term marginal cost is aimed at covering all variable costs of producing a product or delivering a service, where in the case of energy supply, these costs would include

- fuel,
- labor (required number of man-hours), and
- maintenance.

Theoretically, if the marginal cost strategy is implemented, the prices should congregate to a point where the short-term fringe income and the costs are equal and competitive market pricing over the short term is optimal. Marginal cost pricing, however fails to consider the cost of principal investments, as it is presumed to be constant over the long term. If a service expands over the longer term, as is common in services such as energy supply and broadband internet services, the marginal costs could contain the capital price of changing or upgrading older equipment and the initial capital investment is not recovered. To its advantage, the marginal cost pricing strategy can give policymakers a benchmark of current prices and estimated growth even though the current prices are not perfectly matched to the supply and demand of the product or service. It also allows the service provider to allocate costs among different customer categories and allow setting up of the structure of its service offerings accordingly. Bragg [3] expands on the benefits and weaknesses of marginal cost pricing in general. The listed benefits include:

- Customers that are sensitive to pricing would be more prepared to purchase an item or a service from a business if marginal cost pricing is implemented. The business can therefore receive additional augmented profits from these customers.

- Marginal cost pricing is ideal for a company wishing to enter a market and willing to forego large profits; though it is possible to attract *price-sensitive* clients during this stage, which might be detrimental over the longer term.

Bragg [3] listed the following weaknesses of marginal cost pricing:

- Over a extended term, this valuing structure is not feasible, since a company would not be able to cover its initial and ongoing fixed expenses.
- Pricing significantly below the average market price inevitably leads to lower profit margins and a company would forego profits that could have been generated through market-related pricing.
- Price-sensitive customers that initially signed up during the marginal cost phase are likely to abandon the service as the price matures towards market-related margins.
- It is more difficult for companies that routinely implement marginal cost pricing to mature and evolve into a higher-service and higher-quality marketplace.

Apart from companies that wish to enter the market, for example new internet service providers that have the capability to distribute 5G and mm-Wave services on a relatively small scale, marginal cost pricing is also preferred to receive superfluous revenues by exhausting surplus manufacture capability. These two scenarios are far removed from each other, but marginal cost pricing offers advantages for both. In 2006, Biggs and Kelly analyzed the marginal cost to operators of greater number of users and capacity and found it to be zero.

The third pricing approach proposed by the WEC [21] to distribute energy fairly in emerging markets is opportunity cost pricing. This technique is analyzed and related to the broadband internet servicing market in the following section.

6.3.7 Opportunity Cost Pricing

In the context of the energy distribution as analyzed by the WEC [21], "*opportunity cost pricing is based on the value that the energy would have if it could be offered and purchased outside the country, rather than consumed within it*". Following such a strategy allows the setting of a standard on which policymakers can depend. It is therefore a broad measure that the pricing in a country is in line with at least that in neighboring countries. It is also a technique to determine if broadband internet prices in emerging markets, for example, are in line with those in developed countries, notwithstanding additional costs of infrastructure, transmission and distribution that may be required in emerging markets and not in developed countries. In a more general sense, opportunity cost pricing is essentially used when there is a choice or trade-off to consider when pricing a product or a service. An analysis should be made of the potential gains and losses when implementing opportunity pricing. There are two broad categories of opportunity cost pricing, namely

- *"explicit opportunity cost and*
- *implicit opportunity cost"*.

Explicit costs are typically out-of-pocket costs invested in a firm or industry; these costs include the payment of wages and salaries, building or equipment rent, and materials (consumables). Implicit costs are typically the opportunity cost of resources that are already owned by the firm and actively used in the business. Examples of implicit costs are expanding of infrastructure onto land or property that is already owned by the business. Implicit costs may also not be direct costs, and rather a lost opportunity to generate income through resources owned.

In terms of broadband internet products and services in emerging markets, prices could initially be set to market-related values of neighboring countries to recover the costs of infrastructure development. For more mature technologies such as 3G and 4G, the pricing should be a relatively accurate representation of the actual cost of providing the products and services. However, for newer technologies such as 5G and mm-Wave networks, opportunity cost pricing is more difficult to implement because there are fewer countries for comparison and numerous hidden costs for a technology that is not yet mature. As highlighted in Oughton and Frias [17], much work is needed in cost modelling of 5G networks to compare the difference between the data traffic demand and networks costs for alternative deployment scenarios. Oughton and Frias [17] state that cost modelling depends on

- required throughput density; this factor will depend on numerous scenarios, whether the network is deployed in a developed country, emerging market, urban, suburban, or rural area, the income levels of the local population, user demand, types of services used in an area, geographical limitations or weather patterns, among many other factors,
- periodic interest rates in the local economy, and
- the price of the BSs, whether they are locally manufactured or imported, and whether the local populations are skilled to maintain these BSs, which will also add variations to this factor.

From only these three factors as outlined in Oughton and Frias [17], it can be seen that opportunity cost pricing will be more complex to implement, especially in emerging markets that aim to spearhead their participation in Industry 4.0, essentially forcing them to pioneer the pricing strategies of 5G and mm-Wave networking on a per-area basis.

A somewhat related pricing strategy is that of pure market pricing. In emerging markets, there could be factors that prevent market pricing being implemented, as reviewed in the following section.

6.3.8 Market Pricing

Market pricing requires a form of regulation from both national and international bodies to consider explicit market imperfections, together with those that could appear amid the wholesale and trade segments. In some countries, intervention could be beneficial since issues such as

- externalities,
- obstructions to admittance into the country,
- oligopolistic marketplaces,
- non-competitive behavior, and
- corruption

prevent a country from offering products and services at market-related prices. More importantly, as also pointed out by the WEC [21], markets do not all the time offer reasonably priced admittance to products and services such as energy and broadband internet connectivity for their poorest people. Higher levels of regulation and interventions and the role of taxation and subsidies become important considerations to achieve market pricing that is dynamic and fair to boost socioeconomic growth for an entire population.

Subsidies have been used to cross-subsidize numerous sectors in a market to boost economic growth within these sectors, with varying results. A subsidy should, however, be fundable, and there are various ways of achieving this, such as:

- Government-assisted subsidies where transfer payments are made directly to the poor. Through this process, the targeting of the poor is generally well established; however, to improve targeting, an increase in administrative costs will be incurred.
- Funding that is mandated by a local government that requires a utility to sell its product or service at a price that is below its overall costs to manufacture the product or supply the service. The profitability of the service provider will be reduced and could also lead to losses. This technique is not sustainable, however, and unfortunately, according to the WEC [21], this is a common implementation in emerging markets.
- Controlled cross-subsidization as of one customer group to other. Cross-subsidization is an effective means of subsidizing a sector if the need as of the backing client category is not excessively price-elastic and no feasible substitutes exist.
- Finally, progressive tariffs for residential customers where customers that consume more of a service pay a higher fee compared to customers that use less of the service. This sliding scale has been implemented effectively in numerous industries and has the advantage that the variance charges do not have to be extreme, since the sector of customers with high consumption is already large. As discussed in Chap. 2 of this book, 5G network slicing could lead to more progressive tariff structures.

The WEC [21] also lists and describes various issues and challenges in pricing strategies (for energy supply) in emerging markets. These issues and challenges are described in depth by the WEC [21] and briefly listed in this section. According to the WEC [21], the following issues in pricing have been identified.

6.3.9 The Substitution Effect

Distortions can be created through subsidies of a commodity, which are typically as a result of the *substitution effect*. If a product or service is subsidized, the local population need to have an encouragement to purchase additional parts of this product or service and essentially move from any substitutes (alternatives). This would effectively alienate any competing products or services and reduce their market share significantly, and if not subsidized, the profit margin and opportunity to expand would be significantly lower.

However, to its advantage, a well-structured subsidy can also phase out inefficient and obsolete products and services and introduce modern, and sometimes safer, alternatives. In terms of broadband internet connectivity for emerging markets, with the primary goal of boosting socioeconomic growth, subsidizing 5G and mm-Wave networks as well as spectrum will have a detrimental effect on previous generations such as 3G and 4G. Although the incentive to build a 5G and mm-Wave infrastructure exists, current service providers will need to reconsider their long-term vision for the older generations rapidly and push to drive the newer technologies. The transition could be relatively easy; however, smaller 3G and 4G service providers that are still recovering capital investments might not be able to afford such a transition. Careful consideration by policy makers, stakeholders, and governments is therefore required to structure subsidies well enough to avoid an overall negative effect and benefit large corporations and SMEs.

6.3.10 Knock-on Effects

Subsidies can have numerous external effects that might be overlooked at first. The WEC [21] uses an example of subsidizing transport and investigates the increased traffic in terms of knock-on effects. Such knock-on effects will include

- local pollution that will require additional cost to manage and will have an adverse effect on the health of the local population, leading to more health-related expenses,
- increased congestion that will require more investments and expenses in expanding the transportation infrastructure to accommodate higher volumes of traffic,

- an increase in accident rates that will increase the number of health-related expenses and specifically increase the burden on low-income families, and
- a larger impact on climate change; again, with numerous knock-on effects.

More externalities must be accounted for in the example of transport subsidies, and this would essentially mean that these policies must be well-planned and additional financial support must be made available to mitigate an overall negative effect of these factors on the economic wellbeing of the local population. In subsidizing broadband internet, policy makers should therefore investigate any potential knock-on effects for various local industries to ensure that the overall effect of the subsidy does not harm the local economy.

6.3.11 *Larger Effect on the Already Wealthy*

Subsidies might benefit the wealthier population and have limited or no effect on the poor population. According to Ricciardi [20], water subsidies mostly benefit the wealthy, according to a study of 10 countries, namely Ethiopia, Mali, Niger, Nigeria, Uganda, El Salvador, Jamaica, Panama, Bangladesh, and Vietnam. Across all these countries, existing subsidies tended to target networked services; however, the fact is that families that are poor rarely depend on water and sanitary public health access to their homes. These households rely on these services through access from local communal access points. As a result, "*56% of water and sanitation subsidies go to the wealthiest 20% of the population, while only 6% reach the poorest 20%*". Poorer populations therefore benefit very little from these subsidies and their socioeconomic well-being is affected very little. Indirectly, however, the benefits gained by the wealthier population also boost economic growth, which could in turn create opportunities for low-income families, but the goals of these policies and subsidies should be clearly defined.

6.3.12 *Removal of Subsidies*

Subsidies typically have a limited lifetime and are used to boost a market and sector until some form of maturity and competitiveness have been reached. However, removing a subsidy might lead to problems in an industry. These issues could potentially involve social disturbance and even political unrest if dealt with without sufficient planning. Removing a subsidy can only succeed if alternate procedures are effectively realized to overcome the opening amid the price of distributing a product or service and the ability of the local population to pay for these products or services. According to the WEC [21], instances of partial removal of subsidies have led to a rise in non-collection of payable invoiced amounts and additional rises in non-technical damages.

6.3.13 Economies of Scale

Because of economies of scale, cross-subsidies by government in certain markets might only have a significant impact on industrial customers and little to no effect on the general population. Industrial customers typically purchase products and services in high volumes at an already discounted rate. In emerging markets especially, the general population already purchase limited amounts of the product. Cross-subsidies in a specific sector might therefore only have a limited effect on low-income families rather than the economic effects that were initially intended.

In this section, innovative and successful policies and strategies (and warnings against certain techniques) have been presented to price energy competitively in emerging markets and provide access to large populations in unequal markets. Similarities to the broadband internet sector were drawn in this section and it was highlighted where similar strategies could benefit the broadband sector. A similar analogy is proposed in the following section, relating to the pricing of petroleum (fuel) in emerging markets. Petroleum (and related products) is also crucial in ensuring economic growth in any market and its pricing is important to maintain economic stability. In unequal markets, a natural form of discrimination will occur when the poor population cannot afford such products and this will limit its ability to participate in economic development. Pricing strategies are therefore important and in many countries subsidies are needed to provide an entire population with fair-priced petroleum, as reviewed in the following section.

6.4 Pricing of Petroleum in Emerging Markets

In the policy research working paper presented by Kojima [10] published by the World Bank Sustainable Energy Department for Oil, Gas, and Mining, an extensive evaluation on fuel manufactured goods valuing and corresponding guidelines in emerging countries is presented. Again, these policies will be analyzed and reviewed in this section in order to draw some similarities with potential policies for the broadband internet sector in emerging markets.

One of the primary drives to publish the Kojima [10] report emanated from the fact that of the 65 countries reviewed, many showed very little or even reverse progress with reforming pricing of petroleum during the timespan of the study (2009–2012). Many of the problems, challenges, and failing policies that led to this situation outlined in Kojima [10] are outlined and analyzed in this section. From the abstract of Kojima [10], relevant comments on failing policies are made, for example, certain governments in emerging markets have attempted to keep local petroleum prices artificially low through

- controlling the price of the commodity (could be related to the price that consumers pay for broadband internet),
- implementing restrictions on exports and quantities (could be related to the acquisition and distribution of spectrum), or
- putting political pressure on oil companies.

Although some of these efforts have had positive effects on local inflation in the short term, numerous negative effects were also experienced. These included a significant increase in black market activity, smuggling, fuel adulteration, illegal diversion of subsidy funds, financial losses experienced by fuel suppliers, generally deteriorating infrastructure and fuel shortages. As one can imagine, these numerous side-effects have inevitably led to significant damage to the economy rather than growing it. Based on this one example, there should be enough warning signs for any sector, including the broadband internet sector, not to follow a similar path and rather to focus on the policies that introduced positive change in economic performance in these countries. Furthermore, fuel, like energy, medicine, and broadband internet, is essential to growing an economy. If these services disappear from an economy, the effects are typically devastating and have knock-on effects on other sectors in a market, leading to a potential total shutdown in economic activity.

According to Kojima [10], many developing countries control the price of petroleum products. As subsidies on these products increased, an amount of administrations looked at different possibilities and strategies of pricing transformation, especially in the time that lead up to the economic crisis in 2008. In the period after 2008, the prices of petroleum started rising again and these soaring prices forced governments to take action. Many governments attempted to minimize the effect on domestic markets by providing security networks to the underprivileged, a higher lowest wage, emancipating oil tactical services, decreasing taxes, and heavily subsidizing petroleum product prices. Similar to food prices also rising during this time, an expenditure that typically makes up between 20 and 30% of low-income household spending, fuel price reforms were politically challenging. Some emerging markets started protesting against high food and fuel prices and political pressure in these countries also started rising to cumbersome levels [10]. Consequently, prices of petroleum products varied significantly from (developing) country to country, whereas globally the price was broadly uniform. Of the 65 emerging markets studied in Kojima [10], by January 2012 there was a two orders of magnitude difference in fuel prices between all these countries. Predictably, the lowest prices were seen in countries that actively exported fuel to neighboring countries. A complete shutdown and disregard of pricing policies linked international prices towards a global median and some governments had to respond by freezing the prices of these products. Kojima [10] lists several interlinked reasons that affected the costs, availability and consumer-paid prices for petroleum products, which included:

- The deprived economic state of domestic oil corporations in certain emerging markets was exacerbated by price subsidies and led to their incapacity to acquire fuel products duly, which inevitably led to scarcities and excessive black-market costs.

- The fuel price subsidizations in conjunction with high global costs enlarged encouragements to divert to black-market trade that further pushed the domestic market prices far beyond the global median.
- Many emerging markets experience frequent power outages and demanded diesel to generate energy; however, diesel shortages also started emerging, which increased the price of this commodity locally. Shipping delays were experienced owing to piracy in the Gulf of Aden and the Indian Ocean, which led to increases in insurance cost and a further downward spiral in country finance.

As reviewed in this section thus far, it is evident that strict and innovative policies by local governments were needed, especially in emerging markets, as a crisis in the petroleum sector started getting out of hand. These policies were needed to slow down the increase in petroleum product prices and restore the balance in domestic markets towards a price that was globally comparable. Kojima [10] studied these policies and some of the policies are reviewed in this section to determine their feasibility to adapt to telecommunications markets. In recent years, internet connectivity has become an extremely important commodity to conduct business and grow economies; policies are therefore needed to control its price and distribute it ubiquitously. Since the work published in Kojima [10], absolute pricing information has been considered dated and is not directly referred to in this section, but the overall policies and strategies that were put in place to achieve a balance in petroleum product prices are investigated. Great emphasis in the telecommunications industry should be placed on the pricing of spectrum, since this is a section of internet distribution that is especially prone to political interference and unlawful activity, depending on the interests and behavior of a local government.

An important summary presented in Kojima [10] refers to the considerations for petroleum price control and other associated types of price control. Governments can choose where along the supply chain intervention into the price for petroleum products should occur; these stages include

- ex-refinery,
- landed cost,
- wholesale,
- ex-depot, or
- retail.

In the broadband internet sector, the service to the consumer is also part of a larger orchestration of products and services, technologies, spectrum allocation, and government intervention. Price intervention should therefore also not only apply to a single stage, but be considered and analyzed for all stages along the value chain. Furthermore, several mechanisms of price control along the supply chain can be implemented. These mechanisms, as outlined in Kojima [10] include

- "*price ceilings,*
- *price levels,*
- *control at retail,*

- *control at wholesale (or elsewhere upstream of retail),*
- *uniform prices, or*
- *pricing by location*".

Briefly, *price ceilings* are typically used to encourage competition in a struggling market sector. If there is evidence that a provider is selling a product or service below the price ceiling, it gives an indication of competition emerging. A price ceiling should, however, be strategically set. A too low price ceiling will make it difficult to market any product, as it will become impossible to make a profit, or even lead to losses. A too high price ceiling will lead to consumers not being able to purchase a product or service and the providers will suffer from low sales. To its detriment, a price ceiling by definition eliminates competition and if no competition emerges over a long period, it will have little or no effect on economic growth. In terms of *controlling retail prices*, however, in some extreme cases the retail price could be lower than the wholesale price of the product (or service). Kojima [10] found that a large percentage of the emerging markets studied set uniform prices or price ceilings for at least one petroleum product and applied additional considerations to provide a sense of equity. However, even though these strategies can be beneficial over a short term, they inevitably lead to consumers paying higher prices for these commodities, as the suppliers use (sometimes corrupt and illegal) strategies to maximize profits in such an environment. Importantly, as also suggested by Kojima [10], if prices are controlled, it is crucial for local governments to establish and implement criteria for adjusting prices, since the free market and the principle of competition are eliminated. In terms of the petroleum products studied in the 65 emerging markets in Kojima [10], it was found that there are various price adjustment strategies that can be implemented when controlling prices that are relatively sustainable. These mechanisms are reviewed in detail in Kojima [10] and briefly listed below:

- Implement incremental increase of price over systematic phased interims in anticipation of cost-recovery levels being reached.
- De-regulate charges for upper-grade oils.
- Allocate severely funded fuels and ask a greater price outside the set quota.
- Vary prices based on the category (income) of the consumers.
- Shift subsidies between products (that are petroleum-based) based on current supply and demand.
- Introduce temporary stabilization funds.
- Use rule-based pricing if global costs are low.
- Actively regulate prices grounded on changes in the world price to ensure subsidies do not become too costly to maintain.
- Set the subsidy price fiscally and adjust volume.

Among these mechanisms, Kojima [10] also reported that many of the countries studied implemented some form of world price related strategy, adjusting if the world price changed by a specific percentage, in regular intervals to align with the world price, or only if there was a significant change in the world price. In terms of broadband internet, price adjustments would also only be feasible if the government subsidized a significant portion of its distribution. Broadband subsidies are also typically achieved by two primary approaches, namely

- subsidizing usage, therefore on the demand side, and
- subsidizing investment, therefore on the supply side.

Since governmental subsidies in any sector can have certain advantages, broadband subsidies on the supply or demand side will also be reviewed in this chapter, with reference to the work published by Goolsbee [7]. This review will be presented in the following section concerning the pricing of broadband internet in emerging markets. In Kojima [10], an in-depth analysis is presented on the countries studied on how subsidy mechanisms affected the pricing of fuel over the specific period and the impact it had on consumers and the economy. These results are not directly referred to in this section (because of the relatively old statistics); however, a section in Kojima [10] that is specifically related to efforts in these countries to lower the cost of supply (petroleum) has certain similarities that can be adopted in the broadband market, as it is reviewed here. The first technique reviewed in Kojima [10] is that of hedging.

6.4.1 Hedging

Hedging involves insuring the finances of a company against any negative activity that may occur to reduce the impact of such an event. This technique is typically used by portfolio managers, individual investors, as well as corporations to decrease their contact to threats in the marketplace. Hedging is, however, different from the traditional method of insurance and it implicates exhausting monetary mechanisms or market strategies deliberately to counterbalance the threat of any adversative price activities. Essentially, an investor will hedge one investment by trading another that is used as a security.

The most common type of hedge is called a derivative, which is a financial contract that descends its worth from a principal existent asset, for example a stock. The derivative that is most commonly used is an option, the permission to purchase or retail a stock at a listed value inside a timeframe. Another hedging strategy is diversification, where an assortment of assets is owned that track each other's market price. Buying and selling bonds are a typical example of diversification. Privately owned investment funds and hedge funds use many products to hedge reserves and these are not regulated by the government as considerable as mutual funds, whose proprietors are civic businesses.

In Kojima [10], hedging is used as an example to lower the cost of supply of petroleum products. Numerous examples are listed; in many cases, where countries chose options that provided some upside if oil prices remained high but did not protect them against a price collapse, countries and oil companies still suffered significant

Table 6.1 The effect of hedging on oil prices from two scenarios described in Kojima [10]

Market	Absence of a hedging instrument	Presence of a hedging instrument
A global (and domestic) increase in oil prices	Decrease in revenue	Appreciation in revenue due to the change in price of the hedging instrument
No variation in global (and domestic) oil prices	No observed changes	Decrease in revenue resulting from the cost of the hedging investment

losses. Successful hedging also depends on timing, where some examples in Kojima [10] hedged against gasoline and diesel as prices began to rise, and sizable hedging gains paid for fuel subsidies at a later stage when the fuel supply started dwindling. In Table 6.1, a brief summary of the effects of hedging on oil prices is given, based on two different scenarios in the market.

As shown in Table 6.1, hedging can essentially be defined as a mechanism of insurance. If there is any adverse global activity on the oil price, hedging can protect a supplier from a decrease in revenue. If, however, the oil price does not change, or varies only slightly, the cost of the hedging investment will negatively influence the revenue of the supplier. In terms of broadband internet, hedging could have certain local advantages, and where spectrum is acquired from international sources, its significance might be lower when compared to oil price hedging. Oil prices are highly volatile in response to global political activity, and oil companies have become more accustomed to hedging oil prices owing to significant uncertainty in global markets.

Another mechanism presented by Kojima [10] specifically to offset fuel prices in volatile economies is implementing price discounts for fuel import and bulk procurement.

6.4.2 Import Discounts and Bulk Purchasing

According to Kojima [10], certain governments and national oil companies historically negotiated discounts in purchasing products from oil-producing countries when buying (importing) large quantities at a time. Conventionally, bulk purchases reduce unit costs with the scale of the purchase. However, government-mandated bulk procurement could have mixed results, as reported by Kojima [10]. In some instances, in the emerging markets studied by Kojima [10], questions were raised whether bulk purchases actually led to price savings for suppliers and more importantly, consumers. Since oil prices remained high in these countries, the governments were suspected of having manipulated the prices down the value chain, since third parties were required to import the products and it seemed that governments instructed these third parties to manipulate the oil prices as the products entered the countries. These third parties were also only able to participate through a tendering process, which inevitably raised more concerns on which tenders were approved, with possible government intervention based on willingness to manipulate the prices. In emerging markets, unfortunately, corrupt behavior is not uncommon and therefore such direct intervention by a local government in negotiating prices from international sources could be met with unsolicited behavior. An alternative is for private companies to consider bulk purchasing and importing; however, revenue margins for these companies could also lead to smaller gains lower in the value chain. A more viable and sustainable method to lower the cost of supply, also relevant to broadband internet distribution, is strengthening the local infrastructure. In terms of fuel supply, this would involve strengthening the infrastructure of fuel imports, storage, as well as transport. Along all these stages, cost reductions could be indirectly transferred to the end user and have positive effects on economic growth.

6.4.3 Strengthening the Infrastructure

Specific to the petroleum industry, infrastructure enhancements that could lead to overall lower cost of fuel for end users include

- expanding the port capacity to receive higher quantities (bulk) of products; increasing the unloading speed will have additional positive impacts,
- cheaper transport of fuel into the rest of the country, possibly by using a cheaper transport method, for example pipelines as opposed to road transport, which will lower the overheads of distributing fuel across a country, and
- increasing the storage capacity inland, which will reduce the number of trips needed to supply these areas with sufficient stockpile and again reduce transport costs.

These examples of infrastructure development to lower the cost of fuel supply in emerging markets require large capital investment by government, and depending on the current state of the infrastructure, the associated costs might be too high. However, over the long term, the costs could be recovered and other mechanisms such as subsidies and price setting for specific periods can be implemented to recover costs faster. Again, competition is possibly one of the most natural and sustainable means to lower the price of any product, and Kojima [10] also refers to promoting competition in the petroleum sector to reduce the cost of fuel supplies.

6.4.4 Promoting Price Competition

If rates are not regulated by government, or if a price ceiling is not put in place, local governments can (and should) endorse rate competition by being transparent and freely making available data and statistics. All the statistics on the price of fuel (or any product or service, including broadband internet) should be broken down by a provider, even as low down in the value chain as the supplier to the end user. For example, in South Africa the fuel price has been scrutinized in recent years owing to the substantial rise in the price that consumers pay, considering that the global fuel price is significantly less expensive than the price paid. As a result, the local price of fuel was broken down to show all the contributions to the pump-price. The following summary lists these contributions:

- Basic price: The price that the country pays to import fuel from international refineries up to the local port(s).
- Fuel levy: A general tax that is placed on fuel, mandated by the local government and adjusted annually.
- Wholesale margin: A state-regulated and adjusted margin based on external factors such as the efficiency of overseas refineries.
- Demand side administration charge: A charge announced to curtail the expenditure of some petrol variants in inland markets.
- Retail margin: Another state-regulated and adjusted margin based on the operational cost of filling stations.
- Customs and excise: A price that is divided among the local customs unions, a value that has been unchanged for almost a decade.
- Petroleum products charge: A fee taken by producers and traders of fuel merchandises from the controlled fuel costs to be retailed in the country.
- Zone differential in some provinces: The country is separated into zones centered around the distances from the port(s) and an additional levy is incurred based on these zones.
- Storage, handling, and distribution: This includes all associated costs of secondary storage after fuel has been unloaded from the port(s).
- Road accident fund: A compensation fund for third-party accident victims, an amount that is increased annually.
- Fuel pump rounding: A small amount that accounts for price rounding at fuel pumps.

As seen from the number of factors that make up the fuel price in South Africa, the clear, detailed, and transparent information to consumers gives a good indication of how to determine where savings in the local price can occur and what needs to be implemented, for example infrastructure strengthening, to lower the overall cost. This also gives consumers a sense of what they are paying for and if there are any circumstances or improvements through which the local community can endeavor to decrease the price. It is also useful to determine the percentage impact of each factor, as shown in Fig. 6.1, based on the fuel price in South Africa in June 2018.

As shown in Fig. 6.1, the fuel price paid by consumers in South Africa, as of June 2018, consists of numerous factors as listed in this section. Figure 6.1 presents each factor as a percentage of the overall price and such a breakdown gives government, suppliers and consumers an informative view of the taxes, levies, and other costs involved in fuel distribution. A similar strategy could be followed for broadband internet, for example where factors such as profit margins, taxes, and spectrum prices could be outlined and savings implemented where inflated prices are observed. This information could also be used when implementing techniques such as price setting

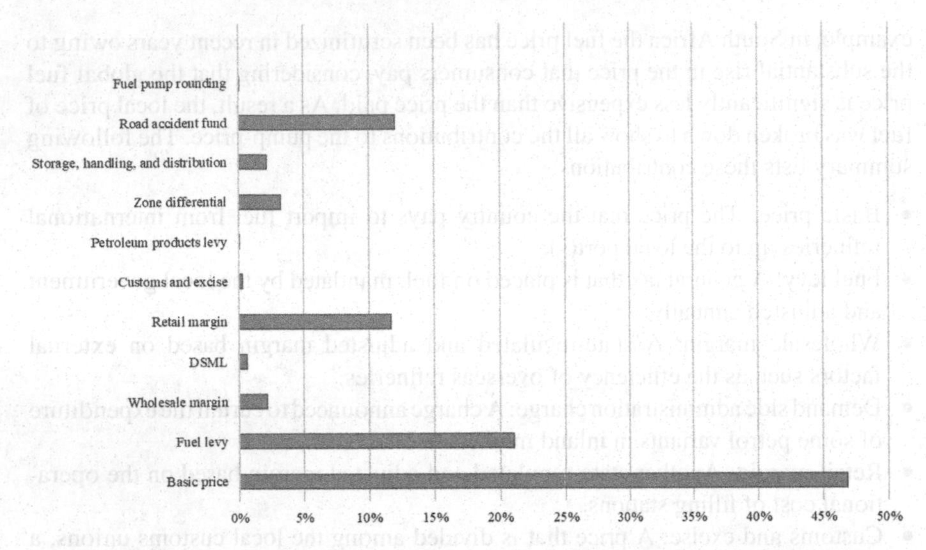

Fig. 6.1 The fuel price in South Africa in June 2018 broken down by all factors that have an influence on the price paid by the consumers. Each factor is represented as a percentage of the price paid by consumers

or price ceilings, and its implementation on only specific factors could be applied as opposed to on the total price. If used for price competition to benefit consumers, effective monitoring and enforcement of standards and policies are important to protect the commodity from exposure to unlawful and corrupt practices that raise prices of certain factors to benefit government or third-party distributors.

Finally, as reported in Kojima [10], numerous negative consequences arose from government control of the downstream petroleum sector and its direct interactions with policies. These consequences were seen in a wide range of countries and show that government intervention, especially in emerging markets, does not always lead to positive stimulation of an economy. These consequences include:

- A shortage in the supply of certain products or services, although only petroleum products are considered in Kojima [10], led to riots, injuries, and death.
- National companies that own a monopoly in a specific sector and in segments downstream can suffer long periods of chronic inefficiency and then have to absorb government subsidies to decrease prices at consumer level.
- Government intervention in the form of state-owned enterprises that are uneconomic can push up domestic prices, increase subsidies and eventually lead to shortages or an additional burden on tax-paying residents.
- Governments often subsidize single entities or sectors in order to exercise greater control over these transactions. As a result, competition is stifled and prices increase, hampering rather than stimulating economic growth.
- Efforts to keep domestic prices artificially low can decapitalize downstream sectors.

- Controlling prices can interfere with improvements in efficiency; based on a cost-plus formula, if incentive is removed, quality and efficiency will inevitably suffer, with detrimental consequences over the long term.

The research published by Kojima [10] therefore indicates that policies and interventions in emerging markets to control the prices of petroleum products, especially during times where global shortages and volatility are experienced, can have mixed results. These results are crucial to consider in other markets, such as in the broadband internet sector, to learn from policies that succeeded and from ones that failed. Implementing subsidies on broadband internet, for example, should not be done to benefit certain entities, but to stimulate economic growth primarily through providing products and services to consumers at cost-effective price points.

In this section, a review is presented of the pricing strategies for energy distribution as researched by the WEC [21] and Kojima [10]. These techniques are somewhat general and can be applied to products and services that require high capital investment and efficient and high-capacity long-term sustainability, such as broadband internet infrastructure and distribution. The energy sector and the broadband internet sector have some similarities identified in this section. It argues that similar approaches could be followed to developing infrastructure for 5G and mm-Wave networks in emerging markets. In the following section, specific works are cited that look at pricing strategies of water in both emerging markets and developed countries. Water is one of the most crucial commodities, more so than medicine, energy, and fuel, and policies and strategies to provide an entire population with water can be analyzed and compared to those regulating other commodities.

6.5 Pricing of Water in Emerging Markets

Clean potable water and sanitation have been acknowledged as human rights by the United Nations (UN) in 2010, and improving access to these services form part of the UN Sustainable Development Goals (SDGs). The water supply and sanitation (WSS) sector is still greatly funded globally, and has been for many decades, according to Andres et al. [1]. The WSS sector is therefore a good sector to review in terms of providing large populations in emerging markets with a critical service, since it has been argued that broadband internet should be a basic human right. Andres et al. [1] maintains that notwithstanding the commonness of grants and the crucial part that active appraising has in suppliers' abilities to provide high-quality amenities; little attention is given to how existing WSS valuing structures and grants actually obstruct advancement on the SDGs. The report therefore explores methods to use scarce public resources in the most effective ways to deliver WSS services universally. In this chapter, reference will be made to existing policies outlined in Andres et al. [1], as well as in an earlier report by Le Blanc [13], and the aim will be to draw a comparison between how these policies affected the WSS sector and which policies would fit the broadband internet sector.

According to Andres et al. [1], there are two primary reasons for a government to subsidize WSS, namely

- the advancement of equitable access to affordable WSS services, therefore helping poor or marginalized families and facilitating either access or consumption, and
- creating a positive environment and harnessing societal benefits through improving the general health and health care for poor populations.

Furthermore, Andres et al. [1] structure the reported work in terms of a few key statements, namely, whether WSS subsidies

- aim to expand access, and
- ensure that a minimal level of consumption is achieved.

Andres et al. [1] propose and analyze two funding mechanisms, namely

- demand-side, which involves straight assignment from the endowment provider to the funded consumer, or
- supply-side, where monies are channeled through a service supplier or other parties, ensuring cost savings to consumers.

Finally, Andres et al. [1] summarize three key messages that led to the study on WSS subsidies:

- Globally, and especially in emerging markets, WSS subsidies have failed to achieve their objectives for various reasons. According to Andres et al. [1], the reasons include poor design in the first place and subsidies that tend to be vague, pervasive or expensive (therefore some entity is gaining from the subsidy), are not targeted efficiently at the parts of the population that need them most, or are nontransparent.
- Design flaws and ineffective subsidy policies can be rectified using the abundance of new knowledge and new technology, including the internet and free access to information.
- A subsidy reform package is needed that includes complementary policy measures, a reassuring governmental alliance, effective and efficient supply service strategies, and a withdrawal policy if required.

Andres et al. [1] provide a detailed review on each of these three key points, specific to the WSS sector. These detailed reviews are not repeated in this chapter. The reviews specifically support the statements (with data) of WSS subsidy policies being pervasive, expensive, poorly targeted, and nontransparent. In a sense, broadband internet distribution to poor countries and low-income families also falls prey to these factors in one way or another. In terms of mobile broadband, allocation of spectrum is one of the primary factors over which governments and third-party entities aim to keep control, since this service/product can be a lucrative investment for its owners. Such details are studied in the next section on the pricing of broadband in emerging markets, and in this section the potential mitigation of these (negative) factors is reviewed. Andres et al. [1] generalize on how to improve the effectiveness and efficiencies of grants through cautious deliberation on five inquiries:

1. *"What is the context of the subsidy?*
2. *What are the policy objectives that the subsidy seeks to achieve?*
3. *Which are the targeted services and/or population(s)?*
4. *How will the subsidy be funded?*
5. *What subsidy design will be most effective and efficient?"*

These guidelines to improve on policies and subsidies are well defended in Andres et al. [1] and could be adapted to the broadband internet sector, specifically mobile broadband for low-income families in rural areas, and emerging markets in general. In terms of the context and objectives of the subsidy, the first and second questions raised by Andres et al. [1], policy makers should initially review the effectiveness and efficiency of current subsidies, before changing them. Specifically, policy makers should comprehend

- *"the magnitude of public resources being expended,*
- *the public perception of the current subsidies,*
- *opportunities for misappropriation, and*
- *the negative effects on sector performance and resource allocation".*

From these listed items, it is clear that an extensive and far-reaching study on current and proposed subsidy policies is required to comprehend the short- and long-term effects they could have. Specifically regarding the negative effects on sector performance and resource allocation, it was seen that in the petroleum sector, as reviewed in this chapter, fuel shortages emanated from certain policies and/or subsidies. The effects down the value chain were not understood (or researched) in those cases and therefore the mention of it in Andres et al. [1] is relevant. The context of the improved policies will also depend heavily on the outcomes of the initial study and dictate its design.

The third question posed by Andres et al. [1] concerns the service(s) and/or populace group(s) that ought to be targeted by the strategies and subsidies. These targets should be very specific, since it is a common mistake to develop policies to support entire sectors. This often leads to ineffectiveness and overlooking consequences down the value chain. For example, in terms of the reviewed WSS subsidies, it should be specified whether the subsidies aim to improve access, or to deliver services to the population directly. These very different mechanisms will structure the subsidies. In broadband internet distribution, a similar question should be asked and the targeted services and population should be clearly outlined. 5G and mm-Wave networks are modular in that they can be designed with tailored QoS based on consumer demand. Therefore, identifying the targeted services and population could prove especially valuable in defining policies that include poor families and rural areas in infrastructure development and sustainability.

An extremely important factor to consider, the fourth question asked by Andres et al. [1], is how the subsidies will be funded. Some costing and pricing strategies have been reviewed in Chap. 5 of this book and are only briefly reiterated in this chapter, but essentially the funding mechanisms include

- government-assisted funding through tax,
- philanthropic funds, or
- cross-subsidization

and the chosen mechanism depends on the fiscal strength of the local government, the availability and opportunities of philanthropic funds, and the opportunities in a market that would allow a form of cross-subsidization. Each mechanism carries its own set of risks and advantages, as reviewed in Chap. 5 of this book.

Finally, Andres et al. [1] ask what subsidy design will be most effective, which is a culmination of the first four questions (and answers). If the first four questions and answers are well defined and researched, the most effective subsidy design should be apparent and it will be ensured that it is well targeted, transparent, and non-distortionary. In 2008, Le Blanc [13] published a proposed theoretical outline for comprehending the primary real difficulties that are related to rates and grants in WSS, specifically in emerging markets. In the work, Le Blanc [13] structured the research into three categories, namely

- the rudimentary commercial concepts pertinent to WSS at the time,
- presentation of a methodical outline to assess the necessity for subsidies, and appraising them, and
- focus on African countries in discussing the features and performance of subsidies in the WSS sector.

The work presented by Le Blanc [13] is therefore relevant in discussing pricing of WSS services in emerging markets and comparing it with the general subsidy guidelines proposed by Andres et al. [1]. The first section presented in Le Blanc [13] examines the finances of overheads and price configurations in water delivery in emerging markets. It highlights the primary factors that influence discrepancies between costs and tariffs in these countries.

As described by Le Blanc [13], the water provision cases researched in this work, in the metropolitan perspective, occur in a lawful and controlling environment and usually combine two levels of intermediation. These two levels are the international and national level. At international level, intervention is typically linked to the UN SDGs, whereas at domestic level, the bodies that have an opinion in describing the provisions (for drinking water in this case) are

- through line ministries in charge of WSS services, the State,
- a controlling organization, which could form portion of the division bureau, or be a self-governing agency,
- intermediary levels of administration that could arbitrate through for instance the carrying out of grants, and
- local metropolises, which are accountable for facility delivery in their respective dominions.

According to Le Blanc [13], other stakeholders that may play some part in policy-making include

- public or private utility companies,
- alternative service providers in communities or public sector entities, and
- various types of consumers in the agricultural, commercial and industrial sectors.

It is therefore vital to take note that there could be a significant amount of stakeholders; all should agree on policies, and this can often become a complex problem. Le Blanc [13] also mentions that at the highest level, the degree of integration of international goals (for example the UN SDGs) into domestic policies should be considered first, but local implementations will vary significantly. Secondly, at the national level, the various stakeholders and actors could have different and potentially conflicting objectives. It is therefore again important to consider, in any sector and including the broadband internet distribution sector, that developing policies and subsidies to grow a sector will often take a long time and will not benefit all associated actors equally (or even be to the detriment of some).

Considering the two primary levels of intervention in developing subsidies, Le Blanc [13] highlights that there are different forms of access and provisions, specifically aimed at WSS in this work. Importantly, there is substantial dissimilarity between emerging markets and developed countries in the number of services and products that are delivered to consumers. For example, in developed countries, delivery of potable water to families is mainly realized through municipalities. Conversely, in emerging markets and especially in rural areas, many mechanisms for drinking water provision exist, each with varying levels of efficiency and sustainability (utilities, water-gathering points, shared access, etc.). In broadband internet distribution, a similar scenario exists, and service providers are encouraged to transform their mechanisms of distribution, cost, and QoS to reach consumers in poor and rural areas effectively. In economic terms and as stated by Le Blanc [13], water provision is a multi-layered product that can be demarcated by three extents, namely

- "*price,*
- *quantity, and*
- *quality*".

Although price and quantity are typically easily defined as the cost per unit and the number of units supplied, quality (of any product) can vary significantly and have effects on the cost. This is also true for broadband internet distribution (varied QoS of 5G and mm-Wave internet services) and will be discussed in the section that covers the pricing of broadband in emerging markets. In terms of water provision, Le Blanc [13] defines various levels of water quality that will affect its price to consumers as

- its composition in terms of toxins and pollutants up to a point where it could be hazardous to the health of the population,
- the extent of confidentiality and accessibility of the amenity, therefore accessibility from a tap in a household, or from a communal tanker (ideally) nearby,
- physical characteristics of the service, for example water flow and pressure, and

- reliability of the service over time, including its predictability, shortage periods, or pressure variations, among other options.

In emerging markets, households might also be dependent on a variety of sources for a service owing to low levels of reliability of each of these sources. This emphasizes the importance of diversity in policies and subsidies that aim to raise an overall level of provision. In the broadband internet distribution sector, a similar argument would be valid and policymakers should account for internet access from a range of sources (comparable to providing electricity through coal/nuclear and various renewable sources). Sources of broadband internet would, for example, include mobile 3G, 4G, or 5G, mm-Wave networks, fiber, copper, satellite, Wi-Fi, Li-Fi, and any other sources that are fit for last mile distribution to local populations, as reviewed in Lambrechts and Sinha [11].

Le Blanc [13] follows up on the various forms of access to a service with an introduction of the considerations of cost. In this section, Le Blanc [13] discusses mainly the structure of costs for water utilities. Water utilities are *usually* responsible for

- "*production,*
- *treatment,*
- *transport, and*
- *distribution*"

of potable water to consumers within their provision zone. In the section on pricing of energy in emerging markets, we saw a similar list of responsibilities of energy providers, being generation, transmission, distribution, and supply. Similarly, these activities have increasing returns to scale. Le Blanc [13] presents a statement that there is a substantial variance in capital investment amid water utilities and telecommunications or energy utilities, with that of water being the largest. Le Blanc [13] argues that water utilities are more capital-intensive, and furthermore, capital assets implemented in water resources is not able to be relocated to an alternative site; they are typically inoperative for any alternative application, and therefore become a great sort of immovable investment related with destroyed expenses. As a result, water utilities for natural monopolies and long-run marginal prices are typically lower than long-run median prices. Water utilities therefore tend to produce structural deficits financed by borrowing and in many cases substantial and vital subsidies. In general, water utilities have long been implementing versions of *price discrimination;* consumers paying different prices for the marginal unit (as also reviewed in the pricing of energy), and using techniques such as

- third-degree value discernment, which comprises asking dissimilar fees for users based on characteristics such as age, gender, location, or time of use,
- second-degree value discernment, where a dissimilar fee is asked for differing capacities consumed,
- a non-linear tariff, which varies the price of the marginal unit based on the amount of consumption by the customer, or
- a combination of these.

These techniques are relatively well-known, at least in terms of energy and water consumption, and some basic forms of price discrimination for broadband internet have been implemented, but not on a comparable scale. For example, in South Africa, relatively recently, price discrimination based on the data validity period was implemented. Therefore, 1 GB of mobile data valid for 30 days is more expensive than 1 GB of data valid for one day, for example. However, the introduction of 5G and mm-Wave networks that allow vast control over QoS as well as available services per customer (as reviewed in Chap. 2 of this book) could lead to increased levels of price discrimination that could have a positive impact in emerging markets and in rural areas. The concepts of varying marginal costs offered specifically by water utilities also have some challenges and limitations, as summarized by Le Blanc [13]:

- On the supply side (the water service provider), short-run and long-run fringe charges of manufacture fluctuate greatly, leading to aggregation of consumer demands, and the pricing mechanisms might not reflect this.
- Tariffs should reflect short-run marginal cost variations, for example in a dry season, and reflect varying opportunity cost. In practice it is challenging to echo all the bases of variation in the cost of manufacture.
- There are (often large) variations in distribution costs and it is challenging to determine who should pay the varying costs. Cases of unfair discrimination will inevitably occur.

From these listed challenges, it becomes clear that marginal price variations are also complex to define and need to account for many actors and stakeholders in the services or products sector. In terms of broadband internet distribution, however, price discrimination might be easier to achieve and customers may initially be more forgiving of the QoS based on the price paid. Finally, in Le Blanc [13], tariffs also form part of water provisions, as briefly reviewed below.

According to Le Blanc [13], if rates were to replicate expenses, owing to the configuration of providers, the dominant kind of tariff configuration might potentially take the arrangement of a two-part tariff, comprised of

- an adjustable-, and
- a fixed-charge.

The variable charge would reflect the marginal costs of, in the case of water utilities, an additional unit of water for the provider. The fixed or immovable fee would be proposed to insure the attributable share of the prices that is not dependent on the expended quantities, therefore the fixed costs of both manufacture and delivery. The fixed charge would also ensure that the utility could break even. The simplest technique to ensure that a provider breaks even is allocating the immovable expenses similarly between all clients and asking the marginal costs on the entire amount of components that are consumed. When assuming that all customers are equal, it could be sensible to offer a facility as soon as the disposable excess of customers is more than zero. For many products and services, including WSS and broadband internet, consumers differ in terms of their net income and requirements (in terms of their demands for quality and quantity). Specifically, a fixed charge inferred even by the dividing of fixed costs could be big likened to the revenue of the most underprivileged household, especially in emerging markets. Research on the dissemination of revenue and appraisal of amenities is needed for a provider to improve its tariff structure. According to Le Blanc [13], forms of second- and third-degree price discrimination can achieve a more dynamic tariff structure that accounts for poorer households. Non-linear charges will be an improvement to put up with the heterogeneity of revenue and service requirements in a specific populace and give families the option of selecting a particular price structure. A simple solution is categorizing low-volume and high-volume households, where the

- "*high-volume consumers have a high fixed charge with a low marginal rate, and*
- *the low-volume consumers have a lower fixed charge with a higher marginal rate*".

Such a solution (often implemented in the energy sector, as reviewed in the pricing of the energy section in this chapter) could potentially mitigate the challenge where a high fixed charge might stop low-income families from even joining the grid. Subsidies can also be implemented to provide certain (quality of) services for free, for example, in the WSS sector, access from public water taps could be free, where access in-house would be a charged service (this would relate to zero-rated internet services). Differentiated subsidies in the WSS sector aim to

- objectify grants to less fortunate families by permitting access to ones that are probable to need subordinate value services, and
- attain larger access coverage by predetermined investment capital.

In Le Blanc [13], a summary of water tariff configurations that are most frequently implemented by providers, according to the research conducted, is provided. This summary is adapted and briefly presented in Fig. 6.3. In terms of broadband internet pricing in emerging economies, some structures could potentially be adopted by the telecommunications sector.

As shown in Fig. 6.2, water utilities typically use four major categories of tariff structures, namely fixed charge, volumetric charge, a two-part fare, and a combination of static rate and volumetric rates. The fixed charge structure, as the name suggests, is a structure where the invoiced amount is not influenced by the number of units expended per set period. Volumetric charges, however, hinge on the quantity of units expended per set period, and in the case of uniform rates, tariffs are equal irrespective of the number of units consumed. Non-uniform rates have different tariffs as a function of the number of units consumed. As a sub-category of non-uniform pricing, block tariffs are used to categorize (or "block") certain bounds into one pricing structure. With increasing block tariffs, the marginal pricing rises with the *block*, whereas with declining *block* tariffs, the fringe pricing decreases. Converse to *block* tariffs, volume-differentiated rates are a scheme where the total amount of components are valued at the equivalent amount, but this amount is dependent on the overall expenditure. Two-part tariffs are comprised of a static rate and an adjustable rate, which is a function of the QoS (in this case the quality of the water), consumed. A uniform approach is adopted where the static portion of the volumetric rates is similar for the total number of connections, and a segregated approach entails a list

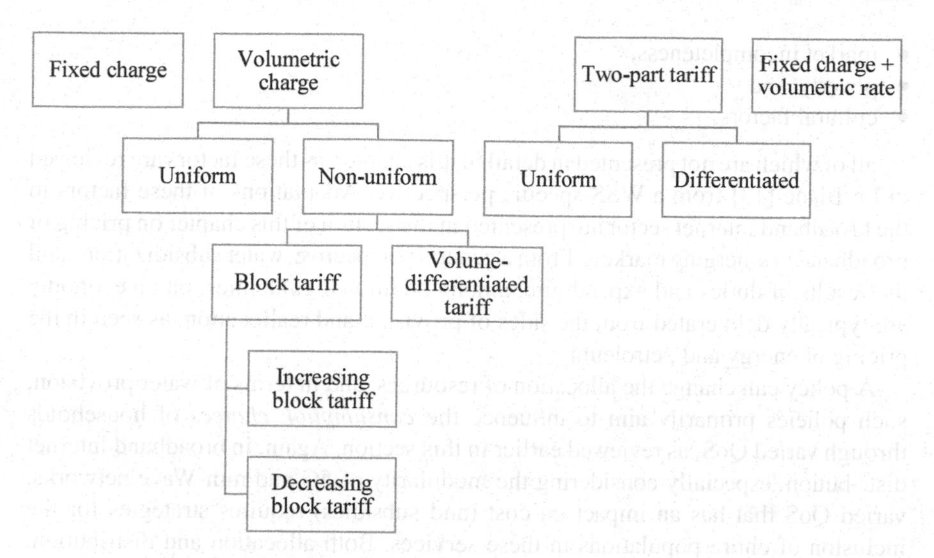

Fig. 6.2 A summary of water fare configurations that are most frequently implemented by providers, according to the research conducted by Le Blanc [13]

of options for amenities in which various sets of static fees and rates are defined. The static charge and volumetric fee are essentially a combination of static charges and differentiated prices.

Finally, with reference to the work published by Le Blanc [13] on water provision in emerging markets, an analysis based on the research conducted is presented, specifically on the effects of subsidies in the WSS sector. It reviews various aspects of water subsidies, outlined to present an overview of the subsidies, their potential effects on an economy, targeting strategies, and the way in which subsidies should be financed. The work provides a good comparison on differences and similarities in water subsidies and energy, petroleum, and medicine subsidies. Le Blanc [13] argues that water, and access to it

- lead to improvements in health,
- save time foraging for water (therefore leaving more time for other activities to build socioeconomic growth),
- result in fewer epidemics, and
- lower the cost of water for poor households.

Similar to many other sectors, including broadband internet distribution, water provision is a function of the economy of scale. Investment in capital-intensive activities is typically introduced where bulk provision of a product or service is required. As a result, rural and poor areas are commonly overlooked during initial provision phases and therefore become a relevant motivation for government intervention. Le Blanc [13] also presents numerous additional reasons that validate the need for intervention in water provision services, such as

- market incompleteness,
- poverty, and
- cultural factors,

all of which are not presented in detail in this chapter, as these factors are reviewed in Le Blanc [13] from a WSS-specific perspective. Adaptations of these factors to the broadband internet sector are presented in the section of this chapter on pricing of broadband in emerging markets. From a general perspective, water subsidizations and the results of duties and expenditure, among additional guidelines, on an economy are typically deliberated from the sides of provision and reallocation, as seen in the pricing of energy and petroleum.

A policy can change the allocation of resources, and in terms of water provision, such policies primarily aim to influence the *consumption choices* of households through varied QoS, as reviewed earlier in this section. Again, in broadband internet distribution, especially considering the modularity of 5G and mm-Wave networks, varied QoS that has an impact on cost (and subsidies) requires strategies for the inclusion of entire populations in these services. Both allocation and distribution, however, require an in-depth economic analysis related to the three criteria proposed by Le Blanc [13], namely

- efficiency,
- equity, and
- coverage.

Also consider reviewing the three world millennium goals relating to energy stipulated by the WEC [21], as reviewed in the pricing of energy section in this chapter: accessibility, availability, and acceptability, and their similarities to goals for water mentioned by Le Blanc [13]. *Efficiency* relates to the allocation of resources in an economy. It can be categorized as

- technical efficiency, the cost of providing a unit of a specific product or service, and
- commercial efficiency, the efficiency of the stages going from distribution to the collection of payments from consumers.

In the context of subsidies, it should be questioned if these give incentives to service suppliers to increase both procedural and service delivery efficiency. In terms of guiding principle assessment, significant results ought to be measurable in the production or consumption changes that occur because of a given policy. These incentives are also directly linked to the sustainability of a service and to its potential to grow an economy over the long term and have an impact on cost recovery.

Again, referring to the three criteria of in-depth economic analysis, *equity* is related to the redistribution of income (explicitly or implicitly) due to a subsidy. Specifically, targeting relates to the degree to which policies and subsidies benefit the part of the population that requires them most.

Table 6.2 The matrix method referred to by Le Blanc [13] to determine the outreach of a subsidy

	Poor	Non-poor
Reached	Ideal case	Leakage occurs
Not reached	Coverage issues	–

Finally, *coverage* denotes the percentage of the goal populace that has efficiently been influenced and extended to by the backing and is often the characteristic most indicative of its effectiveness, performance and efficiency. Coverage and targeting are, however, somewhat antagonistic goals and it may be difficult to achieve success in both simultaneously. Le Blanc [13] refers to a matrix technique to determine the outreach of a subsidy, as presented in Table 6.2.

According to Table 6.2, types of households are considered in terms of the outreach of a subsidy. The households are assumed to be categorized into only poor and non-poor households. It is assumed that the poor households are the primary target of the subsidy. If the subsidy program had been ideal, the diagonal cells (top left and bottom right) should be empty, indicating that only the poor households benefited from the subsidy. If, however, the top right cell requires an "explanation", it indicates a problem of targeting (also referred to as leakage) and households that were not involved in the goal set also benefited from the backing. If the lower left cell requires an explanation, a problem of coverage exists and households that were involved in the goal set have not been extended to by the backing. In practice, however, additional challenges arise from difficulty in measuring poverty and income (and therefore eligibility). Type 1 errors or errors of exclusion might occur, where poor households are classified as non-poor and therefore missing the target group. The opposite, type 2 errors, occur when non-poor households are classified as poor and incorrectly included in the target group.

Finally, with reference to the work published by Le Blanc [13], several other criteria for evaluating subsidies exist, all of which can also be applied to other sectors, such as the broadband internet sector. These criteria are briefly summarized below, and Le Blanc [13] provides more detail on each of these evaluation criteria. The summarized list includes:

- Administrative simplicity. This is applicable from the perspective of the government. The consumer and subsidies should be designed primarily to minimize costs, and importantly, administration costs. In addition, costs, including monitoring, enforcement and legal costs, should be minimized in a well-planned subsidy program.
- Complete transparency. Clarity on the subsidy's eligibility and participation criteria, including effective implementation, is also important to implement a subsidy and reach the goals set initially. Transparency is a qualitative parameter and can be measured by setting and answering a set of questions on, for example, the rules, eligibility, financing, beneficiaries, and management.

- Incentives: It has to be determined whether the encouragements of the various shareholders are well-suited with civic strategy objectives. Assuming the commercial arrangement amid the supervision and the service provider, it should be clear, for example, whether there is an economic incentive for the provider to spread the infrastructure, provide for deprived regions, or intensify metering.
- Political responsibility. If, for example, service is provided by the commercial sector, directive is a fundamental factor to alleviate any probable opposing encouragements to deliver low-quality amenities and take advantage of poor communities.
- Popularity. Has significant research been conducted on the popularity of the subsidy among all stakeholders? Are the incentives equal and fair to stimulate economic growth and achieve accurate targeting over the long term?
- Visibility. Visibility (also referred to as transparency in the section that reviews pricing of fuel in emerging markets) refers to the reflectiveness of all expenses of the funding for domestic duty payers and residents. It also shows how the costs reflect in the government's fiscal budget. High levels of visibility will have advantages for all stakeholders, including improved understanding of the total commercial price of support, improved directing, and less opportunity for exploitation.
- Flexibility. This refers to the ease and related costs of changing or terminating a subsidy if the need arises, or if improvements are to be made during its lifetime. Altering a subsidy is often encouraged as new data and information are gathered that can improve its targeting, efficiency or incentives.
- Sustainability. Related to flexibility, sustainability of a policy asks the question how relevant or essential the subsidy will be a few years down the line of its initial implementation. Issues arise, especially in emerging markets, relating to the compatibility of investment programs with the consequent expenditure on operation and maintenance to sustain the investment.

Measuring the evaluation criteria of a subsidy can also be complex, based on its qualitative or quantitative properties, and certain criteria might have multiple dimensions. It is therefore important to identify and define qualitative or semi-qualitative performance merits early to mitigate two common problems, according to Le Blanc [13], when scrutinizing a subvention, namely

- "*measuring the total amount of subsidies received by households, and*
- *measuring the distributional impact of the subsidy*".

Two methods of achieving these measurements accurately are to approach the problem from

- the demand side, therefore the consumer (recipient of the subsidies and the non-targeted group), and
- the supply side, therefore the perspective of the service provider.

The demand-side approach relies on household survey data and includes both the consumption (units) and the price paid, and typically the subsidy is calculated as "*the difference between what the consumer paid and the price that should have been paid given a normal price*". By aggregating customers by quantiles of revenue dispersal, a degree of dispersal of the grant across these revenue sets can be calculated.

The supply-side method is from the opinion of the government and the component of observation of the service provider. According to Le Blanc [13], this is also referred to as the difference method, since "*subsidies to consumers are calculated as the difference between transfers from the government to the service provider, minus all losses resulting from inefficiencies*". A government will provide a transferal to the service provider to assist it in coping with present outflow necessities. Losses incurred from low efficiency in production, distribution, billing or collection then become an incentive for the service provider to improve. However, as has been seen in emerging markets with very inefficient service providers, much of the government assistance could be wasted and in severe cases consumers might actually subsidize the service provider, the opposite of what would happen if the service and associated costs were efficient.

Up to this point in this chapter, numerous policies and strategies to price essential products and services in emerging markets have been presented, specifically in the medicine, energy, petroleum and water sectors. In each section, the broadband internet sector was referred to and in certain parts, directly compared to the sector reviewed. The following section, pricing of broadband in emerging markets, reviews specific policies and strategies to provide populations and rural communities with broadband internet. Many of these policies can be compared, or directly related, to policies highlighted in previous sections. The goal of this chapter is therefore to review, identify, and highlight as many potential solutions as possible to serve poorer populations with next-generation internet, where 5G and mm-Wave networks are specifically mentioned because of their differences with earlier generation mobile networks (as reviewed in Chaps. 1 and 2 of this book).

6.6 Pricing of Broadband in Emerging Markets

In this section, the pricing of broadband in unequal markets is reviewed. Several methods, policies, strategies, and pricing techniques are available to provide broadband internet access to communities that cannot afford to pay high prices for internet connectivity, or in many cases, are only willing and able to use free internet According to the International Telecommunication Union [8] State of Broadband Report, 66 recommendations on improving the state of broadband globally were put forward between 2012 and 2018, most of them (58) falling into one of ten categories, which include:

- "*general recommendations about information and communications technology (ICT) policies and regulations,*
- *improvements in data, statistics, and monitoring,*
- *improving skills, human capital, and capacity building,*
- *universal service approaches,*
- *taxation,*
- *focusing on local content, language, hosting, and entrepreneurship,*
- *innovative financing and investment,*
- *open access and infrastructure sharing,*
- *spectrum policies, and*
- *national broadband plans*".

The ITU [8] provides an in-depth look at each of these categories and reviews policy and regulatory gaps in ICT to provide global access to the internet. The analysis is focused holistically on improving progress in each of these categories and only used as reference in this section, dedicated to reviewing the pricing strategies of broadband internet. In this section, the pricing strategies of broadband internet will be reviewed from a few perspectives. These perspectives aim to construct a summarized view of past and current pricing strategies (and how these have adapted in line with the internet and its importance). It aims to determine where 5G and mm-Wave will most likely find itself in terms of cost-effective pricing for emerging markets and if a shift in technology will lead to innovative pricing strategies aimed at accommodating the poor. The perspectives include:

- Broadband pricing and strategies in developed countries at a time (in the early 2000s) when "always-on" internet started replacing dial-up connection. During this time, a general model of time spent online was popular, but developments changed the traditional constructs on which the pricing of telecommunication services were based. Time-based internet was relatively expensive and users needed to limit their time spent online to avoid high monthly bills. Rural communities and poor households in emerging markets generally did not take part in the revolution of the internet; however, participation was not yet deemed a socioeconomic benefit but rather a luxury.
- Broadband service pricing and strategies at a time (approximately around the 2010s based on the works referred to in this section) when service providers started monetizing internet traffic growth through value-based pricing as opposed to flat rates. Always-on internet pushed service providers to rethink their pricing strategies and since users were given access to the internet at all times, with no limitations on time spent online, service providers needed to innovate on monetizing their services and attracting new subscribers through pricing.
- Mobile broadband pricing and strategies at a time (approximately around 2016 based on the works referred to in this section) when it became more profitable for service suppliers to offer mobile internet connections, with comparable performance and quality to that of fixed line (asymmetric digital subscriber line or fiber) connections. Mobile internet reached speeds that would give service providers marketing strategies to sell these services where fixed-line infrastructure was not

available or outdated. Mobile data, however, were still much more expensive than fixed-line data and monetizing these services and ensuring good profit margins called for innovations on pricing strategies (such as limiting the availability of data).

- The current milieu of broadband pricing strategies is undergoing change, primarily in response to two factors. First, fiber has been replacing copper lines at a pace, even in emerging markets such as South Africa. The bandwidth gains of fiber compared to copper are significant, and service providers are now in a position to change their pricing strategies and lower their cost (while improving services). Secondly, 5G and mm-Wave are changing the mobile landscape and are allowing operators to enter unserved markets at significantly lower capital investment compared to earlier generation mobile broadband. Pricing strategies can also be adapted based on QoS and give poor households and geographically separated rural areas basic internet at a fraction of the cost of current (one-size-fits-all) implementations.

Furthermore, from the reviews on pricing strategies on medicine, energy, fuel, and water in this chapter, similarities between each of these sectors and the broadband sector should be evident. These similarities will possibly relate to adapting the traditional pricing strategies towards a more inclusive approach that could give all humans access to broadband internet at fair, tailored, and potentially assisted pricing. The first perspective of the review on broadband pricing is on a time when always-on internet started gaining traction.

6.6.1 From Dial-up to Always-on

As far back as 2001, Goolsbee [7] reported that broadband internet would fundamentally change how we use the internet and conduct business (and access entertainment). It is not difficult to argue that this has in fact occurred and with Industry 4.0 looming, even more changes are inevitable. In 2001, Goolsbee [7] mentioned that the USA had already identified fears that some people would be left behind or become the foundation of the digital divide. Many rule makers were already considering ways to accelerate the spread of broadband services. Another prediction from 2000, by Rao [19], was that if there were adequate broadband consumers, merchants of content would produce new methods of content that would entice even further consumers, reminiscent of the snowball effect. Goolsbee [7] listed some ways that the US government proposed to accelerate the spread of broadband at the time, including

- "*subsidies to enable internet adoption in public schools,*
- *subsidized prices of broadband access for all users,*
- *investment tax credits for broadband service providers to expand into rural and under-served markets, and*
- *moratoria on state and local taxation of internet access*".

It is clear from Goolsbee [7] that broadband internet was already at the time earmarked as a revolutionary change and that it was important enough for economic growth to ensure that the greater proportion of the population had access. Looking at the broadband milieu today, there is still a significant digital divide, and the policies proposed in the early 2000s remain relevant. Government subsidies and tax benefits to provide internet access to unserved markets have taken a long time to be realized in many emerging markets, and still today, these policies have not been completely fulfilled.

Goolsbee [7] specifically conducted research in economics on the welfare gains from new products (broadband internet in this case) by means of market value and data capacity by assuming a practical arrangement for the necessity correlation. Goolsbee [7] showed that customer preparedness to pay could be implemented as an alternate method of determining market-level necessity curvatures and computing prosperity improvements. From the earlier statement indicating that content providers will change the form of content if broadband users increase, it should be fair to assume that we are now in this era, and the arguments presented by Goolsbee [7] should have either become a reality or not. An interesting analysis performed by Goolsbee [7], specifically in the San Francisco area in the USA, was to determine the demand curve for broadband, which indicated (non-linearly), that very few people valued broadband internet at the time. Goolsbee [7] continued to propose types of subsidies to increase broadband users (since the demand curvatures formed the foundation of calculating the worth of broadband at the time). Goolsbee [7] proposed their own potential subsidy schemes, still relevant today, which included

- subsidizing each user directly, or
- subsidizing the fixed costs of service.

In terms of subsidizing each user, Goolsbee [7] argued that unless the demand was seamlessly rigid, or there were network externalities, this subvention would spend more than it actually generated. Briefly referring to 5G and mm-Wave that currently offer a new service with a relatively low number of users, a possible similar situation could exist. However, what is different now compared to 2001 is that the economic potential of broadband internet has received much more attention and its importance has increased. Therefore, relooking at such a study, especially in emerging markets and rural areas, would potentially be advantageous (considering the impact that the internet has had and will have) to modernize for 5G and mm-Wave broadband internet. To subsidize the fixed cost of service, as proposed in 2001 by Goolsbee, it has been argued that broadband consists of two discrete mechanisms, namely

- "*a marginal cost component that must be incurred for each additional customer, and*
- *a fixed cost component that needs to be incurred to offer a service in a specific area*".

The fixed costs are pooled by no less than one consumer; the principal component is typically maintenance, upgrading of infrastructure, preliminary advertising, client attainment expenses in a market and the costs related to initiating a billing structure that permits intricate billing and observing. This was true in 2001 and still is, but when compared to modular and tailored 5G and mm-Wave systems, these costs could be much higher and the systems more complex, especially considering tailored QoS. Furthermore, fixed costs related to infrastructure and maintenance are (still) higher in rural areas and many emerging markets with old or non-existent infrastructure, and the customer base would be smaller, leading to difficulty to recover costs. If the anticipated revenues of a market, or area, do not surpass the cost of the initial venture, suppliers will defer or completely avoid these markets.

In 2001, it was already argued that it is important to analyze the effects of taxation of broadband access. Similarly, a subsidy of broadband usage could potentially assist service providers to expand to these markets and areas and recover their cost of investment. Such strategies have no effect on marginal costs and therefore the price of the service would remain the same. However, it might impact the amount of markets that providers select to go into. From a socioeconomic perspective, a similar argument can be made: expanding to poorer and rural areas will provide internet access to these communities, but high marginal costs (or similar to what the wealthier population pays), will lead to very few people signing up and using these services.

Goolsbee [7] therefore presented research on the effects of several guidelines to fund broadband internet acceptance in 2001, by means of approximations of customers' willingness to pay (WTP) for internet in various regions. The comparison was between an appropriation of consumption in present (at the time) broadband markets and an appropriation to invest in markets that had no access. Goolsbee [7] found that "*subsidizing usage generates more adoption than subsidizing fixed costs, but the consumer welfare gains are much smaller and the revenue costs are much higher with a usage subsidy*". By definition, according to Goolsbee [7], funding usage is likely to interest marginal clients who do not rate a new merchandise greatly. This specific result of Goolsbee [7] *could* be different in modern times, since in 2001 the importance of broadband always-on internet was still relatively underestimated. It is therefore likely that a similar study based on 5G and mm-Wave technology could yield different results. Furthermore, and as a generalized statement, an appropriation is likely to total considerably more than it produces in purchaser excess and the appropriation to venture circumvents the issues of little valuation customers by encouraging admission into markets that are not served where new clients value the products highly.

As the Internet changed in the early 2000s to being always on as opposed to dialing in, Biggs and Kelly [2] presented research on broadband pricing strategies and why this technology was so successful in reaching such a great quantity of new users in relatively little time. Specifically, Biggs and Kelly [2] looked at the characteristics of its pricing and how/if this made it possible. Biggs and Kelly [2] conducted their research in 145 countries in terms of their pricing strategies and argued that the work would be beneficial for operators and regulators as the popularity of always-on internet kept growing. Biggs and Kelly [2] initially outlined the primary differences between broadband internet and earlier (dial-up) technology:

- Broadband internet is always on.
- The marginal cost to operators of a greater number of users and capacity is close to zero.
- Pricing is often based on a flat rate with no limitations on data usage (albeit more prevalent in developed countries).
- If data thresholds exist, these are based on content and not time.
- Pricing is independent of distance-pricing and providers do not charge different prices to consumers that are geographically far removed from the service provider.

Important to take away from these factors is how the pricing strategies changed for always-on broadband internet compared to dial-up. These strategies have both positive and negative effects for consumers and providers. Primarily:

- Service providers' incentive to reach markets that are geographically far away has been affected; for example, in rural areas, and with fewer potential consumers, WTP for the service has declined.
- Pricing on time of accessing the internet allowed consumers to have complete control over how much they used the internet and the associated monthly costs, whereas the always-on alternative is typically a fixed cost and therefore requires a consumer to commit to such an investment.
- The pricing structure for service providers was increasingly absorbed by the market and was a function of competition. Revenue per customer depended on various externalities and profit margins were relatively low (per subscriber).

Biggs and Kelly [2] also listed the primary factors that affected broadband pricing at the time, including

- the competitive structure of the broadband market,
- the degree of regulatory intervention,
- existing infrastructure and prior investments in older technology,
- competing technologies, for example, at the time, cable television, and
- indirect competitive pressure from neighboring countries.

Because of the changing scenario in the broadband market, specifically in terms of the pricing structures, service providers only had a few strategies to differentiate their services from the competition; these still hold true today, for example:

- variations and offers based on the installation fee for each new subscriber,
- charging for equipment (such as the broadband modem) or offering fee structures and rental of the equipment,
- the monthly access fees (the flat-rate costs),
- competitive pricing for out-of-bundle data, if a monthly threshold is implemented, and
- additional value-added services such as email addresses.

In the current milieu of internet distribution, specifically in terms of providing next-generation 5G and mm-Wave access, it is a relatively similar situation if compared to the time when always-on internet replaced dial-up (time-based) technologies. 5G and mm-Wave are not completely dependent on current infrastructure, as reviewed in Chap. 2 of this book. Operators are therefore (again) obligated to invest in new, relatively immature, technology infrastructure. Incentive to reach unserved markets, rural areas, and poorer areas in many emerging markets is low, primarily owing to the characteristics of broadband not accounting for distance from the service provider. Recovering costs in these areas will therefore (again) be difficult, and subsidies and other investments are needed to spearhead the technologies for Industry 4.0. However, a major difference in 5G and mm-Wave technology is its modularity in terms of service offerings. Service providers will therefore rely less on the traditional pricing structures and slightly varying access fees, offering additional thresholds, and varying installation fees or equipment fees. Scaling through network slicing and offering subscribers a tailored service at a cost that they are willing to pay could incentivize providers to enter and drastically change internet access in currently unserved (or limited service) markets.

6.6.2 New Ways of Monetizing Always-on Broadband Internet Traffic

However, the market pricing strategies remained relatively constant for some years, and only started changing from approximately the 2010s. Nicosia et al. [16] published a work on how service providers were reconsidering horizontal (flat) rate valuing for broadband facilities and how to monetize internet traffic progress. According to Nicosia et al. [16], telecommunications industries at the time were faced with a fundamental challenge. This is still true today. Nicosia et al. [16] argued that on one side, there were substantially *"increasing requirements for new investments in broadband internet access and transport infrastructure that support continuous growth in broadband traffic. On the other side, there was reduced ability to exercise pricing power with customers to increase revenues"*. The results from Biggs and Kelly [2] were proven to be true and infrastructure investments (because of the complexity and cost of the equipment and technologies to serve large subscriber bases) increased significantly. Nicosia et al. [16] also specifically mentioned that around 2011 and according to the uptake of broadband in Organization for Economic Co-operation and Development countries, broadband internet was approaching maturity. The pricing strategies of broadband at the time therefore required a kind of *revamp* to incentivize operators to reach new markets. At the time, internet rates were reducing, trading capacities were deteriorating and markets were dwindling. The primary offender mentioned by Nicosia et al. [16] was the flat rate ("all-you-can-eat") valuing structure – again, in a sense predicted by Biggs and Kelly [2]. Essentially, the pricing structure introduced when broadband started gaining popularity in the early 2000s did not

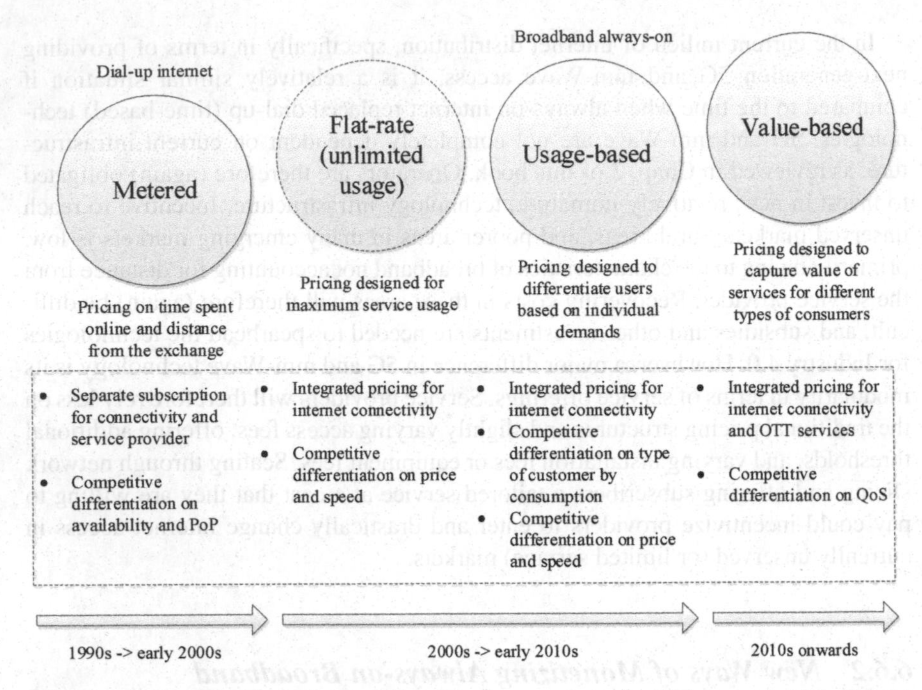

Fig. 6.3 The pricing characteristics of broadband internet from the 1990s up to the time that the work of Nicosia et al. [16] was published, and the predicted (at the time) pricing strategies going forward

account for the large growth in subscribers and data usage of each subscriber, and Nicosia et al. [16] argued that an updated structure was due. Adapted from Nicosia et al. [16], Fig. 6.3 summarizes the pricing characteristics of broadband internet from the 1990s up to the time that the work was published, and the predicted (at the time) pricing strategies going forward.

As shown in Fig. 6.3, originally created by the Cisco Internet Business Solutions Group (IBSG), internet pricing (up to 2012) underwent three waves. During a time when dial-up internet dominated, metered pricing was used. A subscriber paid for two components: "*a subscription fee and a metered fee related to the amount of time that the subscriber was connected to the internet and the distance from the exchange*". The competitive differentiation for service providers was primarily founded on the fee and accessibility of dial-up points of presence.

From the early 2000s, as also presented in for example Biggs and Kelly [2], always-on broadband internet was introduced and changed the pricing strategy significantly. In most countries, specifically in developed countries, operators charged a service fee to consume as much data as a user needed. Variations on the service fee encouraged users to adopt a specific service. Economical variation was founded only on the fee and the connection speed.

As broadband matured around the 2010s, competitive pricing was still founded on fee and connection speed; however, providers started trialing with methods of usage-based valuing and considering traffic levels and traffic limits. Subscribers were, however, not keen on this approach, as they were used to the flat rate, all-you-can-use model. Essentially, traffic tiers and caps (and in some cases bandwidth throttling) raised questions of net neutrality and demands that internet traffic should be treated equally. Operators needed to construct their pricing strategies carefully [4]. (This implied a significant increase in the complexity of billing and monitoring compared to the flat rate model.) The model was, however, a result of the uptake of customers and the data usage per customer, and operators had to convey that data traffic sustainability comes at a cost of infrastructure enhancements.

Nicosia et al. [16] proposed the value-based pricing strategy that is based on access speed and traffic tiers and differentiating service offerings by increasing or decreasing speed, latency, and tiers based on usage and price paid by the subscriber. In terms of "*value-based pricing for business-to-consumer and business-to-business models*", Nicosia et al. [16] reported that the IBSG proposed several techniques to monetize internet traffic (and essentially guard themselves against over-the-top users). These techniques of value-based pricing included:

- Tiers of access speed were implemented to differentiate prices (although this was already commonly used at the time, and still is).
- Guaranteed bandwidth allocated bandwidth to a specific service or application (multimedia streaming or managed backups for example) to enable improved performance.
- The interval for traffic to arrive at its target (latency) was reduced. For example, online gaming typically requires short latency and service providers can offer this as an extra service.
- Time-of-day pricing is a technique to offer price-sensitive customers lower-cost options for accessing the internet at times when traffic is typically low.
- Dynamic or spot pricing links broadband pricing to the available capacity at specific times. Operators therefore need real-time information on the current load of the network. Significant improvements in network utilization result from dynamic or spot pricing. However, costs for monitoring and allocating bandwidth increase for the service provider. As an advantage, this is a resourceful instrument to produce extra need for novel services, proposing enticing fees through specific hours of the day or to consumers in particular area. In terms of 5G and mm-Wave networks, dynamic or spot pricing could be a resourceful pricing strategy, since these networks allow modulatory and tailored service to specific customers.
- All access prioritization at an additional cost, also referred to as the turbo button in some markets, gives customers the ability to receive the best service (bandwidth and latency) for a limited period at a once-off cost. Again, for 5G and mm-Wave networks this could offer customers superior performance for short times along with their monthly subscription that offers lower QoS.

- Shared access between devices. This value-based pricing scheme allows customers to share a total traffic allowance between multiple devices. This model, according to Nicosia et al. [16] was not widely offered at the time and mobile operators were exploring the alternative at the time. In terms of 5G and mm-Wave networks, superior bandwidth and latency compared to previous generation networks could result in successful implementation of such models. Service providers would effectively be able to tailor the bandwidth, latency, and traffic allowance for each device and offer bundles at reduced prices.

Considering the value-based pricing strategies listed above and adapted from Nicosia et al. [16], an evolution on flat rate and usage-based pricing, and reviewing the capabilities and unique characteristics of 5G and mm-Wave networks, at this point it seems that the models proposed by Nicosia et al. [16] could be ideal for the new generation of internet networks. Both "*fixed and mobile broadband operators*" have the capability to possibly increase the value of over-the-top (OTT) facilities such as multimedia content streaming and offer *just-enough* bandwidth for less intensive services such as browsing and email. According to Nicosia et al. [16], broadband providers could package guaranteed QoS for OTT suppliers that ought to comprise of two mechanisms,

- "*content delivery network (CDN) services to host and facilitate the distribution of OTT content over the broadband network, and*
- *guaranteed bandwidth for a specific service*".

Such a bundle would allow a service provider to capture CDN profits at a premium on market CDN tariffs. OTT suppliers ought to then remunerate for the distribution fee of the traffic generated by their own service, and suppliers could relieve this data from their customer's allowance calculation (zero-rating services) and encourage customers to adopt new services without fear of exceeding their quota limits. These proposed value-based pricing models from Nicosia et al. [16] could effectively be adapted to introduce pricing strategies for 5G and mm-Wave internet to unserved markets. As 5G is being rolled out in many developed countries, and its maturity is still relatively low, several works have been published on how to monetize the networks, specifically based on their different architecture when compared to earlier generation networks.

Thought ought to be spared to new and innovative prospects for monetization and valuing policies ought to be established to take into account the distinctive milieu of 5G products. Operators should consider developing new models in three key areas,

- monetizing the network,
- monetizing the slice, and
- monetizing the enterprise market.

5G and mm-Wave give operators the ability to leverage (monetize) their network and infrastructure competences to generate novel monetization prospects, such as

- "*developing and selling premium service offerings charged based on the QoS,*
- *monetizing the mobile network to sell fixed wireless services,*
- *enabling developers to monetize applications that connect to devices and generating variable revenue from these developers (and possibly from the connected devices), and*
- *leveraging infrastructure to create ecosystems with a variety of partnership opportunities and revenue-sharing models*".

Possibly the greatest revenue potential for 5G, as reviewed in Chap. 2 of this book, is network slicing. The technique can support multiple use cases, business models and services, all of which could be adapted to a level where their introduction into emerging markets and rural areas could be cost-effective for all stakeholders. The tailored service on requirements from the end-user in terms of capacity, latency, quality of service, and security could be downscaled for these markets to effectively offer only entry-level internet access for entire communities, and operators can scale the network as subscribers and usage increase.

The enterprise market should also be monetized and 5G and mm-Wave networks could offer opportunities for application in business-to-business and the IoT segment. The enterprise market, as opposed to the consumer market, will be a lucrative area for 5G and mm-Wave to offer premium and once-off services and generate revenue and new innovative strategies that could be passed down to the consumer market. This market typically requires less intervention in terms of subsidies and third-party investments and could generate revenue that will internally subsidize a move towards previously unserved markets.

Meakin et al. [15] also listed the anticipated benefits in a new generation of communications infrastructure (5G and mm-Wave). According to Meakin et al. [15], this can be accomplished by

- unlocking novel income avenues,
- decreasing expenses, and
- cultivating consumer involvement.

According to Meakin et al. [15], operators can unlock new revenue streams through a few techniques, but most importantly, through competing in new markets. 5G fixed wireless access (FWA) and its capacity and latency performance can compete directly against other fixed-line technologies, specifically the more expensive (in terms of infrastructure development) "*fiber to the premises and fiber to the home alternatives*" [15]. Especially considering rural and poor areas in emerging markets, in these markets the 5G FWA could be further feasible, on condition of the availability of spectrum, area concentration (of the local population), and the affordability of customer premises equipment (CPE).

In terms of reducing costs, although initial investment is required in spectrum, network diversification, and equipment modernizations to construct 5G and mm-Wave network infrastructure, these costs will be offset by the sheer capacity and speeds offered by the technology. A large discount in "*unit cost per gigabyte*" or data traffic ought to yield sustainable ecosystems that will assist in maintaining and recovering investments by the operators. Additional cost savings can be accomplished during the rollout phases of 5G and mm-Wave networks through operational simplicity, automation and self-organizing/operating networks. However, it is important to note that much of the costs savings will still rely on policy makers and their investment in building towards Industry 4.0, where spectrum and the ICT sectors (upskilling of local population) will have a significant and long-term influence.

Finally, Meakin et al. [15] identify that improving the customer experience is a method to capitalize on 5G and mm-Wave networks. This factor, however, will have varying degrees and angles of implementation depending on the market served, where urban, developed areas will have significantly different needs compared to rural, emerging areas. Essentially, it is again the scalability and modularity of 5G and mm-Wave networks based on their speed, capacity, and latency that will require extensive research into the requirements and needs of each market.

Meakin et al. [15] also report on recent studies that show that the uptake of 5G, compared to earlier generations, will possibly be slower, for three reasons:

- Consumer awareness of 5G and mm-Wave is relatively low compared to when 4G was launched.
- Devices that have 5G capability built in are not yet commonplace and are expensive, which will lead to consumers waiting to avoid buying first-generation products that could have restricted functionality.
- Customers, according to Meakin et al. [15], are likely to be unwilling to part with large amounts of monies for better service, especially on mobile networks and even less so for home networks. According to Meakin et al. [15], a survey showed that customers are only willing to pay between USD $4.40 (for mobile internet) and USD $5.06 (for home internet) extra for better service. This is therefore an indication for operators of how long it will take to recover costs when investing in 5G and mm-Wave mobile broadband.

Finally, adapted from Meakin et al. [15], selecting the optimal business model for 5G will assist operators and other participants to expand the 5G ecosystem rapidly. Meakin et al. [15] list five criteria for scoring potential use cases, adaptable to both developed countries and emerging markets, as:

- Third-party brand strength: If a third-party service (for example Netflix®) has a stronger brand, specifically for media streaming, a business model advertising the service offerings as opposed to the operator brand would potentially lead to higher uptake.
- Third-party market penetration: In line with the third party's brand strength, its market penetration will also be indicative of the potential success of introducing 5G specifically tailored for the service using the operator's brand, or the third party's customer awareness.
- Alignment with the internal capabilities of the operator: The existing capabilities of an operator, albeit on an earlier generation network, could have established strengths in external departments. These include service delivery, billing and physical presence, and expanding on the current business model could lead to improved

performance (for example in current augmented reality (AR), virtual reality (VR), IoT services) through *only* having to invest in modernized communication technologies such as 5G and mm-Wave networking.

- Brand relevance of the operator: An operator with high brand relevance can choose to collaborate with less known, but more specialized, services to deliver content as opposed to developing these third-party services itself.
- Associated usage intensity or dependence on 5G: As part of market research, the dependence and intensity (need) for 5G would determine the chosen business model. For example, the dependence and intensity for a VR/AR service would be significantly different when compared to the dependence and intensity in a rural community to access basic internet services.

Operators can therefore score a use case against these criteria and determine the optimal business model to implement in a specific market, whether rural, urban, developed, or emerging markets are to be served. The results for each use case will be considerably different and therefore in-depth market analysis and research are encouraged for each use case.

As 5G and mm-Wave are rolled out in developing countries and emerging markets and start exiting the experimental phases in these countries, operators will start adopting new pricing mechanisms for the services offered. Furthermore, technologies such as network slicing and varied QoS are not yet standardized for 5G and mm-Wave deployments and operators are not yet implementing these to their full potential. The technology requires time to mature, and this section highlights the primary areas where innovative pricing of 5G and mm-Wave broadband internet, especially considering unserved and unequal markets, will be implemented.

6.7 Conclusion

In this chapter, a review on pricing mechanisms, policies and strategies, and the failures and successes of these in various sectors is presented. The chapter focuses on the medicine, energy, petroleum, and water sectors in emerging markets and analyzes the policies (such as subsidies) that aim to distribute these products and services even to the poorest communities. Broadband internet shows many similarities to these sectors and is increasingly becoming more similar. For Industry 4.0, broadband internet at affordable prices is crucial to boost economic growth in emerging markets, especially in rural areas. The digital divide has existed for decades, and this book aims to provide relevant reviews on emerging mobile broadband technologies that offer opportunities to bridge the digital divide.

References

1. Andres LA, Thibert M, Cordoba CL, Danilenko AV, Joseph G, Borja-Vega C (2019) Doing more with less. Smarter subsidies for water supply and sanitation. The World Bank Group, CC BY 3.0 IGO
2. Biggs P, Kelly T (2006) Broadband pricing strategies. Info 8(6):3–14
3. Bragg S (2018) Marginal cost pricing. Retrieved Nov 29, 2019 from http://accountingtools.com
4. Calzada J, Martinez-Santos F (2016) Pricing strategies and competition in the mobile broadband market. J Regul Econ 50(1):70–98
5. Couchenery P, Maillé P, Tuffin B (2013) Impact of competition between ISPs on the net neutrality debate. IEEE Trans Netw Serv Manage 10(4):425–433
6. Everard M (2002) Access to medicines in low-income countries. Int J Risk Saf Med 15:137–149
7. Goolsbee A (2002) Subsidies, the value of broadband, and the importance of fixed costs. Broadband: should we regulate high-speed internet access. 278–294
8. ITU (2019) State of broadband report. International Telecommunication Union and United Nations Educational, Scientific and Cultural Organization, Geneva. Licence: CC BY-NC-SA 3.0 IGO
10. Kojima M (2013) Petroleum product pricing and complementary policies: experience of 65 developing countries since 2009. The World Bank, Washington, DC
11. Lambrechts JW, Sinha S (2019) Last mile internet access for emerging economies. Springer International Publishing, Switzerland. ISBN 978-3-030-20956-8
13. Le Blanc D (2008) A framework for analyzing tariffs and subsidies in water provision to urban households in developing countries. DESA working paper no 63, ST/ESA/2008/DWP/63, United Nations
14. Madgavkar A (2019) Emerging markets: why it's tough at the top. McKinsey Quarterly, Mar 2019. Retrieved Nov 28, 2019 from http://www.mckinsey.com
15. Meakin R, Wong S, Zikry K, Shea D (2019) Making 5G pay. Monetizing the impeding revolution in communications infrastructure. PwC. Retrieved Nov 29, 2019 from http://www.pwccn.com
16. Nicosia M, Klemann R, Griffin K, Taylor S, Demuth B, Defour J, Medcalf R, Renger T, Datta P (2012) Rethinking flat rate pricing for broadband services. How service providers can monetize internet traffic growth via value-based pricing. CISCO Internet Business Solutions Group (IBSG). White paper, July 2012
17. Oughton EJ, Frias Z (2018) The cost, coverage and rollout implications of 5G infrastructure in Britain. Telecommun Policy 42(8):636–652
18. Posner RA (1969) Natural monopoly and its regulation. J Reprints Antitrust Law Econ 9:767
19. Rao B (2001) Broadband innovation and the customer experience imperative. JMM 3(11):56–65
20. Ricciardi V (2019) Water subsidies mostly benefit the wealthy. Retrieved December 2019 from http://www.blog.worldbank.org
21. World Energy Council (2001) Pricing energy in developing countries. A report of the World Energy Council. Retrieved Nov 22, 2019 from http://www.regulationbodyofknowledge.org
22. World Energy Council (2019) World energy scenarios | 2019. A report of the World Energy Council. Retrieved Jan, 2020 from http://worldenergy.org

Printed in the United States
by Baker & Taylor Publisher Services

Printed in the United States
by Baker & Taylor Publisher Services